A·N·N·U·A·L E·D

Environment

03/04

Twenty-Second Edition

EDITOR

John L. Allen

University of Wyoming

John L. Allen is professor of geography at the University of Wyoming. He received his bachelor's degree in 1963 and his M.A. in 1964 from the University of Wyoming, and in 1969 he received his Ph.D. from Clark University. His special area of interest is the impact of contemporary human societies on environmental systems.

McGraw-Hill/Dushkin

530 Old Whitfield Street, Guilford, Connecticut 06437

Visit us on the Internet

http://www.dushkin.com

Credits

1. **The Global Environment: An Emerging World View**
   Unit photo—© 2003 by PhotoDisc, Inc.

2. **The World's People: Population and Economy**
   Unit photo—United Nations photo by Ray Witlin.

3. **Energy: Present and Future Problems**
   Unit photo—© 2003 by Sweet By & By/Cindy Brown.

4. **Biosphere: Endangered Species**
   Unit photo—United Nations photo by M. Gonzalez.

5. **Resources: Land, Water, and Air**
   Unit photo—United Nations photo.

6. **Pollution: The Hazards of Growth**
   Unit photo—United Nations photo by Rick Grunbaum.

Copyright

Cataloging in Publication Data
Main entry under title: Annual Editions: Environment. 2003/2004.
1. Environment—Periodicals. I. Allen, John L., *comp.* II. Title: Environment.
ISBN 0–07–283851–5   658'.05   ISSN 1092–4876

Twenty-Second Edition

Cover image © 2003 PhotoDisc, Inc.
Printed in the United States of America   1234567890BAHBAH543   Printed on Recycled Paper

Editors/Advisory Board

Members of the Advisory Board are instrumental in the final selection of articles for each edition of ANNUAL EDITIONS. Their review of articles for content, level, currentness, and appropriateness provides critical direction to the editor and staff. We think that you will find their careful consideration well reflected in this volume.

EDITOR

John L. Allen
University of Wyoming

ADVISORY BOARD

Daniel Agley
Towson University

Matthew R. Auer
Indiana University

Robert V. Bartlett
Purdue University

Susan W. Beatty
University of Colorado, Boulder

William P. Cunningham
University of Minnesota - St. Paul

Lisa Danko
Mercyhurst College

Dianne Draper
University of Calgary

Juris Dreifelds
Brock University

Debbie R. Folkerts
Auburn University - Main

Gian Gupta
University of Maryland

George N. Huppert
University of Wisconsin, La Crosse

Jeffrey W. Jacobs
National Research Council

Vishnu R. Khade
Eastern Connecticut State University

Adil Najam
Boston University

David J. Nemeth
University of Toledo

Shannon O'Lear
University of Illinois, Urbana Champaign

David Padgett
Tennessee State University

John H. Parker
Florida International University

Joseph David Shorthouse
Laurentian University

Bradley F. Smith
Western Washington University

Nicholas J. Smith-Sebasto
Montclair State University

Ben E. Wodi
SUNY at Cortland

Staff

Jeffrey L. Hahn, Vice President/Publisher

To the Reader

In publishing ANNUAL EDITIONS we recognize the enormous role played by the magazines, newspapers, and journals of the public press in providing current, first-rate educational information in a broad spectrum of interest areas. Many of these articles are appropriate for students, researchers, and professionals seeking accurate, current material to help bridge the gap between principles and theories and the real world. These articles, however, become more useful for study when those of lasting value are carefully collected, organized, indexed, and reproduced in a low-cost format, which provides easy and permanent access when the material is needed. That is the role played by ANNUAL EDITIONS.

At the beginning of our new millennium, environmental dilemmas long foreseen by natural and social scientists began to emerge in a number of guises: regional imbalances in numbers of people and the food required to feed them, international environmental crime, energy scarcity, acid rain, build-up of toxic and hazardous wastes, ozone depletion, water shortages, massive soil erosion, global atmospheric pollution and possible climate change, forest dieback and tropical deforestation, and the highest rates of plant and animal extinction the world has known in 65 million years.

These and other environmental problems continue to worsen in spite of an increasing amount of national and international attention to the issues surrounding them and increased environmental awareness and legislation at both global and national levels. The problems have resulted from centuries of exploitation and unwise use of resources, accelerated recently by the shortsighted public policies that have favored the short-term, expedient approach to problem solving over longer-term economic and ecological good sense. In Africa, for example, the drive to produce enough food to support a growing population has caused the use of increasingly fragile and marginal resources, resulting in the dryland deterioration that brings famine to that troubled continent. Similar social and economic problems have contributed to massive deforestation in middle and South America and in Southeast Asia.

Part of the problem is that efforts to deal with environmental issues have been intermittent. During the decade of the 1980s, economic problems generated by resource scarcity caused the relaxation of environmental quality standards and contributed to the refusal of many of the world's governments and international organizations to develop environmentally sound protective measures, which were viewed as too costly. More recently, in the 1990s, as environmental protection policies were adopted, they were often cosmetic, designed for good press but little else. Even with these public relations policies, governments often lacked either the will or the means to implement them properly. The absence of effective environmental policy has been particularly apparent in those countries that are striving to become economically developed. But even in the more highly developed nations, economic concerns tend to favor a loosening of environmental controls. In the United States, for example, the interests of maintaining jobs for the timber industry imperil many of the last areas of old-growth forests, and the desire to maintain agricul-

tural productivity at all costs causes the continued use of destructive and toxic chemicals on the nation's farmlands. In addition, concerns over energy availability have created the need for foreign policy and military action to protect the developed nations' access to cheap oil and have prompted increasing reliance on technological quick fixes, as well as the development of environmentally sensitive areas to new energy resource exploration and exploitation.

Despite the recent tendency of the U.S. government to turn its back on environmental issues and refuse to participate in international environmental accords, particularly those related to global warming, there is some reason to hope that, globally, a new environmental consciousness is awakening. Unfortunately, increasing globalization of the economy has meant globalization of other things as well, such as internal conflict and disease transmission. The emergence of terrorism as an instrument of national or quasi-national policy—particularly where terrorism may employ environmental contamination as a weapon—has the potential to produce future environmental problems that are almost too frightening to think about.

In *Annual Editions: Environment 03/04* every effort has been made to choose articles that encourage an understanding of the nature of the environmental problems that beset us and how, with wisdom and knowledge and the proper perspective, they can be solved or at least mitigated. Accordingly, the selections in this book have been chosen more for their intellectual content than for their emotional tone. They have been arranged into an order of topics—the global environment, population and economy, energy, the biosphere, resources, and pollution—that lends itself to a progressive understanding of the causes and effects of human modifications of Earth's environmental systems. We will not be protected against the ecological consequences of human actions by remaining ignorant of them.

Readers can have input into the next edition of *Annual Editions: Environment* by completing and returning the postpaid *article rating form* at the back of the book.

John L. Allen
Editor

Contents

UNIT 1
The Global Environment: An Emerging World View

Three selections provide information on the current state of Earth and the changes we will face.

UNIT 2
The World's People: Population and Economy

Six unit selections examine the problems the world will have in feeding and caring for its ever-increasing population.

The concepts in bold italics are developed in the article. For further expansion, please refer to the Topic Guide and the Index.

UNIT 3
Energy: Present and Future Problems

Five articles in this unit consider the problems of meeting present and future energy needs. Alternative energy sources are also examined.

Unit Overview 80

The concepts in bold italics are developed in the article. For further expansion, please refer to the Topic Guide and the Index.

UNIT 4
Biosphere: Endangered Species

Three unit articles examine the problems in the world's biosphere that include eco-nomic issues, natural ecosystems, and bioinvasion.

The concepts in bold italics are developed in the article. For further expansion, please refer to the Topic Guide and the Index.

UNIT 5
Resources: Land, Water, and Air

In this unit, five selections discuss the environmental problems affecting our land, water, and air resources.

Unit Overview   **148**

The concepts in bold italics are developed in the article. For further expansion, please refer to the Topic Guide and the Index.

UNIT 6
Pollution: The Hazards of Growth

The four selections in this unit weigh the environmental impacts of the growth of human population.

The concepts in bold italics are developed in the article. For further expansion, please refer to the Topic Guide and the Index.

Topic Guide

This topic guide suggests how the selections in this book relate to the subjects covered in your course. You may want to use the topics listed on these pages to search the Web more easily.

On the following pages a number of Web sites have been gathered specifically for this book. They are arranged to reflect the units of this *Annual Edition*. You can link to these sites by going to the DUSHKIN ONLINE support site at *http://www.dushkin.com/online/*.

ALL THE ARTICLES THAT RELATE TO EACH TOPIC ARE LISTED BELOW THE BOLD-FACED TERM.

Agribusiness
18. Where Have All the Farmers Gone?

Agriculture
18. Where Have All the Farmers Gone?
20. Growing More Food With Less Water

Air pollution
23. Three Pollutants and an Emission

Alternative energy
11. Beyond Oil: The Future of Energy

Aquifers
24. Groundwater Shock: The Polluting of the World's Major Freshwater Stores

Atmosphere
22. Feeling the Heat: Life in the Greenhouse

Biodiversity
15. What Is Nature Worth?
16. A Fragile Cornucopia: Assessing the Status of U.S. Biodiversity

Bioinvasion
17. Invasive Species: Pathogens of Globalization

Biosphere
15. What Is Nature Worth?

Carbon dioxide emissions
23. Three Pollutants and an Emission

Clean Water Act
25. Water Quality: The Issues

Climate change
1. How Many Planets? A Survey of the Global Environment
21. Oceans Are on the Critical List
22. Feeling the Heat: Life in the Greenhouse
26. Statehouse and Greenhouse: The States Are Taking the Lead on Climate Change

Common resource problems
7. The Eco-Economic Revolution: Getting the Market in Sync With Nature

Conservation of energy
13. Fossil Fuels and Energy Independence

Cultural customs
18. Where Have All the Farmers Gone?

Cultural values
18. Where Have All the Farmers Gone?

Dams and reservoirs
19. All the Wild Rivers

Developing countries
4. Population Control Today—and Tomorrow?

Development, economic
18. Where Have All the Farmers Gone?

Eco-economy
7. The Eco-Economic Revolution: Getting the Market in Sync With Nature

Ecology
15. What Is Nature Worth?
17. Invasive Species: Pathogens of Globalization
18. Where Have All the Farmers Gone?

Economics
6. An Economy for the Earth
18. Where Have All the Farmers Gone?

Ecosystem
2. Forget Nature. Even Eden Is Engineered
15. What Is Nature Worth?
16. A Fragile Cornucopia: Assessing the Status of U.S. Biodiversity

Endangered species
3. Crimes of (a) Global Nature

Energy resources
10. Energy: A Brighter Future?

Energy system
7. The Eco-Economic Revolution: Getting the Market in Sync With Nature
11. Beyond Oil: The Future of Energy
12. Renewable Energy: A Viable Choice
14. Power Struggle: California's Engineered Energy Crisis and the Potential of Public Power

Environmental crises
5. Population and Consumption: What We Know, What We Need to Know

Environmental deterioration
6. An Economy for the Earth

Environmental laws
26. Statehouse and Greenhouse: The States Are Taking the Lead on Climate Change

World Wide Web Sites

The following World Wide Web sites have been carefully researched and selected to support the articles found in this reader. The easiest way to access these selected sites is to go to our DUSHKIN ONLINE support site at *http://www.dushkin.com/online/*.

AE: Environment 03/04

The following sites were available at the time of publication. Visit our Web site—we update DUSHKIN ONLINE regularly to reflect any changes.

General Sources

Britannica's Internet Guide
http://www.britannica.com

This site presents extensive links to material on world geography and culture, encompassing material on wildlife, human lifestyles, and the environment.

EnviroLink
http://envirolink.netforchange.com

One of the world's largest environmental information clearinghouses, EnviroLink is a grassroots nonprofit organization that unites organizations and volunteers around the world and provides up-to-date information and resources.

Library of Congress
http://www.loc.gov

Examine this extensive Web site to learn about resource tools, library services/resources, exhibitions, and databases in many different subfields of environmental studies.

The New York Times
http://www.nytimes.com

Browsing through the archives of the New York Times will provide a wide array of articles and information related to the different subfields of the environment.

SocioSite: Sociological Subject Areas
http://www.pscw.uva.nl/sociosite/TOPICS/

This huge sociological site from the University of Amsterdam provides many discussions and references of interest to students of the environment, such as the links to information on ecology and consumerism.

U.S. Geological Survey
http://www.usgs.gov

This site and its many links are replete with information and resources in environmental studies, from explanations of El Niño to discussion of concerns about water resources.

UNIT 1: The Global Environment: An Emerging World View

Alternative Energy Institute (AEI)
http://www.altenergy.org

The AEI will continue to monitor the transition from today's energy forms to the future in a "surprising journey of twists and turns." This site is the beginning of an incredible journey.

Earth Science Enterprise
http://www.earth.nasa.gov

Information about NASA's Mission to Planet Earth program and its Science of the Earth System can be found here. Surf to learn about satellites, El Niño, and even "strategic visions" of interest to environmentalists.

IISDnet
http://iisd.ca

The International Institute for Sustainable Development, a Canadian organization, presents information through gateways entitled Business and Sustainable Development, Develping Ideas, and Hot Topics. Linkages is its mutlimedia resource for environment and development policymakers.

National Geographic Society
http://www.nationalgeographic.com

Links to *National Geographic*'s huge archive are provided here. There is a great deal of material related to the atmosphere, the oceans, and other environmental topics.

Research and Reference (Library of Congress)
http://lcweb.loc.gov/rr/

This research and reference site of the Library of Congress will lead to invaluable information on different countries. It provides links to numerous publication, bibliographies, and guides in area studies that can be of great help to environmentalists.

Santa Fe Institute
http://acoma.santafe.edu

This home page of the Santa Fe Institute—a nonprofit, multidisciplinary research and education center—will lead to many interesting links related to its primary goal: to create a new kind of scientific research community, pursuing emerging science.

Solstice: Documents and Databases
http://solstice.crest.org/docndata.shtml

In this online source for sustainable energy information, the Center for Renewable Energy and Sustainable Technology (CREST) offers documents and databases on renewable energy, energy efficiency, and sustainable living. The site also offers related Web sites, case studies, and policy issues. Solstice also connects to CREST's Web presence.

United Nations
http://www.unsystem.org

Visit this official Web site Locator for the United Nations System of Organizations to get a sense of the scope of international environmental inquiry today. Various UN organizations concern themselves with everything from maritime law to habitat protection to agriculture.

United Nations Environment Programme (UNEP)
http://www.unep.ch

Consult this home page of UNEP for links to critical topics of concern to environmentalists, including desertification, migratory species, and the impact of trade on the environment. The site will direct you to useful databases and global resource information.

UNIT 2: The World's People: Population and Economy

The Hunger Project
http://www.thp.org

Browse through this nonprofit organization's site to explore the ways in which it attempts to achieve its goal: the sustainable end to global hunger through leadership at all levels of society. The

Hunger Project contends that the persistence of hunger is at the heart of the major security issues that are threatening our planet.

Poverty Mapping

http://www.povertymap.net

Poverty maps can quickly provide information on the spatial distribution of poverty. This site provides maps, graphics, data, publications, news, and links that provide the public with poverty mapping from the global to the subnational level.

World Health Organization

http://www.who.int

The home page of the World Health Organization provides links to a wealth of statistical and analytical information about health and the environment in the developing world.

World Population and Demographic Data

http://geography.about.com/cs/worldpopulation/

On this site, information about world population and additional demographic data for all the countries of the world is provided.

WWW Virtual Library: Demography & Population Studies

http://demography.anu.edu.au/VirtualLibrary/

This is a definitive guide to demography and population studies. A multitude of important links to information about global poverty and hunger can be found here.

UNIT 3: Energy: Present and Future Problems

Alliance for Global Sustainability (AGS)

http://www.global-sustainability (AGS)

The AGS is a cooperative venture seeking solutions to today's urgent and complex environmental problems. Research teams from four research universities study large-scale, multidisciplinary environmental problems that are faced by the world's ecosystems, economies, and societies.

Alternative Energy Institute, Inc.

http://www.altenergy.org

On this site created by a nonprofit organization, learn about the impacts of the use of conventional fuels on the environment. Also learn about research work on new forms of energy.

Communications for a Sustainable Future

http://csf.colorado.edu

This site will lead to information on topics in international environmental sustainability. It pays particular attention to the political economics of protecting the environment.

Energy and the Environment: Resources for a Networked World

http://zebu.uoregon.edu/energy.html

An extensive array of materials having to do with energy sources—both renewable and nonrenewable—as well as other topics of interest to students of the environment is found on this site.

Institute for Global Communication/EcoNet

http://www.igc.org/igc/gateway/

This environmentally friendly site provides links to dozens of governmental, organizational, and commercial sites having to do with energy sources. Resources address energy efficiency, renewable generating sources, global warming, and more.

Nuclear Power Introduction

http://library.thinkquest.org/17658/pdfs/nucintro.pdf

Information regarding alternative energy forms can be accessed here. There is a brief introduction to nuclear power and a link to maps that show where nuclear power plants exist.

U.S. Department of Energy

http://www.energy.gov

Scrolling through the links provided by this Department of Energy home page will lead to information about fossil fuels and a variety of sustainable/renewable energy sources.

UNIT 4: Biosphere: Endangered Species

Endangered Species

http://www.endangeredspecie.com/

This site provides a wealth of information on endangered species anywhere in the world. Links providing data on the causes, interesting facts, law issues, case studies, and other issues on endangered species are available.

Friends of the Earth

http://www.foe.co.uk/index.html

Friends of the Earth, a nonprofit organization based in the United Kingdom, pursues a number of campaigns to protect the Earth and its living creatures. This site has links to many important environmental sites, covering such broad topics as ozone depletion, soil erosion, and biodiversity.

Smithsonian Institution Web Site

http://www.si.edu

Looking through this site, which will provide access to many of the enormous resources of the Smithsonian, offers a sense of the biological diversity that is threatened by humans' unsound environmental policies and practices.

World Wildlife Federation (WWF)

http://www.wwf.org

This home page of the WWF leads to an extensive array of information links about endangered species, wildlife management and preservation, and more. It provides many suggestions for how to take an active part in protecting the biosphere.

UNIT 5: Resources: Land, Water, and Air

Agriculture Production Statistics

http://www.wri.org/statistics/fao-prd.html

The Food and Agriculture Organizatin of the UN (FAO) provides annual statistics, on a world-wide basis, on all important data of crop and livestock production. Coverage includes land use, irrigation, human population, index numbers of agriculture production, major crops, livestock numbers, livestock products, food supply, and means of production for individual countries, continents, and the world. Web links to the FAOSTAT Database Gateway are provided.

Global Climate Change

http://www.puc.state.oh.us/consumer/gcc/index.html

The goal of this PUCO (Public Utilities Commission of Ohio) site is to serve as a clearinghouse of information related to global climate change. Its extensive links provide an explanation of the science and chronology of global climate change, acronyms, definitions, and more.

National Oceanic and Atmospheric Administration (NOAA)

http://www.noaa.gov

Through this home page of NOAA, you can find information about coastal issues, fisheries, climate, and more.

National Operational Hydrologic Remote Sensing Center (NOHRSC)

http://www.nohrsc.nws.gov

Flood images are available at this site of the NOHRSC, which works with the U.S. National Weather Service to track weather-related information.

Virtual Seminar in Global Political Economy/Global Cities & Social Movements
http://csf.colorado.edu/gpe/gpe95b/resources.html

Links to subjects of interest in regional environmental studies, covering topics such as sustainable cities, megacities, and urban planning are available here. Many international nongovernmental organizations are included.

Websurfers Biweekly Earth Science Review
http://www.mindspring.com/~michaelg2/weeksreviews.html

This is a biweekly compilation of Internet sites devoted to the terrestrial and planetary sciences. It includes a list of hyperlinks to related earth science sites and news items.

UNIT 6: Pollution: The Hazards of Growth

IISDnet
http://www.iisd.org/default.asp

The International Institute for Sustainable Development's site presents information through links on business and sustainable development, developing ideas, and Hot Topics.

Persistant Organic Pollutants (POP)
http://irptc.unep.ch/pops/

Visit this site to learn more about persistant organic pollutants (POPs) and the issues and concerns surrounding them.

School of Labor and Industrial Relations (SLIR): Hot Links
http://www.lir.msu.edu/hotlinks/

Michigan State University's SLIR page connects to industrial relations sites throughout the world. It has links to U.S. government statistics, newspapers and libraries, international intergovernmental organizations, and more.

Space Research Institute
http://arc.iki.rssi.ru/Welcome.html

For a change of pace, browse through this home page of Russia's Space Research Institute for information on its Environment Monitoring Information Systems, the IKI Satellite Situation Center, and its Data Archive.

Worldwatch Institute
http://www.worldwatch.org

The Worldwatch Institute, dedicated to fostering the evolution of an environmentally sustainable society, presents this site with access to *World Watch Magazine* and *State of the World 2000*. Click on Alerts and Press Briefings for discussions of current problems.

We highly recommend that you review our Web site for expanded information and our other product lines. We are continually updating and adding links to our Web site in order to offer you the most usable and useful information that will support and expand the value of your Annual Editions. You can reach us at: *http://www.dushkin.com/annualeditions/*.

UNIT 1
The Global Environment: An Emerging World View

Unit Selections

1. **How Many Planets? A Survey of the Global Environment**, *The Economist*
2. **Forget Nature. Even Eden Is Engineered**, Andrew C. Revkin
3. **Crimes of (a) Global Nature**, Lisa Mastny and Hilary French

Key Points to Consider

- What are the connections between the attempts to develop sustainable systems and the quantity and quality of environmental data? Are there also relationships between data and the role of technology and economic systems in shaping the environmental future?

- In what ways are environmental changes engineered by humans, and how can human planning systems develop mechanisms to ensure that the human-designed environment will be a sustainable one?

- How well do international agreements work in controlling "environmental crimes" such as the taking of endangered species, hazardous waste dumping, or emissions of harmful pollutants? Are there ways in which international environmental accords could be made more enforceable?

 **Links: www.dushkin.com/online/**
These sites are annotated in the World Wide Web pages.

Alternative Energy Institute (AEI)
http://www.altenergy.org

Earth Science Enterprise
http://www.earth.nasa.gov

IISDnet
http://iisd.ca

National Geographic Society
http://www.nationalgeographic.com

Research and Reference (Library of Congress)
http://lcweb.loc.gov/rr/

Santa Fe Institute
http://acoma.santafe.edu

Solstice: Documents and Databases
http://solstice.crest.org/docndata.shtml

United Nations
http://www.unsystem.org

United Nations Environment Programme (UNEP)
http://www.unep.ch

More than three decades after the celebration of the first Earth Day in 1970, public apprehension over the environmental future of the planet has reached levels unprecedented even during the late 1960s and early 1970s "Age of Aquarius." No longer are those concerned about the environment dismissed as "ecofreaks" and "tree-huggers." Many serious scientists have joined the rising clamor for environmental protection, as have the more traditional environmentally conscious public-interest groups. There are a number of reasons for this increased environmental awareness. Some of these reasons arise from environmental events; it is, for example, becoming increasingly difficult to deny the effects of global warming. But more arise simply from the process of globalization: the increasing unity of the world's economic, social, and information systems. Hailed by many as the salvation of the future, globalization has done little to make the world a better or safer place. Diseases once confined to specific regions now have the capacity for widespread dissemination. Increasing human mobility has allowed human-caused disruptions to political, cultural, and economic systems to spread, and acts of terrorism now take place in locations once thought safe from such manifestations of hatred and despair. On the more positive side, the expansion of global information systems has fostered a maturation of concepts about the global nature of environmental processes.

Much of what has been learned through this increased information flow, particularly by American observers, has been of the environmentally ravaged world behind the old Iron Curtain—a chilling forecast of what other industrialized regions as well as the developing countries can become in the near future unless strict international environmental measures are put in place. For perhaps the first time ever, countries are beginning to recognize that environmental problems have no boundaries and that international cooperation is the only way to solve them.

The subtitle of this first unit, "An Emerging World View," is an optimistic assessment of the future: a future in which less money is spent on defense and more on environmental protection and cleanup—a new world order in which political influence might be based more on leadership in environmental and economic issues than on military might. It is probably far too early to make such optimistic predictions, to conclude that the world's nations—developed and underdeveloped—will begin to recognize that Earth's environment is a single unit. Thus far those nations have shown no tendency to recognize that humankind is a single unit and that what harms one harms all. The recent emergence of wide-scale terrorism as an instrument of political and social policy is evidence of such a failure of recognition. Nevertheless, there is a growing international realization—aided by the information superhighway—that we are all, as environmental activists have been saying for decades, inhabitants of Spaceship Earth and will survive or succumb together.

The articles selected for this unit have been chosen to illustrate the increasingly global perspective on environmental problems and the degree to which their solutions must be linked to political, economic, and social problems and solutions. In the lead piece of the unit, "How Many Planets?" the editors of *The Economist* attempt an analysis of what they admit is a very slippery subject by beginning with the observation that "it comes as a shock to discover how little information there is on the environment." They note the lip service paid everywhere to the concept of sustainability and acknowledge that economic growth and environmental health are not mutually inconsistent but that a great deal more work is necessary to make them compatible. They also conclude that governments, corporations, and individuals are more prepared now to think about how to use the planet than they were even 10 years ago.

Planning for the wise use of the planet is the subject of the next selection in the unit. In "Forget Nature. Even Eden Is Engineered," science writer Andrew Revkin directs his attention toward the interconnectedness of environmental systems and toward new concepts and ways of thinking about the environment and human impact. Science has finally recognized, says Revkin, the degree to which people have significantly altered global atmospheric systems and the biosphere. With that recog-

nition comes the first real chance to begin balancing economic development with a sustainable future on Earth for human beings. Whatever we make of that chance, the world of the future will be one in which, through technology and sheer numbers, the forces of human society rather than the forces of nature will be the chief architect.

Some of the impediments in making the human-engineered Earth one in which we would want to live are discussed in the final article in this unit. Lisa Mastny and Hilary French of the Worldwatch Institute describe the difficulties of developing and enforcing international environmental treaties and other agreements. In "Crimes of (a) Global Nature," Mastny and French focus on three types of international environmental agreements: those related to the taking and sale of endangered terrestrial and aquatic species; those governing the disposal of toxic and hazardous waste materials; and those mandating restrictions on the manufacture and use of certain chemicals that damage portions of the atmosphere. It is easy enough for countries to agree to international environmental accords. It is much more difficult for officials to enforce those same accords and some of the authors' statistics are staggering—for example, illegally cut wood accounted for 65 percent of the world's supply in 2000. The future of the environment depends not just upon international agreement but upon enforcement.

How many planets?

A survey of the global environment

The great race

Growth need not be the enemy of greenery. But much more effort is required to make the two compatible, says Vijay Vaitheeswaran

SUSTAINABLE development is a dangerously slippery concept. Who could possibly be against something that invokes such alluring images of untouched wildernesses and happy creatures? The difficulty comes in trying to reconcile the "development" with the "sustainable" bit: look more closely, and you will notice that there are no people in the picture.

That seems unlikely to stop a contingent of some of 60,000 world leaders, businessmen, activists, bureaucrats and journalists from travelling to South Africa next month for the UN-sponsored World Summit on Sustainable Development in Johannesburg. Whether the summit achieves anything remains to be seen, but at least it is asking the right questions. This survey will argue that sustainable development cuts to the heart of mankind's relationship with nature—or, as Paul Portney of Resources for the Future, an American think-tank, puts it, "the great race between development and degradation". It will also explain why there is reason for hope about the planet's future.

The best way known to help the poor today—economic growth—has to be handled with care, or it can leave a degraded or even devastated natural environment for the future. That explains why ecologists and economists have long held diametrically opposed views on development. The difficult part is to work out what we owe future generations, and how to reconcile that moral obligation with what we owe the poorest among us today.

It is worth recalling some of the arguments fielded in the run-up to the big Earth Summit in Rio de Janeiro a decade ago. A publication from UNESCO, a United Nations agency, offered the following vision of the future: "Every generation should leave water, air and soil resources as pure and unpolluted as when it came on earth. Each generation should leave undiminished all the species of animals it found existing on earth." Man, that suggests, is but a strand in the web of life, and the natural order is fixed and supreme. Put earth first, it seems to say.

Robert Solow, an economist at the Massachusetts Institute of Technology, replied at the time that this was "fundamentally the wrong way to go", arguing that the obligation to the future is "not to leave the world as we found it in detail, but rather to leave the option or the capacity to be as well off as we are." Implicit in that argument is the seemingly hard-hearted notion of "fungibility": that natural resources, whether petroleum or giant pandas, are substitutable.

Rio's fatal flaw

Champions of development and defenders of the environment have been locked in battle ever since a UN summit in Stockholm launched the sustainable-development debate three decades ago. Over the years, this debate often pitted indignant politicians and social activists from the poor world against equally indignant politicians and greens from the rich world. But by the time the Rio summit came along, it seemed they had reached a truce. With the help of a committee of grandees led by Gro Harlem Brundtland, a former Norwegian prime minister, the interested parties struck a deal in 1987: development and the environment, they declared, were inextricably linked. That compromise generated a good deal of euphoria. Green groups grew concerned over poverty, and development charities waxed lyrical about greenery. Even the World Bank joined in. Its World Development Report in 1992 gushed about "win-win" strategies, such as ending environmentally harmful subsidies, that would help both the economy and the environment.

By nearly universal agreement, those grand aspirations have fallen flat in the decade since that summit. Little headway has been made with environmental problems such as climate change and loss of biodiversity. Such progress as has been achieved has been largely due to three factors that this survey will explore in later sections: more decision-making at local level, technological innovation, and the rise of market forces in environmental matters.

The main explanation for the disappointment—and the chief lesson for those about to gather in South Africa—is that Rio overreached itself. Its participants were so anxious to reach a political consensus that they agreed to the Brundtland definition of sustainable development, which Daniel Esty of Yale University thinks has turned into "a buzz-word largely devoid of content". The biggest mistake, he reckons, is that it slides over the difficult trade-offs between environment and development in the real world. He is careful to note that there are plenty of cases where those goals are linked—but also many where they are not: "Environmental and economic policy goals are distinct, and the actions needed to achieve them are not the same."

No such thing as win-win

To insist that the two are "impossible to separate", as the Brundtland commission claimed, is nonsense. Even the World Bank now accepts that its much-trumpeted 1992 report was much too optimistic. Kristalina Georgieva, the Bank's director for the environment, echoes comments from various colleagues when she says: "I've never seen a real win-win in my life. There's always somebody, usually an elite group grabbing rents, that loses. And we've learned in the past decade that those losers fight hard to make sure that technically elegant win-win policies do not get very fat."

So would it be better to ditch the concept of sustainable development altogether? Probably not. Even people with their feet firmly planted on the ground think one aspect of it is worth salvaging: the emphasis on the future.

Nobody would accuse John Graham of jumping on green bandwagons. As an official in President George Bush's Office of Management and Budget, and previously as head of Harvard University's Centre for Risk Analysis, he has built a reputation for evidence-based policymaking. Yet he insists sustainable development is a worthwhile concept: "It's good therapy for the tunnel vision common in government ministries, as it forces integrated policymaking. In practical terms, it means that you have to take economic cost-benefit trade-offs into account in environmental laws, and keep environmental trade-offs in mind with economic development."

Jose Maria Figueres, a former president of Costa Rica, takes a similar view. "As a politician, I saw at first hand how often policies were dictated by short-term considerations such as elections or partisan pressure. Sustainability is a useful template to align short-term policies with medium- to long-term goals."

It is not only politicians who see value in saving the sensible aspects of sustainable development. Achim Steiner,

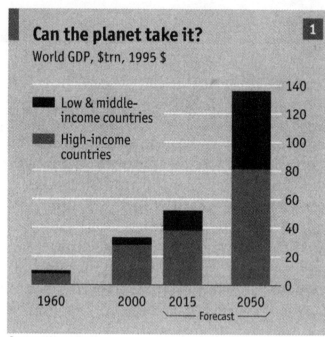

Can the planet take it?
World GDP, $trn, 1995 $

Source: World Bank

head of the International Union for the Conservation of Nature, the world's biggest conservation group, puts it this way: "Let's be honest: greens and businesses do not have the same objective, but they can find common ground. We look for pragmatic ways to save species. From our own work on the ground on poverty, our members—be they bird watchers or passionate ecologists—have learned that 'sustainable use' is a better way to conserve."

Sir Robert Wilson, boss of Rio Tinto, a mining giant, agrees. He and other business leaders say it forces hard choices about the future out into the open: "I like this concept because it frames the trade-offs inherent in a business like ours. It means that single-issue activism is simply not as viable."

Kenneth Arrow and Larry Goulder, two economists at Stanford University, suggest that the old ideological enemies are converging: "Many economists now accept the idea that natural capital has to be valued, and that we need to account for ecosystem services. Many ecologists now accept that prohibiting everything in the name of protecting nature is not useful, and so are being selective." They think the debate is narrowing to the more empirical question of how far it is possible to substitute natural capital with the man-made sort, and specific forms of natural capital for one another.

The job for Johannesburg

So what can the Johannesburg summit contribute? The prospects are limited. There are no big, set-piece political treaties to be signed as there were at Rio. America's acrimonious departure from the Kyoto Protocol, a UN treaty on climate change, has left a bitter taste in many mouths. And the final pre-summit gathering, held in early June in

4

Indonesia, broke up in disarray. Still, the gathered worthies could usefully concentrate on a handful of areas where international co-operation can help deal with environmental problems. Those include improving access for the poor to cleaner energy and to safe drinking water, two areas where concerns about human health and the environmental overlap. If rich countries want to make progress, they must agree on firm targets and offer the money needed to meet them. Only if they do so will poor countries be willing to cooperate on problems such as global warming that rich countries care about.

That seems like a modest goal, but it just might get the world thinking seriously about sustainability once again. If the Johannesburg summit helps rebuild a bit of faith in international environmental cooperation, then it will have been worthwhile. Minimising the harm that future economic growth does to the environment will require the rich world to work hand in glove with the poor world—which seems nearly unimaginable in today's atmosphere poisoned by the shortcomings of Rio and Kyoto.

To understand why this matters, recall that great race between development and degradation. Mankind has stayed comfortably ahead in that race so far, but can it go on doing so? The sheer magnitude of the economic growth that is hoped for in the coming decades (see chart 1) makes it seem inevitable that the clashes between man-

kind and nature will grow worse. Some are now asking whether all this economic growth is really necessary or useful in the first place, citing past advocates of the simple life.

"God forbid that India should ever take to industrialism after the manner of the West... It took Britain half the resources of the planet to achieve this prosperity. How many planets will a country like India require?", Mahatma Gandhi asked half a century ago. That question encapsulated the bundle of worries that haunts the sustainable-development debate to this day. Today, the vast majority of Gandhi's countrymen are still living the simple life—full of simple misery, malnourishment and material want. Grinding poverty, it turns out, is pretty sustainable.

If Gandhi were alive today, he might look at China next door and find that the country, once as poor as India, has been transformed beyond recognition by two decades of roaring economic growth. Vast numbers of people have been lifted out of poverty and into middle-class comfort. That could prompt him to reframe his question: how many planets will it take to satisfy China's needs if it ever achieves profligate America's affluence? One green group reckons the answer is three. The next section looks at the environmental data that might underpin such claims. It makes for alarming reading—though not for the reason that first springs to mind.

Flying blind

It comes as a shock to discover how little information there is on the environment

WHAT is the true state of the planet? It depends from which side you are peering at it. "Things are really looking up," comes the cry from one corner (usually overflowing with economists and technologists), pointing to a set of rosy statistics. "Disaster is nigh," shouts the other corner (usually full of ecologists and environmental lobbyists), holding up a rival set of troubling indicators.

According to the optimists, the 20th century marked a period of unprecedented economic growth that lifted masses of people out of abject poverty. It also brought technological innovations such as vaccines and other advances in public health that tackled many preventable diseases. The result has been a breath-taking enhancement of human welfare and longer, better lives for people everywhere on earth (see chart 2).

At this point, the pessimists interject: "Ah, but at what ecological cost?" They note that the economic growth which made all these gains possible sprang from the rapid spread of industrialisation and its resource-guzzling cousins, urbanisation, motorisation and electrification. The earth provided the necessary raw materials, ranging from coal to pulp to iron. Its ecosystems—rivers,

seas, the atmosphere—also absorbed much of the noxious fallout from that process. The sheer magnitude of ecological change resulting directly from the past century's economic activity is remarkable (see table 3).

To answer that Gandhian question about how many planets it would take if everybody lived like the West, we need to know how much—or how little—damage the West's transformation from poverty to plenty has done to the planet to date. Economists point to the remarkable improvement in local air and water pollution in the rich world in recent decades. "It's Getting Better All the Time", a cheerful tract co-written by the late Julian Simon, insists that: "One of the greatest trends of the past 100 years has been the astonishing rate of progress in reducing almost every form of pollution." The conclusion seems unavoidable: "Relax! If we keep growing as usual, we'll inevitably grow greener."

The ecologically minded crowd takes a different view. "GEO3", a new report from the United Nations Environment Programme, looks back at the past few decades and sees much reason for concern. Its thoughtful boss, Klaus Töpfer (a former German environment minister),

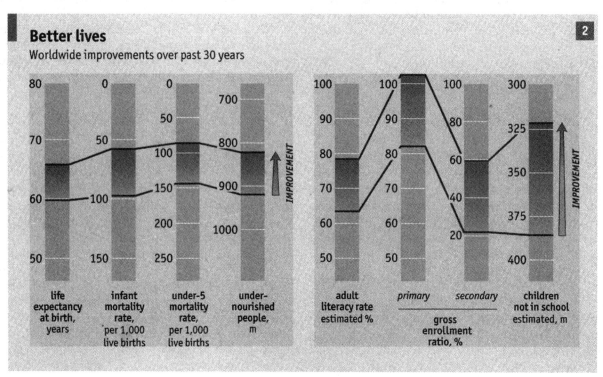

Better lives
Worldwide improvements over past 30 years

2

Source: UNEP, GEO3

insists that his report is not "a document of doom and gloom". Yet, in summing it up, UNEP decries "the declining environmental quality of planet earth", and wags a finger at economic prosperity: "Currently, one-fifth of the world's population enjoys high, some would say excessive, levels of affluence." The conclusion seems unavoidable: "Panic! If we keep growing as usual, we'll inevitably choke the planet to death."

"People and Ecosystems", a collaboration between the World Resources Institute, the World Bank and the United Nations, tried to gauge the condition of ecosystems by examining the goods and services they produce—food, fibre, clean water, carbon storage and so on—and their capacity to continue producing them. The authors explain why ecosystems matter: half of all jobs worldwide are in agriculture, forestry and fishing, and the output from those three commodity businesses still dominates the economies of a quarter of the world's countries.

The report reached two chief conclusions after surveying the best available environmental data. First, a number of ecosystems are "fraying" under the impact of human activity. Second, ecosystems in future will be less able than in the past to deliver the goods and services human life depends upon, which points to unsustainability. But it took care to say: "It's hard, of course, to know what will be truly sustainable." The reason this collection of leading experts could not reach a firm conclusion was that, remarkably, much of the information they needed was in-

complete or missing altogether: "Our knowledge of ecosystems has increased dramatically, but it simply has not kept pace with our ability to alter them."

Another group of experts, this time organised by the World Economic Forum, found itself similarly frustrated. The leader of that project, Daniel Esty of Yale, exclaims, throwing his arms in the air: "Why hasn't anyone done careful environmental measurement before? Businessmen always say, 'what matters gets measured.' Social scientists started quantitative measurement 30 years ago, and even political science turned to hard numbers 15 years ago. Yet look at environmental policy, and the data are lousy."

Gaping holes

At long last, efforts are under way to improve environmental data collection. The most ambitious of these is the Millennium Ecosystem Assessment, a joint effort among leading development agencies and environmental groups. This four-year effort is billed as an attempt to establish systematic data sets on all environmental matters across the world. But one of the researchers involved grouses that it "has very, very little new money to collect or analyse new data". It seems astonishing that governments have been making sweeping decisions on environmental policy for decades without such a baseline in the first place.

One positive sign is the growing interest of the private sector in collecting environmental data. It seems plain that leaving the task to the public sector has not worked.

A century that changed the world   3

Change between 1890 (=1) and 1990s

Industrial output	40
Marine fish catch	35
Carbon dioxide emissions	17
Energy use	16
World economy	14
World urban population	13
Coal production	7
Air pollution	5
Irrigated area	5
World population	4
Horse population	1.1
Bird and mammal species	0.99
Forest area	0.8
Blue-whale population	0.0025

Source: "Something New Under the Sun" by John McNeill

Information on the environment comes far lower on the bureaucratic pecking order than data on education or social affairs, which tend to be overseen by ministries with bigger budgets and more political clout. A number of countries, ranging from New Zealand to Austria, are now looking to the private sector to help collect and manage data in areas such as climate. Development banks are also considering using private contractors to monitor urban air quality, in part to get around the corruption and apathy in some city governments.

"I see a revolution in environmental data collection coming because of computing power, satellite mapping, remote sensing and other such information technologies," says Mr Esty. The arrival of hard data in this notoriously fuzzy area could cut down on environmental disputes by reducing uncertainty. One example is the long-running squabble between America's mid-western states, which rely heavily on coal, and the north-eastern states, which suffer from acid rain. Technology helped disprove claims by the mid-western states that New York's problems all resulted from home-grown pollution.

The arrival of good data would have other benefits as well, such as helping markets to work more robustly: witness America's pioneering scheme to trade emissions of sulphur dioxide, made possible by fancy equipment capable of monitoring emissions in real time. Mr Esty raises an even more intriguing possibility: "Like in the American West a hundred years ago, when barbed wire helped establish rights and prevent overgrazing, information technology can help establish 'virtual barbed wire' that secures property rights and so prevents over-exploitation of the commons." He points to fishing in the waters between Australia and New Zealand, where tracking and monitoring devices have reduced over-exploitation.

Best of all, there are signs that the use of such fancy technology will not be confined to rich countries. Calestous Juma of Harvard University shares Mr Esty's excitement about the possibility of such a technology-driven revolution even in Africa: "In the past, the only environmental 'database' we had in Africa was our grandmothers. Now, with global information systems and such, the potential is enormous." Conservationists in Namibia, for example, already use satellite tracking to keep count of their elephants. Farmers in Mali receive satellite updates about impending storms on hand-wound radios. Mr Juma thinks the day is not far off when such technology, combined with ground-based monitoring, will help Africans measure trends in deforestation, soil erosion and climate change, and assess the effects on their local environment.

Make a start

That is at once a sweeping vision and a modest one. Sweeping, because it will require heavy investment in both sophisticated hardware and nuts-and-bolts information infrastructure on the ground to make sense of all these new data. As the poor world clearly cannot afford to pay for all this, the rich world must help—partly for altruistic reasons, partly with the selfish aim of discovering in good time whether any global environmental calamities are in the making. A number of multilateral agencies now say they are willing to invest in this area as a "neglected global public good"—neglected especially by those agencies themselves. Even President Bush's administration has recently indicated that it will give environmental satellite data free to poor countries.

But that vision is also quite a modest one. Assuming that this data "revolution" does take place, all it will deliver is a reliable assessment of the health of the planet today. We will still not be able to answer the broader question of whether current trends are sustainable or not.

To do that, we need to look more closely at two very different sorts of environmental problems: global crises and local troubles. The global sort is hard to pin down, but can involve irreversible changes. The local kind is common and can have a big effect on the qualify of life, but is usually reversible. Data on both are predictably inadequate. We turn first to the most elusive environmental problem of all, global warming.

Blowing hot and cold

Climate change may be slow and uncertain, but that is no excuse for inaction

WHAT would Winston Churchill have done about climate change? Imagine that Britain's visionary wartime leader had been presented with a potential time bomb capable of wreaking global havoc, although not certain to do so. Warding it off would require concerted global action and economic sacrifice on the home front. Would he have done nothing?

Not if you put it that way. After all, Churchill did not dismiss the Nazi threat for lack of conclusive evidence of Hitler's evil intentions. But the answer might be less straightforward if the following provisos had been added: evidence of this problem would remain cloudy for decades; the worst effects might not be felt for a century; but the costs of tackling the problem would start biting immediately. That, in a nutshell, is the dilemma of climate change. It is asking a great deal of politicians to take action on behalf of voters who have not even been born yet.

One reason why uncertainty over climate looks to be with us for a long time is that the oceans, which absorb carbon from the atmosphere, act as a time-delay mechanism. Their massive thermal inertia means that the climate system responds only very slowly to changes in the composition of the atmosphere. Another complication arises from the relationship between carbon dioxide (CO_2), the principal greenhouse gas (GHG), and sulphur dioxide (SO_2), a common pollutant. Efforts to reduce man-made emissions of GHGs by cutting down on fossil-fuel use will reduce emissions of both gases. The reduction in CO_2 will cut warming, but the concurrent SO_2 cut may mask that effect by contributing to the warming.

There are so many such fuzzy factors—ranging from aerosol particles to clouds to cosmic radiation—that we are likely to see disruptions to familiar climate patterns for many years without knowing why they are happening or what to do about them. Tom Wigley, a leading climate scientist and member of the UN's Intergovernmental Panel on Climate Change (IPCC), goes further. He argues in an excellent book published by the Aspen Institute, "US Policies on Climate Change: What Next?", that whatever policy changes governments pursue, scientific uncertainties will "make it difficult to detect the effects of such changes, probably for many decades."

As evidence, he points to the negligible short- to medium-term difference in temperature resulting from an array of emissions "pathways" on which the world could choose to embark if it decided to tackle climate change (see chart 4). He plots various strategies for reducing GHGs (including the Kyoto one) that will lead in the next century to the stabilisation of atmospheric concentrations of CO_2 at 550 parts per million (ppm). That is roughly double the level which prevailed in pre-industrial times, and

is often mooted by climate scientists as a reasonable target. But even by 2040, the temperature differences between the various options will still be tiny—and certainly within the magnitude of natural climatic variance. In short, in another four decades we will probably still not know if we have over- or undershot.

Ignorance is not bliss

However, that does not mean we know nothing. We do know, for a start, that the "greenhouse effect" is real: without the heat-trapping effect of water vapour, CO_2, methane and other naturally occurring GHGs, our planet would be a lifeless 30°C or so colder. Some of these GHG emissions are captured and stored by "sinks", such as the oceans, forests and agricultural land, as part of nature's carbon cycle.

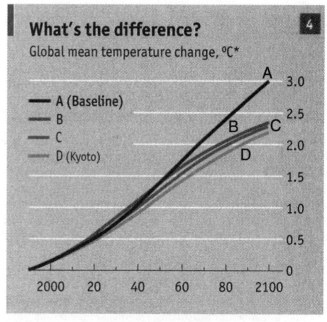

What's the difference?
Global mean temperature change, °C*

A (Baseline)
B
C
D (Kyoto)

*Various plausible scenarios for stabilising atmospheric concentration of CO_2 at 550 parts per million

Source: Tom Wigley

We also know that since the industrial revolution began, mankind's actions have contributed significantly to that greenhouse effect. Atmospheric concentrations of GHGs have risen from around 280ppm two centuries ago to around 370ppm today, thanks chiefly to mankind's use of fossil fuels and, to a lesser degree, to deforestation and other land-use changes. Both surface temperatures and sea levels have been rising for some time.

There are good reasons to think temperatures will continue rising. The IPCC has estimated a likely range for that increase of 1.4°C–5.8°C over the next century, although the lower end of that range is more likely. Since what matters is not just the absolute temperature level but the rate of change as well, it makes sense to try to slow down the increase.

The worry is that a rapid rise in temperatures would lead to climate changes that could be devastating for many (though not all) parts of the world. Central America, most of Africa, much of south Asia and northern China could all be hit by droughts, storms and floods and otherwise made miserable. Because they are poor and have the misfortune to live near the tropics, those most likely to be affected will be least able to adapt.

The colder parts of the world may benefit from warming, but they too face perils. One is the conceivable collapse of the Atlantic "conveyor belt", a system of currents that gives much of Europe its relatively mild climate; if temperatures climb too high, say scientists, the system may undergo radical changes that damage both Europe and America. That points to the biggest fear: warming may trigger irreversible changes that transform the earth into a largely uninhabitable environment.

Given that possibility, extremely remote though it is, it is no comfort to know that any attempts to stabilise atmospheric concentrations of GHGs at a particular level will take a very long time. Because of the oceans' thermal inertia, explains Mr Wigley, even once atmospheric concentrations of GHGs are stabilised, it will take decades or centuries for the climate to follow suit. And even then the sea level will continue to rise, perhaps for millennia.

This is a vast challenge, and it is worth bearing in mind that mankind's contribution to warming is the only factor that can be controlled. So the sooner we start drawing up a long-term strategy for climate change, the better.

What should such a grand plan look like? First and foremost, it must be global. Since CO_2 lingers in the atmosphere for a century or more, any plan must also extend across several generations.

The plan must recognise, too, that climate change is nothing new: the climate has fluctuated through history, and mankind has adapted to those changes—and must continue doing so. In the rich world, some of the more obvious measures will include building bigger dykes and flood defences. But since the most vulnerable people are those in poor countries, they too have to be helped to adapt to rising seas and unpredictable storms. Infrastructure improvements will be useful, but the best investment will probably be to help the developing world get wealthier.

It is essential to be clear about the plan's long-term objective. A growing chorus of scientists now argues that we need to keep temperatures from rising by much more than 2–3°C in all. That will require the stabilisation of atmospheric concentrations of GHGs. James Edmonds of the University of Maryland points out that because of the long life of CO_2, stabilisation of CO_2 concentrations is not at all the same thing as stabilisation of CO_2 emissions. That, says Mr Edmonds, points to an unavoidable conclusion: "In the very long term, global net CO_2 emissions must eventually peak and gradually decline toward zero, regardless of whether we go for a target of 350ppm or 1,000ppm."

A low-carbon world

That is why the long-term objective for climate policy must be a transition to a low-carbon energy system. Such a transition can be very gradual and need not necessarily lead to a world powered only by bicycles and windmills, for two reasons that are often overlooked.

One involves the precise form in which the carbon in the ground is distributed. According to Michael Grubb of the Carbon Trust, a British quasi-governmental body, the long-term problem is coal. In theory, we can burn all of the conventional oil and natural gas in the ground and still meet the most ambitious goals for tackling climate change. If we do that, we must ensure that the far greater amounts of carbon trapped as coal (and unconventional resources like tar sands) never enter the atmosphere.

The snag is that poor countries are likely to continue burning cheap domestic reserves of coal for decades. That suggests the rich world should speed the development and diffusion of "low carbon" technologies using the energy content of coal without releasing its carbon into the atmosphere. This could be far off, so it still makes sense to keep a watchful eye on the soaring carbon emissions from oil and gas.

The other reason, as Mr Edmonds took care to point out, is that it is net emissions of CO_2 that need to peak and decline. That leaves scope for the continued use of fossil fuels as the main source of modern energy if only some magical way can be found to capture and dispose of the associated CO_2. Happily, scientists already have some magic in the works.

One option is the biological "sequestration" of carbon in forests and agricultural land. Another promising idea is capturing and storing CO_2—underground, as a solid or even at the bottom of the ocean. Planting "energy crops" such as switch-grass and using them in conjunction with sequestration techniques could even result in negative net CO_2 emissions, because such plants use carbon from the atmosphere. If sequestration is combined with techniques for stripping the hydrogen out of this hydrocarbon, then coal could even offer a way to sustainable hydrogen energy.

But is anyone going to pay attention to these long-term principles? After all, over the past couple of years all participants in the Kyoto debate have excelled at producing short-sighted, selfish and disingenuous arguments. And the political rift continues: the EU and Japan pushed ahead with ratification of the Kyoto treaty a month ago, whereas President Bush reaffirmed his opposition.

However, go back a decade and you will find precisely those principles enshrined in a treaty approved by the elder George Bush and since reaffirmed by his son: the UN Framework Convention on Climate Change (FCCC). This treaty was perhaps the most important outcome of the Rio summit, and it remains the basis for the international climate-policy regime, including Kyoto.

The treaty is global in nature and long-term in perspective. It commits signatories to pursuing "the stabilisation of GHG concentrations in the atmosphere at a level that would prevent dangerous interference with the climate system." Note that the agreement covers GHG concentrations, not merely emissions. In effect, this commits even gas-guzzling America to the goal of declining emissions.

Better than Kyoto

Crucially, the FCCC treaty not only lays down the ends but also specifies the means: any strategy to achieve stabilisation of GHG concentrations, it insists, "must not be disruptive of the global economy". That was the stumbling block for the Kyoto treaty, which is built upon the FCCC agreement: its targets and timetables proved unrealistic.

Any revised Kyoto treaty or follow-up accord (which must include the United States and the big developing countries) should rest on the three basic pillars. First, governments everywhere (but especially in Europe) must understand that a reduction in emissions has to start modestly. That is because the capital stock involved in the global energy system is vast and long-lived, so a dash to scrap fossil-fuel production would be hugely expensive. However, as Mr Grubb points out, that pragmatism must be flanked by policies that encourage a switch to low-carbon technologies when replacing existing plants.

Second, governments everywhere (but especially in America) must send a powerful signal that carbon is going out of fashion. The best way to do this is to levy a carbon tax. However, whether it is done through taxes, mandated restrictions on GHG emissions or market mechanisms is less important than that the signal is sent clearly, forcefully and unambiguously. This is where President Bush's mixed signals have done a lot of harm: America's industry, unlike Europe's, has little incentive to invest in low-carbon technology. The irony is that even some coal-fired utilities in America are now clamouring for CO_2 regulation so that they can invest in new plants with confidence.

The third pillar is to promote science and technology. That means encouraging basic climate and energy research, and giving incentives for spreading the results.

Rich countries and aid agencies must also find ways to help the poor world adapt to climate change. This is especially important if the world starts off with small cuts in emissions, leaving deeper cuts for later. That, observes Mr Wigley, means that by mid-century "very large investments would have to have been made—and yet the 'return' on these investments would not be visible. Continued investment is going to require more faith in climate science than currently appears to be the case."

Even a visionary like Churchill might have lost heart in the face of all this uncertainty. Nevertheless, there is a glimmer of hope that today's peacetime politicians may rise to the occasion.

Miracles sometimes happen

Two decades ago, the world faced a similar dilemma: evidence of a hole in the ozone layer. Some inconclusive signs suggested that it was man-made, caused by the use of chlorofluorocarbons (CFCs). There was the distant threat of disaster, and the knowledge about a concerted global response was required. Industry was reluctant at first, yet with leadership from Britain and America the Montreal Protocol was signed in 1987. That deal has proved surprisingly successful. The manufacture of CFCs is nearly phased out, and there are already signs that the ozone layer is on the way to recovery.

This story holds several lessons for the admittedly far more complex climate problem. First, it is the rich world which has caused the problem and which must lead the way in solving it. Second, the poor world must agree to help, but is right to insist on being given time—as well as money and technology—to help it adjust. Third, industry holds the key: in the ozone-depletion story, it was only after DuPont and ICI broke ranks with the rest of the CFC manufacturers that a deal became possible. On the climate issue, BP and Shell have similarly broken ranks with Big Oil, but the American energy industry—especially the coal sector—remains hostile.

The final lesson is the most important: that the uncertainty surrounding a threat such as climate change is no excuse for inaction. New scientific evidence shows that the threat from ozone depletion had been much deadlier than was thought at the time when the world decided to act. Churchill would surely have approved.

Local difficulties

Greenery is for the poor too, particularly on their own doorstep

WHY should we care about the environment? Ask a European, and he will probably point to global warming. Ask the two little boys playing outside a newsstand in Da Shilan, a shabby neighbourhood in the heart of Beijing,

and they will tell you about the city's notoriously foul air: "It's bad—like a virus!"

Given all the media coverage in the rich world, people there might believe that global scares are the chief envi-

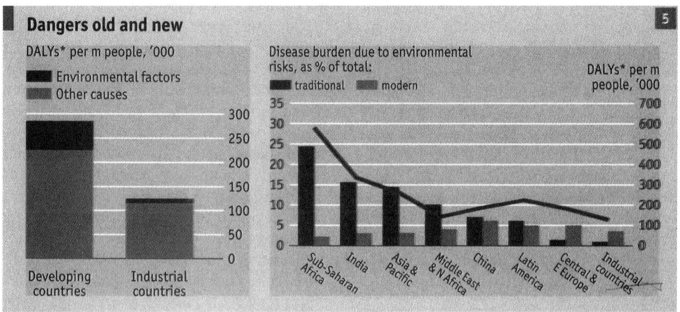

Dangers old and new

DALYs* per m people, '000
- ■ Environmental factors
- ■ Other causes

Disease burden due to environmental risks, as % of total:
- ■ traditional ■ modern

DALYs* per m people, '000

*Disability Adjusted Life Years—number of years lived with disability and years lost to premature death
Source: World Bank

ronmental problems facing humanity today. They would be wrong. Partha Dasgupta, an economics professor at Cambridge University, thinks the current interest in global, future-oriented problems has "drawn attention away from the economic misery and ecological degradation endemic in large parts of the world today. Disaster is not something for which the poorest have to wait; it is a frequent occurrence."

Every year in developing countries, a million people die from urban air pollution and twice that number from exposure to stove smoke inside their homes. Another 3m unfortunates die prematurely every year from water-related diseases. All told, premature deaths and illnesses arising from environmental factors account for about a fifth of all diseases in poor countries, bigger than any other preventable factor, including malnutrition. The problem is so serious that Ian Johnson, the World Bank's vice-president for the environment, tells his colleagues, with a touch of irony, that he is really the bank's vice-president for health: "I say tackling the underlying environmental causes of health problems will do a lot more good than just more hospitals and drugs."

The link between environment and poverty is central to that great race for sustainability. It is a pity, then, that several powerful fallacies keep getting in the way of sensible debate. One popular myth is that trade and economic growth make poor countries' environmental problems worse. Growth, it is said, brings with it urbanisation, higher energy consumption and industrialisation—all factors that contribute to pollution and pose health risks.

In a static world, that would be true, because every new factory causes extra pollution. But in the real world, economic growth unleashes many dynamic forces that, in the longer run, more than offset that extra pollution. As

chart 5 makes clear, traditional environmental risks (such as water-borne diseases) cause far more health problems in poor countries than modern environmental risks (such as industrial pollution).

Rigged rules

However, this is not to say that trade and economic growth will solve all environmental problems. Among the reasons for doubt are the "perverse" conditions under which world trade is carried on, argues Oxfam. The British charity thinks the rules of trade are "unfairly rigged against the poor", and cites in evidence the enormous subsidies lavished by rich countries on industries such as agriculture, as well as trade protection offered to manufacturing industries such as textiles. These measurements hurt the environment because they force the world's poorest countries to rely heavily on commodities—a particularly energy-intensive and ungreen sector.

Mr Dasgupta argues that this distortion of trade amounts to a massive subsidy of rich-world consumption paid by the world's poorest people. The most persuasive critique of all goes as follows: "Economic growth is not sufficient for turning environmental degradation around. If economic incentives facing producers and consumers do not change with higher incomes, pollution will continue to grow unabated with the growing scale of economic activity." Those words come not from some anti-globalist green group, but from the World Trade Organisation.

Another common view is that poor countries, being unable to afford greenery, should pollute now and clean up later. Certainly poor countries should not be made to adopt American or European environmental standards. But there is evidence to suggest that poor countries can and should try to tackle some environmental problems now, rather than wait till they have become richer.

This so-called "smart growth" strategy contradicts conventional wisdom. For many years, economists have observed that as agrarian societies industrialised, pollution increased at first, but as the societies grew wealthier it declined again. The trouble is that this applies only to some pollutants, such as sulphur dioxide, but not to others, such as carbon dioxide. Even more troublesome, those smooth curves going up, then down, turn out to be misleading. They are what you get when you plot data for poor and rich countries together at a given moment in time, but actual levels of various pollutants in any individual country plotted over time wiggle around a lot more. This suggests that the familiar bell-shaped curve reflects no immutable law, and that intelligent government policies might well help to reduce pollution levels even while countries are still relatively poor.

Developing countries are getting the message. From Mexico to the Philippines, they are now trying to curb the worst of the air and water pollution that typically accompanies industrialisation. China, for example, was persuaded by outside experts that it was losing so much potential economic output through health troubles caused by pollution (according to one World Bank study, somewhere between 3.5% and 7.7% of GDP) that tackling it was cheaper than ignoring it.

One powerful—and until recently ignored—weapon in the fight for a better environment is local people. Old-fashioned paternalists in the capitals of developing countries used to argue that poor villagers could not be relied on to look after natural resources. In fact, much academic research has shown that the poor are more often victims than perpetrators of resource depletion: it tends to be rich locals or outsiders who are responsible for the worst exploitation.

Local people usually have a better knowledge of local ecological conditions than experts in faraway capitals, as well as a direct interest in improving the quality of life in their village. A good example of this comes from the bone-dry state of Rajasthan in India, where local activism and indigenous know-how about rainwater "harvesting" provided the people with reliable water supplies—something the government had failed to do. In Bangladesh, villages with active community groups or concerned mullahs proved greener than less active neighbouring villages.

Community-based forestry initiatives from Bolivia to Nepal have shown that local people can be good custodians of nature. Several hundred million of the world's poorest people live in and around forests. Giving those villagers an incentive to preserve forests by allowing sustainable levels of harvesting, it turns out, is a far better way to save those forests than erecting tall fences around them.

To harness local energies effectively, it is particularly important to give local people secure property rights, argues Mr Dasgupta. In most parts of the developing world, control over resources at the village level is ill-defined. This often means that local elites usurp a disproportionate share of those resources, and that individuals have little incentive to maintain and upgrade forests or agricultural land. Authorities in Thailand tried to remedy this problem by distributing 5.5m land titles over a 20-year period. Agricultural output increased, access to credit improved and the value of the land shot up.

Name and shame

Another powerful tool for improving the local environment is the free flow of information. As local democracy flourishes, ordinary people are pressing for greater environmental disclosure by companies. In some countries, such as Indonesia, governments have adopted a "sunshine" policy that involves naming and shaming companies that do not meet environmental regulations. It seems to achieve results.

Bringing greenery to the grass roots is good, but on its own it will not avert perceived threats to global "public goods" such as the climate or biodiversity. Paul Portney of Resources for the Future explains: "Brazilian villagers may think very carefully and unselfishly about their future descendants, but there's no reason for them to care about and protect species or habitats that no future generation of Brazilians will care about."

That is why rich countries must do more than make pious noises about global threats to the environment. If they believe that scientific evidence suggests a credible threat, they must be willing to pay poor countries to protect such things as their tropical forests. Rather than thinking of this as charity, they should see it as payment for environmental services (say, for carbon storage) or as a form of insurance.

In the case of biodiversity, such payments could even be seen as a trade in luxury goods: rich countries would pay poor countries to look after creatures that only the rich care about. Indeed, private green groups are already buying up biodiversity "hot spots" to protect them. One such initiative, led by Conservation International and the International Union for the Conservation of Nature (IUCN), put the cost of buying and preserving 25 hot spots exceptionally rich in species diversity at less than $30 billion. Sceptics say it will cost more, as hot spots will need buffer zones of "sustainable harvesting" around them. Whatever the right figure, such creative approaches are more likely to achieve results than bullying the poor into conservation.

It is not that the poor do not have green concerns, but that those concerns are very different from those of the rich. In Beijing's Da Shilan, for instance, the air is full of soot from the many tiny coal boilers. Unlike most of the neighbouring districts, which have recently converted from coal to natural gas, this area has been considered too poor to make the transition. Yet ask Liu Shihua, a shopkeeper who has lived in the same spot for over 20 years, and he insists he would readily pay a bit more for the cleaner air that would come from using natural gas. So would his neighbours.

To discover the best reason why poor countries should not ignore pollution, ask those two little boys outside Mr Liu's shop what colour the sky is. "Grey!" says one tyke, as if it were the most obvious thing in the world. "No, stupid, it's blue!" retorts the other. The children deserve blue skies and clean air. And now there is reason to think they will see them in their lifetime.

Working miracles

Can technology save the planet?

"Nothing endures but change." That observation by Heraclitus often seems lost on modern environmental thinkers. Many invoke scary scenarios assuming that resources—both natural ones, like oil, and man-made ones, like knowledge—are fixed. Yet in real life man and nature are entwined in a dynamic dance of development, scarcity, degradation, innovation and substitution.

The nightmare about China turning into a resource-guzzling America raises two questions: will the world run out of resources? And even if it does not, could the growing affluence of developing nations lead to global environmental disaster?

The first fear is the easier to refute; indeed, history has done so time and again. Malthus, Ricardo and Mill all worried that scarcity of resources would snuff out growth. It did not. A few decades ago, the limits-to-growth camp raised worries that the world might soon run out of oil, and that it might not be able to feed the world's exploding population. Yet there are now more proven reserves of petroleum than three decades ago; there is more food produced than ever; and the past decade has seen history's greatest economic boom.

What made these miracles possible? Fears of oil scarcity prompted investment that led to better ways of producing oil, and to more efficient engines. In food production, technological advances have sharply reduced the amount of land required to feed a person in the past 50 years. Jesse Ausubel of Rockefeller University calculates that if in the next 60 to 70 years the world's average farmer reaches the yield of today's average (not best) American maize grower, then feeding 10 billion people will require just half of today's cropland. All farmers need to do is maintain the 2%-a-year productivity gain that has been the global norm since 1960.

"Scarcity and Growth", a book published by Resources for the Future, sums it up brilliantly: "Decades ago Vermont granite was only building and tombstone material; now it is a potential fuel, each ton of which has a usable energy content (uranium) equal to 150 tons of coal. The notion of an absolute limit to natural resource availability is untenable when the definition of resources changes drastically and unpredictably over time." Those words

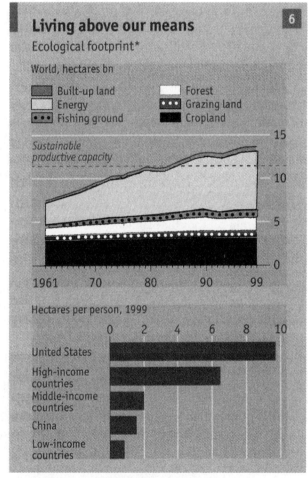

*Land needed to meet human needs

Source: WWF, Living Planet Report 2002

were written by Harold Barnett and Chandler Morse in 1963, long before the limits-to-growth bandwagon got rolling.

Giant footprint

Not so fast, argue greens. Even if we are not going to run out of resources, guzzling ever more resources could still do irreversible damage to fragile ecosystems.

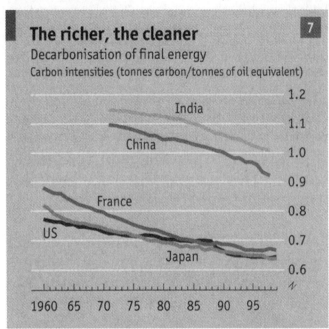

The richer, the cleaner 7

Decarbonisation of final energy

Carbon intensities (tonnes carbon/tonnes of oil equivalent)

Source: Nebojsa Nakicenovic and Arnulf Gruebler, International Institute for Applied Systems Analysis

WWF, an environmental group, regularly calculates mankind's "ecological footprint", which it defines as the "biologically productive land and water areas required to produce the resources consumed and assimilate the wastes generated by a given population using prevailing technology." The group reckons the planet has around 11.4 billion "biologically productive" hectares of land available to meet continuing human needs. As chart 6 overleaf shows, WWF thinks mankind has recently been using more than that. This is possible because a forest harvested at twice its regeneration rate, for example, appears in the footprint accounts at twice its area—an unsustainable practice which the group calls "ecological overshoot."

Any analysis of this sort must be viewed with scepticism. Everyone knows that environmental data are incomplete. What is more, the biggest factor by far is the land required to absorb CO_2 emissions of fossil fuels. If that problem could be managed some other way, then mankind's ecological footprint would look much more sustainable.

Even so, the WWF analysis makes an important point: if China's economy were transformed overnight into a clone of America's, an ecological nightmare could ensue. If a billion eager new consumers were suddenly to produce CO_2 emissions at American rates, they would be bound to accelerate global warming. And if the whole of the developing world were to adopt an American lifestyle tomorrow, local environmental crises such as desertification, aquifer depletion and topsoil loss could make humans miserable.

So is this cause for concern? Yes, but not for panic. The global ecological footprint is determined by three factors: population size, average consumption per person and

technology. Fortunately, global population growth now appears to be moderating. Consumption per person in poor countries is rising as they become better off, but there are signs that the rich world is reducing the footprint of its consumption (as this survey's final section explains). The most powerful reason for hope—innovation—was foreshadowed by WWF's own definition. Today's "prevailing technologies" will, in time, be displaced by tomorrow's greener ones.

"The rest of the world will not live like America," insists Mr Ausubel. Of course poor people around the world covet the creature comforts that Americans enjoy, but they know full well that the economic growth needed to improve their lot will take time. Ask Wu Chengjian, an environmental official in booming Shanghai, what he thinks of the popular notion that his city might become as rich as today's Hong Kong by 2020: "Impossible—that's just not enough time." And that is Shanghai, not the impoverished countryside.

Leaps of faith

This extra time will allow poor countries to embrace new technologies that are more efficient and less environmentally damaging. That still does not guarantee a smaller ecological footprint for China in a few decades' time than for America now, but it greatly improves the chances. To see why, consider the history of "dematerialisation" and "decarbonisation" (see chart 7). Viewed across very long spans of time, productivity improvements allow economies to use ever fewer material inputs—and to emit ever fewer pollutants—per unit of economic output. Mr Ausubel concludes: "When China has today's American mobility, it will not have today's American cars," but the cleaner and more efficient cars of tomorrow.

The snag is that consumers in developing countries want to drive cars not tomorrow but today. The resulting emissions have led many to despair that technology (in the form of vehicles) is making matters worse, not better.

Can they really hope to "leapfrog" ahead to cleaner air? The evidence from Los Angeles—a pioneer in the fight against air pollution—suggests the answer is yes. "When I moved to Los Angeles in the 1960s, there was so much soot in the air that it felt like there was a man standing on your chest most of the time," says Ron Loveridge, the mayor of Riverside, a city to the east of LA that suffers the worst of the region's pollution. But, he says, "We have come an extraordinary distance in LA."

Four decades ago, the city had the worst air quality in America. The main problem was the city's infamous "smog" (an amalgam of "smoke" and "fog"). It took a while to figure out that this unhealthy ozone soup developed as a result of complex chemical reactions between nitrogen oxides and volatile organic compounds that need sunlight to trigger them off.

Arthur Winer, an atmospheric chemist at the University of California at Los Angeles, explains that tackling smog required tremendous perseverance and political will. Early regulatory efforts met stiff resistance from business interests, and began to falter when they failed to show dramatic results.

Clean-air advocates like Mr Loveridge began to despair: "We used to say that we needed a 'London fog' [a reference to an air-pollution episode in 1952 that may have killed 12,000 people in that city] here to force change." Even so, Californian officials forged ahead with an ambitious plan that combined regional regulation with stiff mandates for cleaner air. Despite uncertainties about the cause of the problem, the authorities introduced a sequence of controversial measures: unleaded and low-sulphur petrol, on-board diagnostics for cars to minimise emissions, three-way catalytic converters, vapour-recovery attachments for petrol nozzles and so on.

As a result, the city that two decades ago hardly ever met federal ozone standards has not had to issue a single alert in the past three years. Peak ozone levels are down by 50% since the 1960s. Though the population has shot up in recent years, and the vehicle-miles driven by car-crazy Angelenos have tripled, ozone levels have fallen by two-thirds. The city's air is much cleaner than it was two decades ago.

"California, in solving its air-quality problem, has solved it for the rest of the United States and the world—but it doesn't get credit for it," says Joe Norbeck of the University of California at Riverside. He is adamant that the poor world's cities can indeed leapfrog ahead by embracing some of the cleaner technologies developed specifically for the Californian market. He points to China's vehicle fleet as an example: "China's typical car has the emissions of a 1974 Ford Pinto, but the new Buicks sold there use 1990s emissions technology." The typical car sold today produces less than a tenth of the local pollution of a comparable model from the 1970s.

That suggests one lesson for poor cities such as Beijing that are keen to clean up: they can order polluters to meet high emissions standards. Indeed, from Beijing to Mexico city, regulators are now imposing rich-world rules, mandating new, cleaner technologies. In China's cities, where pollution from sooty coal fires in homes and industrial boilers had been a particular hazard, officials are keen to switch to natural-gas furnaces.

However, there are several reasons why such mandates—which worked wonders in LA—may be trickier to achieve in impoverished or politically weak cities. For a start, city officials must be willing to pay the political price of reforms that raise prices for voters. Besides, higher standards for new cars, useful though they are, cannot do the trick on their own. Often, clean technologies such as catalytic converters will require cleaner grades of petrol too. Introducing cleaner fuels, say experts, is an essential lesson from LA for poor countries. This will not come free either.

There is another reason why merely ordering cleaner new cars is inadequate: it does nothing about the vast stock of dirty old ones already on the streets. In most cities of the developing world, the oldest fifth of the vehicles on the road is likely to produce over half of the total pollution caused by all vehicles taken together. Policies that encourage a speedier turnover of the fleet therefore make more sense than "zero emissions" mandates.

Policy matters

In sum, there is hope that the poor can leapfrog at least some environmental problems, but they need more than just technology. Luisa and Mario Molina of the Massachusetts Institute of Technology, who have studied such questions closely, reckon that technology is less important than the institutional capacity, legal safeguards and financial resources to back it up: "The most important underlying factor is political will." And even a techno-optimist such as Mr Ausubel accepts that: "There is nothing automatic about technological innovation and adoption; in fact, at the micro level, it's bloody."

Clearly innovation is a powerful force, but government policy still matters. That suggests two rules for policymakers. First, don't do stupid things that inhibit innovation. Second, do sensible things that reward the development and adoption of technologies that enhance, rather than degrade, the environment.

The greatest threat to sustainability may well be the rejection of science. Consider Britain's hysterical reaction to genetically modified crops, and the European Commission's recent embrace of a woolly "precautionary principle". Precaution applied case-by-case is undoubtedly a good thing, but applying any such principle across the board could prove disastrous.

Explaining how not to stifle innovation that could help the environment is a lot easier than finding ways to encourage it. Technological change often goes hand-in-hand with greenery by saving resources, as the long history of dematerialisation shows—but not always. Sports utility vehicles, for instance, are technologically innovative, but hardly green. Yet if those SUVs were to come with hydrogen-powered fuel cells that emit little pollution, the picture would be transformed.

The best way to encourage such green innovations is to send powerful signals to the market that the environment matters. And there is no more powerful signal than price, as the next section explains.

The invisible green hand

Markets could be a potent force for greenery—if only greens could learn to love them

"MANDATE, regulate and litigate." That has been the environmentalists' rallying cry for ages. Nowhere in the green manifesto has there been much mention of the market. And, oddly, it was market-minded America that led the dirigiste trend. Three decades ago, Congress passed a sequence of laws, including the Clean Air Act, which set lofty goals and generally set rigid technological standards. Much of the world followed America's lead.

This top-down approach to greenery has long been a point of pride for groups such as the Natural Resources Defence Council (NRDC), one of America's most influential environmental outfits. And with some reason, for it has had its successes: the air and water in the developed world is undoubtedly cleaner than it was three decades ago, even though the rich world's economies have grown by leaps and bounds. This has convinced such groups stoutly to defend the green status quo.

But times may be changing. Gus Speth, now head of Yale University's environment school and formerly head of the World Resources Institute and the UNDP, as well as one of the founders of the NRDC, recently explained how he was converted to market economics: "Thirty years ago, the economists at Resources for the Future were pushing the idea of pollution taxes. We lawyers at NRDC thought they were nuts, and feared that they would derail command-and-control measures like the Clean Air Act, so we opposed them. Looking back, I'd have to say this was the single biggest failure in environmental management—not getting the prices right."

A remarkable mea culpa; but in truth, the command-and-control approach was never as successful as its advocates claimed. For example, although it has cleaned up the air and water in rich countries, it has notably failed in dealing with waste management, hazardous emissions and fisheries depletion. Also, the gains achieved have come at a needlessly high price. That is because technology mandates and bureaucratic edicts stifle innovation and ignore local realities, such as varying costs of abatement. They also fail to use cost-benefit analysis to judge trade-offs.

Command-and-control methods will also be ill-suited to the problems of the future, which are getting trickier. One reason is that the obvious issues—like dirty air and water—have been tackled already. Another is increasing technological complexity: future problems are more likely to involve subtle linkages—like those involved in ozone depletion and global warming—that will require sophisticated responses. The most important factor may be society's ever-rising expectations; as countries grow

wealthier, their people start clamouring for an ever-cleaner environment. But because the cheap and simple things have been done, that is proving increasingly expensive. Hence the greens' new interest in the market.

Carrots, not just sticks

In recent years, market-based greenery has taken off in several ways. With emissions trading, officials decide on a pollution target and then allocate tradable credits to companies based on that target. Those that find it expensive to cut emissions can buy credits from those that find it cheaper, so the target is achieved at the minimum cost and disruption.

The greatest green success story of the past decade is probably America's innovative scheme to cut emissions of sulphur dioxide (SO_2). Dan Dudek of Environmental Defence, a most unusual green group, and his market-minded colleagues persuaded the elder George Bush to agree to an amendment to the sacred Clean Air Act that would introduce an emissions-trading system to achieve sharp cuts in SO_2. At the time, this was hugely controversial: America's power industry insisted the cuts were prohibitively costly, while nearly every other green group decried the measure as a sham. In the event, ED has been vindicated. America's scheme has surpassed its initial objectives, and at far lower cost than expected. So great is the interest worldwide in trading that ED is now advising groups ranging from hard-nosed oilmen at BP to bureaucrats in China and Russia.

Europe, meanwhile, is forging ahead with another sort of market-based instrument: pollution taxes. The idea is to levy charges on goods and services so that their price reflects their "externalities"—jargon for how much harm they do to the environment and human health. Sweden introduced a sulphur tax a decade ago, and found that the sulphur content of fuels dropped 50% below legal requirements.

Though "tax" still remains a dirty word in America, other parts of the world are beginning to embrace green tax reform by shifting taxes from employment to pollution. Robert Williams of Princeton University has looked at energy use (especially the terrible effects on health of particulate pollution) and concluded that such externalities are comparable in size to the direct economic costs of producing that energy.

Externalities are only half the battle in fixing market distortions. The other half involves scrapping environmentally harmful subsidies. These range from prices below market levels for electricity and water to shameless cash handouts for industries such as coal. The boffins at the OECD reckon that stripping away harmful subsidies, along with introducing taxes on carbon-based fuels and chemicals use, would result in dramatically lower emissions by 2020 than current policies would be able to achieve. If the revenues raised were then used to reduce other taxes, the cost of these virtuous policies would be less than 1% of the OECD's economic output in 2020.

Such subsidies are nothing short of perverse, in the words of Norman Myers of Oxford University. They do double damage, by distorting markets and by encouraging behaviour that harms the environment. Development banks say such subsidies add up to $700 billion a year, but Mr Myers reckons the true sum is closer to $2 trillion a year. Moreover, the numbers do not fully reflect the harm done. For example, EU countries subsidise their fishing fleets to the tune of $1 billion a year, but that has encouraged enough overfishing to drive many North Atlantic fishing grounds to near-collapse.

Fishing is an example of the "tragedy of the commons", which pops up frequently in the environmental debate. A resource such as the ocean is common to many, but an individual "free rider" can benefit from plundering that commons or dumping waste into it, knowing that the costs of his actions will probably be distributed among many neighbours. In the case of shared fishing grounds, the absence of individual ownership drives each fisherman to snatch as many fish as he can—to the detriment of all.

Of rights and wrongs

Assigning property rights can help, because providing secure rights (set at a sustainable level) aligns the interests of the individual with the wider good of preserving nature. This is what sceptical conservationists have observed in New Zealand and Iceland, where schemes for tradable quotas have helped revive fishing stocks. Similar rights-based approaches have led to revivals in stocks of African elephants in southern Africa, for example, where the authorities stress property rights and private conservation.

All this talk of property rights and markets makes many mainstream environmentalists nervous. Carl Pope, the boss of the Sierra Club, one of America's biggest green groups, does not reject market forces out of hand, but expresses deep scepticism about their scope. Pointing to the difficult problem of climate change, he asks: "Who has property rights over the commons?"

Even so, some greens have become converts. Achim Steiner of the IUCN reckons that the only way forward is rights-based conservation, allowing poor people "sustainable use" of their local environment. Paul Faeth of the World Resources Institute goes further. He says he is convinced that market forces could deliver that holy grail of environmentalism, sustainability—"but only if we get prices right."

The limits to markets

Economic liberals argue that the market itself is the greatest price-discovery mechanism known to man. Allow it to function freely and without government meddling, goes the argument, and prices are discovered and internalised automatically. Jerry Taylor of the Cato Institute, a libertarian think-tank, insists that "The world today is already sustainable—except those parts where western capitalism doesn't exist." He notes that countries that have relied on central planning, such as the Soviet Union, China and India, have invariably misallocated investment, stifled innovation and fouled their environment far more than the prosperous market economies of the world have done.

All true. Even so, markets are currently not very good at valuing environmental goods. Noble attempts are under way to help them do better. For example, the Katoomba Group, a collection of financial and energy companies that have linked up with environmental outfits, is trying to speed the development of markets for some of forestry's ignored "co-benefits" such as carbon storage and watershed management, thereby producing new revenue flows for forest owners. This approach shows promise: water consumers ranging from officials in New York City to private hydro-electric operators in Costa Rica are now paying people upstream to manage their forests and agricultural land better. Paying for greenery upstream turns out to be cheaper than cleaning up water downstream after it has been fouled.

Economists too are getting into the game of helping capitalism "get prices right." The World Bank's Ian Johnson argues that conventional economic measures such as gross domestic product are not measuring wealth creation properly because they ignore the effects of environmental degradation. He points to the positive contribution to China's GDP from the logging industry, arguing that such a calculation completely ignores the billions of dollars-worth of damage from devastating floods caused by over-logging. He advocates a more comprehensive measure the Bank is working on, dubbed "genuine GDP", that tries (imperfectly, he accepts) to measure depletion of natural resources.

That could make a dramatic difference to how the welfare of the poor is assessed. Using conventional market measures, nearly the whole of the developing world save Africa has grown wealthier in the past couple of decades. But when the degradation of nature is properly accounted for, argues Mr Dasgupta at Cambridge, the countries of Africa and south Asia are actually much worse off today than they were a few decades ago—and even China, whose economic "miracle" has been much trumpeted, comes out barely ahead.

The explanation, he reckons, lies in a particularly perverse form of market distortion: "Countries that are exporting resource-based products (often among the poorest) may be subsidising the consumption of countries that are doing the importing (often among the richest).” As evidence, he points to the common practice in poor countries of encouraging resource extraction. Whether through licenses granted at below-market rates, heavily subsidised exports or corrupt officials tolerating illegal exploitation, he reckons the result is the same: “The cruel paradox we face may well be that contemporary economic development is unsustainable in poor countries because it is sustainable in rich countries.”

One does not have to agree with Mr Dasgupta's conclusion to acknowledge that markets have their limits. That should not dissuade the world from attempting to get prices right—or at least to stop getting them so wrong. For grotesque subsidies, the direction of change should be obvious. In other areas, the market itself may not provide enough information to value nature adequately. This

is true of threats to essential assets, such as nature's ability to absorb and “recycle” CO_2, that have no substitute at any price. That is when governments must step in, ensuring that an informed public debate takes place.

Robert Stavins of Harvard University argues that the thorny notion of sustainable development can be reduced to two simple ideas: efficiency and intergenerational equity. The first is about making the economic pie as large as possible; he reckons that economists are well equipped to handle it, and that market-based policies can be used to achieve it. On the second (the subject of the next section), he is convinced that markets must yield to public discourse and government policy: “Markets can be efficient, but nobody ever said they're fair. The question is, what do we owe the future?”

Insuring a brighter future

How to hedge against tomorrow's environmental risks

So WHAT do we owe the future? A precise definition for sustainable development is likely to remain elusive but, as this survey has argued, the hazy outline of a useful one is emerging from the experience of the past decade.

For a start, we cannot hope to turn back the clock and return nature to a pristine state. Nor must we freeze nature in the state it is today, for that gift to the future would impose an unacceptable burden on the poorest alive today. Besides, we cannot forecast the tastes, demands or concerns of future generations. Recall that the overwhelming pollution problem a century ago was horse manure clogging up city streets: a century hence, many of today's problems will surely seem equally irrelevant. We should therefore think of our debt to the future as including not just natural resources but also technology, institutions and especially the capacity to innovate. Robert Solow got it mostly right a decade ago: the most important thing to leave future generations, he said, is the capacity to live as well as we do today.

However, as the past decade has made clear, there is a limit to that argument. If we really care about the “sustainable” part of sustainable development, we must be much more watchful about environmental problems with critical thresholds. Most local problems are reversible and hence no cause for alarm. Not all, however: the depletion of aquifers and the loss of topsoil could trigger irreversible changes that would leave future generations worse off. And global or long-term threats, where victims are far removed in time and space, are easy to brush aside.

In areas such as biodiversity, where there is little evidence of a sustainability problem, a voluntary approach is best. Those in the rich world who wish to preserve pandas, or hunt for miracle drugs in the rainforest, should pay for their predilections. However, where there are strong scientific indications of unsustainability, we must

act on behalf of the future—even at the price of today's development. That may be expensive, so it is prudent to try to minimise those risks in the first place.

A riskier world

Human ingenuity and a bit of luck have helped mankind stay a few steps ahead of the forces degrading the environment this past century, the first full one in which the planet has been exposed to industrialisation. In the century ahead, the great race between development and degradation could well become a closer call.

On one hand, the demands of development seem sure to grow at a cracking pace in the next few decades as the Chinas, Indias and Brazils of this world grow wealthy enough to start enjoying not only the necessities but also some of the luxuries of life. On the other hand, we seem to be entering a period of huge technological advances in emerging fields such as biotechnology that could greatly increase resource productivity and more than offset the effect of growth on the environment. The trouble is, nobody knows for sure.

Since uncertainty will define the coming era, it makes sense to invest in ways that reduce that risk at relatively low cost. Governments must think seriously about the future implications of today's policies. Their best bet is to encourage the three powerful forces for sustainability outlined in this survey: the empowerment of local people to manage local resources and adapt to environmental change; the encouragement of science and technology, especially innovations that reduce the ecological footprint of consumption; and the greening of markets to get prices right.

To advocate these interventions is not to call for a return to the hubris of yesteryear's central planners. These

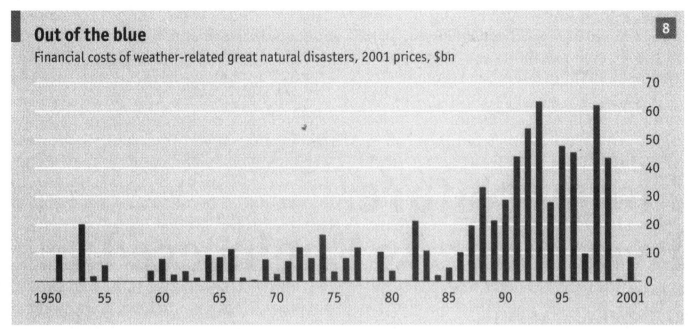

Out of the blue ⬛8

Financial costs of weather-related great natural disasters, 2001 prices, $bn

Source: Munich Re

measures would merely give individuals the power to make greener choices if they care to. In practice, argues Chris Heady of the OECD, this may still not add up to sustainability "because we might still decide to be greedy, and leave less for our children."

Happily, there are signs of an emerging bottom-up push for greenery. Even such icons of western consumerism as Unilever and Procter & Gamble now sing the virtues of "sustainable consumption." Unilever has vowed that by 2005 it will be buying fish only from sustainable sources, and P&G is coming up with innovative products such as detergents that require less water, heat and packaging. It would be naive to label such actions as expressions of "corporate social responsibility": in the long run, firms will embrace greenery only if they see profit in it. And that, in turn, will depend on choices made by individuals.

Such interventions should really be thought of as a kind of insurance that tilts the odds of winning that great race just a little in humanity's favour. Indeed, even some of the world's most conservative insurance firms increasingly see things this way. As losses from weather-related disasters have risen of late (see chart 8), the industry is getting more involved in policy debates on long-term environmental issues such as climate change.

Bruno Porro, chief risk officer at Swiss Re, argues that: "The world is entering a future in which risks are more concentrated and more complex. That is why we are pressing for policies that reduce those risks through preparation, adaptation and mitigation. That will be cheaper than covering tomorrow's losses after disaster strikes."

Jeffrey Sachs of Columbia University agrees: "When you think about the scale of risk that the world faces, it is clear that we grossly underinvest in knowledge... we have enough income to live very comfortably in the developed world and to prevent dire need in the developing world. So we should have the confidence to invest in longer-term issues like the environment. Let's help insure the sustainability of this wonderful situation."

He is right. After all, we have only one planet, now and in the future. We need to think harder about how to use it wisely.

Acknowledgements

In addition to those cited in the text, the author would like to thank Robert Socolow, David Victor, Geoffrey Heal, and experts at Tsinghua University, Friends of the Earth, the European Commission, the World Business Council for Sustainable Development, the International Energy Agency, the OECD and the UN for sharing their ideas with him. A list of sources can be found on *The Economist's* website.

MANAGING PLANET EARTH

Forget Nature. Even Eden Is Engineered

By ANDREW C. REVKIN

Nearly 70 years ago, a Soviet geochemist, reflecting on his world, made a startling observation: through technology and sheer numbers, he wrote, people were becoming a geological force, shaping the planet's future just as rivers and earthquakes had shaped its past.

Eventually, wrote the scientist, Vladimir I. Vernadsky, global society, guided by science, would soften the human environmental impact, and earth would become a "noosphere," a planet of the mind, "life's domain ruled by reason."

Today, a broad range of scientists say, part of Vernadsky's thinking has already been proved right: people have significantly altered the atmosphere and are the dominant influence on ecosystems and natural selection. The question now, scientists say, is whether the rest of his vision will come to pass. Choices made in the next few years will determine the answer.

Aided by satellites and supercomputers, and mobilized by the evident environmental damage of the last century, humans have a real chance to begin balancing economic development with sustaining earth's ecological webs, said Dr. William C. Clark, a biologist at Harvard who heads an international effort to build a scientific foundation for such a shift.

"We've come through a period of finally understanding the nature and magnitude of humanity's transformation of the earth," Dr. Clark said. "Having realized it, can we become clever enough at a big enough scale to be able to maintain the rates of progress? I think we can."

Some scientists say it is anthropocentric hubris to think people understand the living planet well enough to know how to manage it. But that prospect is attracting more than 100 world leaders and thousands of other participants to the United Nations' World Summit on Sustainable Development, which starts on Monday in Johannesburg.

No matter what they come up with, ice ages, volcanoes and shifting tectonic plates will dwarf human activities in the long run. But communities and countries face concrete choices in the next decade that are likely to determine the quality of human life and the environment well into the 22nd century.

Human activity is such a pervasive influence on the planet's ecological framework that it is no longer possible to separate people and nature.

Emissions of heat-trapping carbon dioxide, whether from an Ohio power plant or a Bangkok taxicab, contribute to global warming.

Seafood lovers dining in Manhattan bistros prompt fishing vessels to sweep Antarctic waters for slow-growing Chilean sea bass. Shoppers in Tokyo seeking inexpensive picture frames send loggers deep into Indonesian forests.

In a new book "Great Transition: The Promise and Lure of the Times Ahead," published by the Stockholm Environment Institute, a group of top geographers, economists, engineers and other experts concludes that the same inventiveness that accelerated the human ascent can be harnessed to soften human impact.

The need for a new approach is urgent, the researchers say, because a surge of growth in quickly industrializing regions of Asia and the Americas could have environmental effects that exceed those of the industrialization of the West. More pressure for change comes from southern Africa and other pockets of extreme poverty where the brutal calculus of Malthus still holds sway.

Even in the industrialized north, after generations of prosperity, people are hemmed in by concrete, seeing commuting times grow and starting to question their definition of progress.

As a result, countless communities, from the charred fringes of the Amazon to the spreading suburbs of Seattle, are balancing growing needs and limited resources.

If development does not change course, the new book concludes, "the nightmare of an impoverished, mean, destructive future looms."

The World Remade By Human Hands

When delegates convene in Johannesburg next week for a United Nations environment summit, they will confront the problems human interference has brought to the world's ecosystems. But some trends once seen as problems may not be so bad after all. And some seemingly beneficial trends have serious unexpected side effects. Here are some examples.

MEGACITIES

By 2015, demographers predict, the number of cities with more than 10 million people—so-called megacities— will rise to 36, from about 20 today. Megacities are notorious for pollution and sprawling slums. But people who live in them tend to have smaller families and greater opportunities for jobs and educations. And environmental improvements can come quickly to cities. For example, in Mexico City, the percentage of days violating pollution standards has fallen to less than 20 percent, from 50 percent. Now the city is considering a mass transit program that would bring further improvements.

FOOD SUPPLIES

Demographers are lowering their estimates for how much the population of the world will increase before it stabilizes, but even so they predict 50 percent growth by the end of the century. Demand for food will grow even faster, as people in developing countries start consuming more meat and fish. Aquaculture, once thought of as a solution, has produced problems of its own. In Thailand, and elsewhere, shrimp farms have replaced valuable mangrove ecosystems. In China, however, a recent experiment in mixing rice varieties in a single field greatly increased yields, without added chemicals.

FALLING FORESTS

In the tropics, forests continue to fall before farmers, loggers and timber processors like the saw mill, on a branch of the Amazon River in Brazil. But new surveys show that in the 1990's the rate of cutting was 23 percent less than earlier estimates. Some timber companies have gained certification from environmental groups as managing forests with care. Companies that buy their wood, like the Gibson guitar company, promote their products as environmentally friendly. Environmental groups are working with growers of coca, bananas, coffee and other commodities to adopt similar programs. And some conservationists now say that judicious logging can actually benefit the plants and animals that live in the forest.

Signs of Improvement

Unexpected Cleanups Generate Optimism

Over the past 30 years, "sustainability" has become the mantra of many private groups, government officials, scientists and, even, a growing number of businesses. Most define the notion as advancing human endeavors without diminishing prospects for future generations. The Johannesburg summit will be the third global conclave in that span chasing this elusive goal.

But movement toward concrete action has been slow. The first meeting, in Stockholm in 1972, rang an alarm about despoiling the earth. Wealthy nations began cleaning air and water, but continued to assault forests and other resources elsewhere to fuel growth.

In 1992 came the second meeting, in Rio de Janeiro, called the Earth Summit. There, diplomats forged ambitious agreements aimed at holding back deserts and protecting the atmosphere, forests and pockets of biological richness.

But the agreements were vague, relying more on good will than on concrete obligations. Developing countries refused to take on obligations, saying the north should step first.

After Rio, population continued to grow, poverty persisted, forests retreated, soils eroded, fish stocks shrank, and concentrations of heat-trapping greenhouse gases rose, despite a treaty in which industrialized countries pledged to "strive" to reduce them.

Now, a host of satellites provides streams of data that powerful computers sift and disseminate on the Web. Communities can track forest loss in Indonesia, sprawl in Indiana and the flow of pollution from state to state, country to country.

After disasters like the chemical release in Bhopal, India, in 1984 and the grounding of the Exxon Valdez in Alaska in 1989, many companies have shifted practices to avoid environmental damage, shareholder wrath and consumer boycotts.

Fast-growing developing countries including China and Mexico are rapidly cutting urban air pollution..

They have been spurred both by community pressure and awareness of the high costs of treating illnesses caused by pollution.

Indonesia, China and other countries are posting factories' chemical emissions on the Web. The technique, pioneered in the United States, is prompting cleanups.

No one expects that people will be able to manage the planet like some giant corporation—the real Big Blue.

"If you mean making the thousands of little decisions that need to be made, we can no more effectively manage the world than the Soviet Union could manage its centrally planned economy," said Dr. Robert W. Kates, a geographer who headed a National Academy of Sciences committee on sustainable development and is an author of "Great Transition."

But Dr. Kates says the potential exists to make informed choices that spread the benefits of development to an impoverished majority while not depleting vital assets.

One impediment to such a transition is the change itself, the environmental and societal turbulence created by explosive

Look What We've Done...

Experts from the World Bank and elsewhere say the availability of environmental information can, in itself, lead to environmental improvements. Satellites are producing a wealth of data.

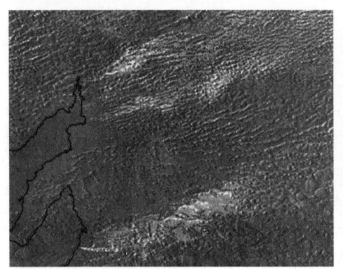

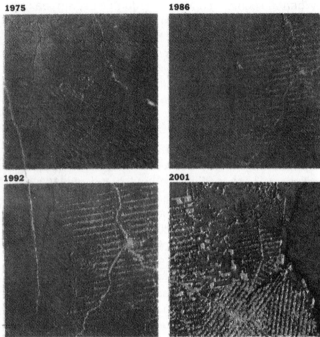

1975   1986

1992   2001

Top **POLLUTION PLUMES OVER AUSTRALIA** In this image, the clouds near Adelaide on the south coast of Australia are color coded to show their density. The light streaks indicate pollution emanating from smokestacks.

Right **DEFORESTATION IN RONDÔNIA, BRAZIL** Brazil completed the Cuiaba-Porto Velho highway in 1960 but growth in the region was slow until after the World Bank paid to pave the highway in 1980. The region, in northwest Brazil, has since grown to become a substantial farming community.

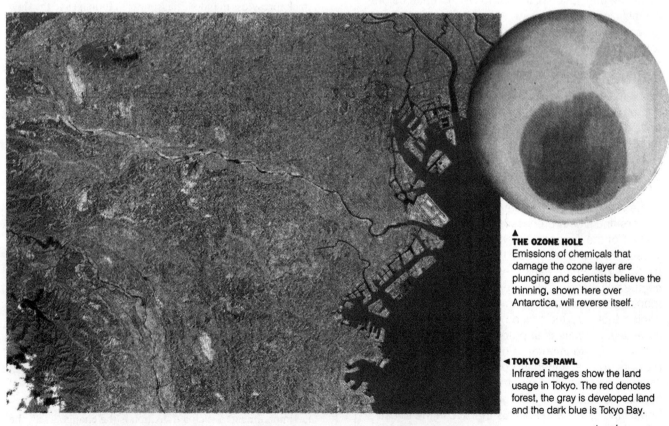

▲
THE OZONE HOLE
Emissions of chemicals that damage the ozone layer are plunging and scientists believe the thinning, shown here over Antarctica, will reverse itself.

◄ **TOKYO SPRAWL**
Infrared images show the land usage in Tokyo. The red denotes forest, the gray is developed land and the dark blue is Tokyo Bay.

continued on next page

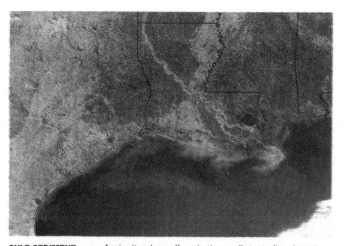

GULF SEDIMENT
- Sediment
- Phytoplankton

Agricultural runoff and other pollutants flow into the Mississippi River and into the Gulf of Mexico where they nourish phytoplankton, creating, in effect, a dead zone for fish.

INDONESIA SMOG
- Smoke
- Smog

Wildfires in Indonesia in 1997 left a plume of smoke and smog over Southeast Asia. Researchers are using satellite data to track how different types of pollution move through the atmosphere.

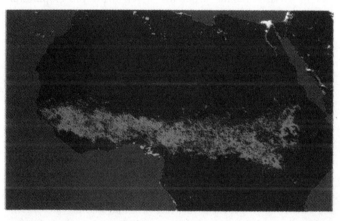

LIGHT IN AFRICA
- Gas flares
- City lights
- Fires

Using a light-sensing satellite, scientists tracked light sources over the course of six months to study land usage. The data will help researchers gauge global change in the environment and population.

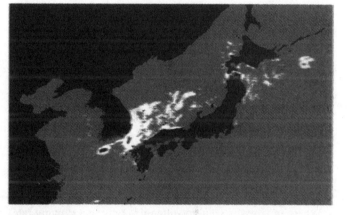

SQUID FISHING
- Fishing lights
- Land
- Ocean

Fishing boats use lights to attract squid. This image shows fishing activity near Japan and Korea.

Sources: NASA; National Oceanic & Atmospheric Administration; U.S. Geological Survey; Daniel Rosenfeld, The Hebrew University of Jerusalem; Michael King, Goddard Space Flight Center; University of Colorado; Colorado State University. Images by Daniel Rosenfeld, The Hebrew University of Jerusalem (pollution over Australia); Advanced Spaceborne Thermal Emission and Reflection Radiometer (Tokyo sprawl) Landsat (deforestation in Rondônia, Brazil); Terra satellite (Gulf sediment); Total Ozone Mapping Spectrometer (ozone hole, Indonesia smog); Defense Meterological Satellite Program (lights in Africa, squid fishing)

Brett Taylor/The New York Times

human growth, technological advance and the planetwide linkup of disparate cultures, Dr. Kates and other experts say.

Another barrier, they add, is the enormous growth of population and consumption. Although global population appears headed for a 50 percent increase in the next 50 years, for example, demand for food will likely double, as prosperity raises the per capita consumption of calories.

There is another roadblock. Not every problem of consequence comes with a Bhopal-style wake-up call. Global warming and species extinction are examples of potential catastrophes that are hiding in plain sight, experts say.

Scientists are helping identify problems and opportunities. But communities will make choices guided only in part by what makes sense for the long haul.

For one thing, big gaps persist in the basic information needed to measure progress. When a team from Yale and Co-

lumbia studied dozens of trends in 142 countries to rank their sustainability, the members had to leave 40 percent of their spreadsheet blank, said Daniel C. Esty, a Yale law professor, who was a leader of the project.

Nonetheless, optimists say they see signs of hope. Not the least of them is the intensifying dialogue on the problem, which includes parties as disparate as multinational companies and tribal bands.

In essence, the human capacity for understanding the world is catching up with the human capacity to change it, Dr. Clark at Harvard said. "It really is a plausible case that we're coming on a key stage now, with the cold war under control, with globalization happening," he said.

The hard part, he said, will be for societies to overcome a habit of focusing on present needs. "Do we move beyond simply being a big bull in a china shop, having impacts, to becoming a reflective capacity on the planet?" Dr. Clark asked. "Or do we simply bungle ahead?"

An Altered World

Human Imprints From Pole to Pole

Evidence abounds now that the world is a human-dominated place.

By flooding the atmosphere with synthetic chemicals and heat-trapping carbon dioxide and other greenhouse gases, for example, people damaged the protective ozone layer and contributed to a warming climate, scientists have said. The ozone depletion became vividly and unexpectedly evident in the 1980's, when a gaping hole was detected over Antarctica.

The hole will shrink in the next 50 years because of a ban on ozone-eating chlorofluorocarbons. Other damage will not be so easy to repair.

Long before they are cataloged, thousands of plant and animal species are likely to be driven to extinction as forests, wetlands, mountain slopes and other habitats are exploited or harmed by climate change.

Satellites that map vegetation and the nighttime signature of human activity—fire and light—show that people have altered more than one-third of the terrestrial landscape. Once it is changed, it is usually changed forever, Dr. G. David Tilman, an ecologist at the University of Minnesota, said. "When you add another 1,000 acres of shopping mall or another highway, far into the future those are probably close to permanent acts," Dr. Tilman said.

Where progress is seen, too often it is only in a slowing rate of destruction, ecologists say. For example, new satellite surveys show that forest loss in the tropics through the 1990's occurred at a rate 23 percent less than previous estimates. But losses still add up to some 14 million acres a year, with 5 million more acres visibly damaged.

The human imprint is evident almost everywhere. In the South Atlantic, fleets illuminate so much of the ocean to attract squid that the illuminated area dwarfs the megalopolis of São Paulo. The squid harvest has in part grown because commercial fish stocks have been overfished.

Altogether, scientists have found that two-thirds of commercial marine fish species are fully exploited or diminishing, prompting companies to move down the food chain.

Aquaculture is a fast-growing alternative, but often causes damage like the destruction of coastal mangroves in southern Asia to make way for shrimp farms.

Also, in many cases, farmed species are fed fish meal made of other fish. So the cultivation still indirectly depletes the oceans.

Hydrologists estimate that people appropriate half the world's flowing fresh water. Across the American West in the last 20 years, circular patches made by great rotating irrigation rigs have peppered the land like an expanding checkerboard, marking the draining of aquifers under the plains.

Scientists have concluded that humans not only now dominate the planet, but have also become the dominant driver of natural selection, the machinery of evolution.

The main influence, experts say, is the continuing chemical arms race against germs and pests, which kills most, but leaves a resistant minority behind.

Also, by wittingly and unwittingly moving myriad species around the globe, humans have become a biological blender, carrying West Nile virus to America and overrunning the Bordeaux countryside in France with American bullfrogs that residents say do not even taste good.

Troubling Trends

An S.U.V. Culture Shifts to Third World

Projections for the next two generations do not bode well for easing environmental problems. Even with the population bomb predicted in the 1960's substantially defused, the human population is likely headed for at least nine billion before leveling off.

Most of the growth will be in poor countries, and as people there pursue prosperity, consumption of natural resources will rapidly increase.

Half the world's 17,000 major wildlife refuges are already being heavily used for agriculture, recent studies showed.

Car companies are racing to build factories to assemble sport utility vehicles in India, even as its once-legendary rail system, plagued by mismanagement, is deteriorating and losing freight and passengers, said Dr. Rajendra K. Pachauri, director general of the Tata Energy Research Institute of India. The private organization assesses energy and environmental problems in India.

"In the last 10 years, every major manufacturer has set up facilities in India," Dr. Pachauri said. "They see it as a major market. They have the buzz of the people. This is something that should cause real concern."

Depending on how power is generated, bringing electricity to the two billion people in the world who still lack it could greatly increase emissions of greenhouse gases. But, experts note, the options available to those who remain off the grid also cause harm. The two billion people who cook on wood or dung fires live in acrid clouds of toxic smoke and deplete forests.

In India, Dr. Pachauri said, millions of people light their homes with kerosene, using government-subsidized fuel. Together, cooking fires and sputtering lanterns create indoor pollution that causes asthma and other ailments and that, in India alone, is estimated to kill 600,000 women a year.

Dr. Pachauri's group has experimented with distributing solar-powered lanterns to rural communities. Other projects push cleaner ovens that use less-polluting fuels.

But the question of how to take good ideas from pilot projects to the new norm remains largely unsolved.

Then there are the costs. For example, about a billion people have no clean water. More than twice that number live where raw sewage flows unchecked. So water is a prime focus of the delegates in Johannesburg.

When even countries like the United States lag tens of billions of dollars in improving their sewage systems, experts say the prospects of big investments elsewhere are dim.

New Strategies

Learning to Harvest More From Less

The hardest part of meshing economic and environmental progress, experts say, is that this shift cannot be engineered with top-down directives.

It will be a result of 10,000 decisions, large and small, by countries, communities, companies and individuals, said Dr. Kates, the geographer.

Action will have to be focused where the human imprint is most intense, in forests, on the farm and in the fast-expanding cities, experts say.

An analysis by the World Wildlife Fund found that 20 percent of the existing forest area could provide all the world's future needs for wood and pulp if it was all managed according to environmentally sound practices that a few big companies have already adopted.

Some forestry companies and wood and paper buyers, including Home Depot and Ikea, participate in a program in which wood is certified by the Forest Stewardship Council, a private group. The council monitors forest holdings and products to ensure that wood marketed as environmentally friendly is produced with limited damage.

Other organizations run similar certification programs to encourage growers of other crops like bananas and cocoa to preserve habitat or limit pesticide use.

Those efforts remain a tiny fringe of the markets for those commodities. The Forest Stewardship Council has certified 70 million acres of forests—just 4 percent of the total acreage controlled by timber companies.

If farming does not change drastically in the next few decades, enormous ecological damage will result, many scientists say.

Dr. Tilman of Minnesota notes that farming has already produced the biggest global imprint of humanity, affecting half the earth's habitable land. The challenge now, he said, will be to double agricultural productivity without using substantially more land.

There are signs this can be achieved. In an extraordinary experiment several years ago, in Yunnan Province, China, rice farmers were recruited to intersperse two varieties on 8,000 acres instead of planting one. The yield nearly doubled, and the occurrence of rice blast, the most harmful disease in the world's biggest crop, fell 94 percent.

Recognizing a good thing, China has expanded the work to 250,000 acres, said Dr. Christopher C. Mundt, a plant pathologist at Oregon State University who helped conduct the research.

"Perhaps more important," he added, "the Chinese are taking the general concept of diversification into related approaches" to other crops.

The key to the next green revolution, Dr. Mundt said, will be abandoning most of the industrial model of agriculture of the 20th century and shifting to a "biological model based on management of ecological processes" like applying fertilizer and water only as needed.

Such efforts are being made in agriculture, forestry, fisheries and other fields around the world. But, once again, experts say the challenge lies in moving to the scales needed to avert widespread harm.

The Role of the City

The Megalopolis As Eco-Strategy

Another focal point for experts who envision a managed earth is cities. In many ways, they are where the battle will be won or lost.

Cities are where almost all remaining population growth will occur, demographers say. The roster of megacities, those with populations exceeding 10 million, is widely expected to climb, from 20 today to 36 by 2015.

These vast metropolises have been widely characterized as a nightmarish element of the new century, sprawling and chaotic and spawning waste and illness.

But increasingly, demographers and other experts say that cities may actually be a critical means of limiting environmental damage. Most significantly, they say, family size drops sharply in urban areas.

"The city is perhaps the most effective device for reducing the birthrate," said Dr. George Bugliarello, chancellor of the Polytechnic University in Brooklyn and an expert on urban trends.

For the poor, access to health care, schools and other basic services is generally greater in the city than in the countryside. Energy is used more efficiently, and drinking and wastewater systems, although lacking now, can be built relatively easily.

And for every person who moves to a city, that is one person fewer chopping firewood or poaching game.

Still, many cities face decisions now that may permanently alter the quality of human lives and the environment.

Dr. Kates said the pivotal nature of these times is perfectly illustrated by Mexico City, which is just behind Tokyo atop the list of megacities. The sprawling megalopolis, where traffic is paralyzed, is about to choose in a referendum between double-decking its downtown highways or expanding its subway system.

One course could encourage sprawl and pollution; the other would conserve energy, experts say.

Chances of Change

'Pernicious Fad' Or Real Prospect?

Some environmentalists say the whole notion of sustainable development is an oxymoron, that the Western industrial model of endless growth, however packaged, cannot possibly persist without grievous environmental damage. At the same time, some business leaders still scoff at the effort.

In a new PricewaterhouseCoopers survey of Fortune 1000 executives' attitudes toward sustainable development, one corporate vice president for environmental affairs called the concept "a pernicious fad."

Some skeptics note that even if cleaner, less destructive industries take hold everywhere, the sheer volume of economic growth could still cause big problems.

But a durable line of pragmatic optimists from science, business and environmental groups, for now, holds center stage.

They say that cleaning the environment and reducing poverty are not only required from an ethical standpoint, but also because they are in humanity's self-interest.

Without improvements, said Nitin Desai, the United Nations official running the Johannesburg meeting, "you create societies that live in a state of perpetual hopelessness."

"That is obviously going to hit back at you at some stage," Mr. Desai said.

Hundreds of businesses, though still a minority, have added sustainability managers to their executive roster.

Prof. Jeffrey D. Sachs, director of the Columbia University Earth Institute, said the world was quickly shifting from a model in which wealth was derived mainly through exploiting resources.

"Most growth now comes from increased knowledge, not from the mining of nature," Professor Sachs said. "And knowledge isn't limited in the way that, say, soil fertility is."

Dr. Kates agrees that economic advancement is vital and says evidence is emerging in many places that it can occur without too much environmental harm. Despite federal inaction on climate change, Dr. Kates said, 129 American cities have programs to cut greenhouse-gas emissions, and California has moved to reduce car emissions.

"This fits with history," he said. "States and cities have always been a major set of social experimenters. This was true on disability insurance and child-labor laws, on antimonopoly laws."

It may well end up being the case that local communities, here and abroad, lead the way in harmonizing people and the planet, he said.

"That ferment," Dr. Kates said, "is the most encouraging sign."

From the *New York Times*, August 20, 2002, pp. F1, F5. © 2002 by The New York Times Company. Reprinted by permission.

Crimes of (a) Global Nature

FORGING ENVIRONMENTAL TREATIES IS DIFFICULT.
ENFORCING THEM IS EVEN TOUGHER.

by Lisa Mastny and Hilary French

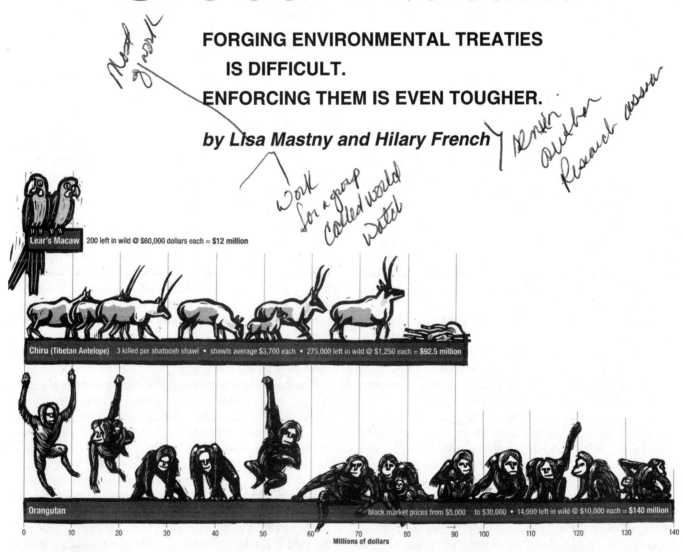

Lear's Macaw   200 left in wild @ $60,000 dollars each = **$12 million**

Chiru (Tibetan Antelope)   3 killed per shatoosh shawl • shawls average $3,700 each • 275,000 left in wild @ $1,250 each = **$92.5 million**

Orangutan   black market prices from $5,000   to $30,000 • 14,000 left in wild @ $10,000 each = **$140 million**

0   10   20   30   40   50   60   70   80   90   100   110   120   130   140
Millions of dollars

Extinction's Payoff
Pets, aphrodisiacs, distinctive clothing: these are a few of our favorite things—even if having them drives a species or two to extinction. The staggering prices some threatened animals fetch on the black market create powerful incentives for illegal trafficking and help increase the risk of extinction. The prices in the graph are probably conservative, since prices would tend to rise with increasing scarcity.

Last February, armed troops and fisheries officials on two Australian navy ships and a helicopter boarded and seized the Volga and the Lena, Russian-flagged fishing vessels operating near Heard Island, some 2,200 nautical miles southwest of Perth. The two ships were found to be carrying about 200 tons of illegally caught Patagonian toothfish in their holds. This bounty, valued at an estimated $1.25 million, had been taken in violation of conservation agreements negotiated under the auspices of the Commission for the Conservation of Antarctic Marine Living Resources.

Few casual seafood lovers have heard of Patagonian toothfish, but many are familiar with Chilean sea bass, a different name for the same fish. Chilean sea bass began appearing on menus in the early 1990s, and consumption of the flaky, white fish took off fast, quickly endangering the health of the fishery. Large-scale commercial fishing of the species began only a decade ago, but scientists estimate that at current rates of plunder the fish could become commercially extinct in less than five years.

A few months after the drama in the Southern Ocean, a different front in the same battle opened up thousands of miles away in Washington, D.C., where nearly 60 restaurants and caterers pledged to keep the fish off their menus. More than 90 restaurants in the Los Angeles area did the same a few weeks later, following similar promises by chefs in Northern California, Chicago, and Houston. Thus the fight to save the Patagonian toothfish is beginning to hit close to home. But many Chilean sea bass fans remain unaware that they may be accomplices to a growing phenomenon known as international environmental crime.

Although variously defined, in this article international environmental crime means an activity that violates the letter or the spirit of an international environmental treaty and the implementing national legislation. Trade in endangered species, illegal fishing, CFC smuggling, and the illicit dumping of wastes are all cases in point that are explored below. Illegal logging is another major category of environmental crime, although environmental treaties currently impose few specific constraints on logging (see Box, *Logging Illogic*). The rapidly growing illegal trade in these environmentally sensitive products stems from strong demand, low risk, and other factors (see Box, *Variable Crimes, Constant Incentives*).

The number of international environmental accords has exploded as countries awaken to the seriousness of transboundary and global ecological threats. The UN Environment Programme (UNEP) estimates that there are now more than 500 international treaties and other agreements related to the environment, more than 300 of them negotiated in the last 30 years.

But reaching such agreements is only the first step. The larger challenge is seeing that the ideals expressed in them become reality. What is needed is not necessarily more agreements, but a commitment to breathe life into the hundreds of existing accords by implementing and enforcing them.

Here the genteel world of diplomacy often runs into hard-nosed domestic politics. Countries that ratify treaties are responsible for upholding them by enacting and enforcing the necessary domestic laws. This requires the backing of businesses, consumers, and other constituencies, which may not be easily secured. Countries with strong fossil fuel industries, for instance, may meet staunch resistance to international rules to mitigate climate change. And countries where natural resource industries are politically powerful will probably find it difficult to adequately enforce environmental treaties designed to regulate resource-related activity. The effect has been to expand trafficking in a number of restricted substances, an increasingly urgent problem that is beginning to stimulate a stronger international response.

Trading in Wildlife

Undercover Russian police officers in the port city of Vladivostok recently trailed two investigators from environmental groups posing as eager purchasers of Siberian tiger skins from a corrupt official. When the deal went down, the officers arrested the wildlife trader on the spot. Russian investigators earlier had infiltrated the wildlife trade crime ring and determined that it was raking in some $5 million a year from smuggling wild ginseng, tiger skins, and bear paws and gallbladders across the Russian border.

Trade in these wildlife products is restricted under the terms of the 1973 Convention on International Trade in Endangered Species of Wild Fauna and Flora (CITES), which bans international trade in some 900 animal and plant species in danger of extinction, including all tigers, great apes, and sea turtles, and many species of elephants, orchids, and crocodiles. CITES also restricts trade in some 29,000 additional species that are threatened by commerce; among them birdwing butterflies, parrots, black and stony corals, and some hummingbirds.

CITES has shrunk the trade in many threatened species, including cheetahs, chimpanzees, crocodiles, and elephants. But the trafficking in these and other species continues, earning smugglers profits of $8 billion to $12 billion annually. Among the most coveted black market items are tigers and other large cats, rhinos, reptiles, rare birds, and botanical specimens.

Most illegally traded wildlife originates in developing countries, home to most of the world's biological diversity. Brazil alone supplies some 10 percent of the global black market, and its nonprofit wildlife-trade monitoring body, RENCTAS, estimates that poachers steal some 38 million animals a year from the country's Amazon forests, Pantanal wetlands, and other important habitats, generating annual revenues of $1 billion. Southeast Asian wildlife has also been plundered: the Gibbon Foundation reports that in a single recent year, traders smuggled out some 2,000 orangutans from Indonesia—at an average street price of $10,000 apiece.

The demand for illegal wildlife—for food and medicine, as clothing and ornamentation, for display in zoo collections and horticulture, and as pets—comes primarily from wealthy collectors and other consumers in Europe, North America, Asia, and the Middle East. In the United States (the world's largest market for reptile trafficking) exotic pets such as the Komodo dragon of Indonesia, the plowshare tortoise of northeast Madagascar, and the tuatara (a small lizard-like reptile from New Zealand) reportedly sell for as much as $30,000 each on the black market. Wildlife smuggling is also a growing concern in the United Kingdom, where in 1999 alone, customs officials confiscated some 1,600 live animal and birds, 1,800 plants, 52,000 parts and derivatives of endangered species, and 388,000 grams of smuggled caviar.

The multibillion-dollar traditional Asian medicine industry has also been a strong source of demand for illegal wildlife, with adherents from Beijing to New York purchasing potions made from ground tiger bone, rhino horn, and other wildlife derivatives for their alleged effect on ailments from impotence to asthma. In parts of Asia, bile from bear gall bladders (used to

VARIABLE CRIMES, CONSTANT INCENTIVES

International environmental crime involves many different kinds of activities and contraband, ranging from illegal fishing and trading in endangered wildlife to smuggling ozone-depleting chlorofluorocarbons (CFCs) across borders. But in most cases there are several common elements:

• **Booming demand, minimal investment, high profits**. Such crime is generally very attractive to traders and smugglers. Demand for increasingly rare plants, fish, or CFCs is strong. Investment can be minimal and the rewards great, with prices climbing substantially as the items change hands. For instance, a Senegalese wholesaler may buy a grey parrot in Gabon for $16–20 and sell it to a European wholesaler for $300–360, who then gets $600–1,200 for it.

• **Low risk, low penalties**. Environmental crimes normally carry low risk, with domestic penalties often light or nonexistent. A 2000 study by the secretariat of the Convention on International Trade in Endangered Species of Wild Fauna and Flora found that roughly half of the treaty's 158 parties failed to implement the treaty adequately. A common deficiency is a lack of appropriate penalties to deter treaty violations.

• **Creative smuggling methods**. Smugglers have devised clever ways to evade controls on restricted items, from using unauthorized crossing points to shipping illegal items along with legal consignments.

• **Links to organized crime**. Environmental criminals sometimes launder or funnel the proceeds from their trafficking to other illicit activities, such as buying drugs or weapons. The National Network for Combating the Traffic of Wild Animals (RENCTAS), a Brazilian NGO, reports that as much as 40 percent of the 300 to 400 criminal groups that control that country's wildlife trade also have links with drug trafficking.

• **Weak domestic treaty enforcement**. Because treaty secretariats have little centralized authority, signatory countries are expected to designate their own permitting authorities and train local customs inspectors and police to detect and penalize illegal activity. But many countries simply don't have the resources to do this.

• **Porous borders**. Many countries lack the money, equipment, or political will to effectively monitor illegal activity at border crossings. In the developing world in particular, customs offices are chronically understaffed and underfunded. Agents are often untrained to spot crimes or overwhelmed by the sheer numbers of items they must track.

• **Permitting challenges**. Local authorities may unwittingly issue permits for animals, seafood, chemicals, or other items they don't know are restricted, or let consignments slip through because they can't identify a fake or altered permit. Corrupt local officials may issue false permits or overlook illicit consignments in return for bribes or kickbacks.

Scarcity=Value
The tuatara has a third eye on top of its skull and can live to be 100 years old. Its cousins in the reptilian order Rhynchocephalia all died out 139 million years ago. These oddities may help explain why collectors will pay up to $30,000 apiece. At least one species is nominally protected under CITES.

treat cancers, asthma, eye disease, and other afflictions) can be worth more than narcotics. The New York-based Wildlife Conservation Society reports that the illegal hunting and trafficking of animals for medicine, aphrodisiacs, and gourmet food is now the single greatest threat to endangered species in Asia.

The scale of the illegal wildlife trade reflects the serious obstacles to enforcement under CITES. Smugglers may conceal the items on their persons or in vehicles, baggage, or postal and courier shipments, resulting in fatality rates as high as 90 percent for many live species. They may also alter the required CITES permits to indicate a different quantity, type, or destination of species, or to change the appearance of items so they appear ordinary. In February 2000, the U.S. Fish and Wildlife Service arrested a Cote d'Ivoire man for smuggling 72 elephant ivory carvings, valued at $200,000, through New York's John F. Kennedy airport. Many of them had been painted to resemble stone.

As wildlife smuggling grows in sophistication, it accounts for a rapidly rising share of international criminal activity. RENCTAS now ranks the illicit wildlife trade as the third largest illegal cross-border activity, after the arms and drug trades. Wildlife smugglers commonly rely on the same international trafficking routes as dealers of other contraband goods, such as gems and drugs. In the United States, consignments of snakes have been found stuffed with cocaine, and illegally traded turtles have entered on the same boats as marijuana. The U.S.-Mexico border has become a significant transfer point for environmental contraband: in the first eight months of 2001, Mexican authorities reported seizing more than 50,000 smuggled animals en route to the U.S. border.

As the scope of international wildlife trafficking becomes clearer, authorities are beginning to take action on the domestic, regional, and global fronts. In Europe, officials are working together to improve regional cooperation in enforcement of CITES and to strengthen legal mechanisms for prosecuting violators. The United Kingdom has established a new national police unit to combat wildlife crime and announced tougher penalties for persistent offenders. And in the first program of its kind in East Asia, a team at South Korea's Seoul airport now uses specially trained dogs to detect tiger bone, musk, bear gall bladders, and other illegal wildlife derivatives smuggled in luggage and freight.

In general, developing countries have had greater difficulty in controlling wildlife smuggling and implementing CITES. Many lack the political will, money, or equipment to effectively monitor their wildlife populations, much less the wildlife trade. In Kenya, where elephant deaths declined dramatically following a CITES-imposed ban on the international ivory trade in the late 1980s, poaching is again surging as funds to hire additional game wardens have run dry. In two separate incidents in April 2002, poachers armed with automatic weapons slaughtered 25 elephants in the country's wildlife parks and removed the tusks for sale on the black market.

At the international level, one tool that has proved successful in enticing (some would say coercing) countries to uphold their obligations under CITES is the use of trade sanctions. CITES is empowered to recommend that its members temporarily suspend all wildlife trade with noncomplying countries. Within the past year, such sanctions have been levied against the United Arab Emirates for not taking strong enough measures to combat the illicit trade in falcons, against Russia for not cracking down on the illegal caviar trade, and against Fiji and Vietnam for not enacting adequate wildlife trade legislation by the required deadline. In most instances, the governments scrambled to strengthen their legislation and enforcement, and the sanctions were soon lifted.

High Seas, High Crimes

Although CITES focuses mainly on terrestrial animals and plants, it protects several highly endangered fish species as well. In February 2002, the Philippine navy arrested 95 Chinese fishermen near a national marine park in the Sulu Sea and charged them with multiple counts of poaching fish, harvesting endangered species, and using illegal fishing methods like poison and explosives. Among the species found aboard their four vessels were endangered sea turtles and giant clams, both of which are prohibited from trade under CITES.

Countries have negotiated a wide range of other agreements to oversee and regulate the world's fisheries. Like CITES, many of these are poorly enforced, resulting in illegal fishing in all types of fisheries and in all the world's oceans, including national waters, regionally managed fisheries, and the high seas. In some of the world's most important fisheries, as much as 30 percent of the catch is illegal, according to the UN Food and Agriculture Organization (FAO).

As with other forms of wildlife trade, booming consumer demand is an important driver behind the rise in this activity. Big profits can be made selling black market seafood to selective buyers, who are willing to pay a premium for increasingly rare items. In Japan, species such as the threatened Southern bluefin tuna now fetch up to $50,000 per fish.

One major form of illegal fishing occurs when foreign vessels fish without authorization in the waters of other countries, often developing nations unable to patrol their shores adequately. In early 2000, the Tanzanian government estimated that more than 70 vessels, most of them from Mediterranean countries and the Far East, were fishing illegally in its waters. Tanzania has few police boats of its own and has had to rely on assistance from France and the United Kingdom to crack down on offenders. Mozambique, Somalia, and other African countries also report increased illegal fishing, often by heavily armed foreign vessels, and worry that the continued poaching will damage national economies and deprive coastal villagers of the healthy fisheries they depend upon for their livelihoods.

Commercial fishers are also turning to the largely unmonitored high seas, including the Mediterranean Sea and the Indian, South Atlantic, and Southern Oceans. Their ships illegally penetrate the borders of regional fishing grounds that are closed or restricted under international law, and deliberately hide their flags or other markings to avoid recognition. Some offload their illicit catch to other vessels to further disguise its origin and to minimize the penalties if discovered. Often these fishers do not report their activity, or if they do, they may falsify the equipment used, the fishing area frequented, or the species or amount caught.

TRAFFIC, a UK-based nonprofit group that monitors the international wildlife trade, reports that illegal Russian trawlers are accelerating the collapse of once productive fisheries in the Bering Sea. Backed by the Russian mafia, the fishers remove billions of dollars worth of pollack, cod, herring, flounder, halibut, and other species from the ecosystem each year, often trawling in prohibited areas and using illegal nets and other gear. They then transfer the catches to ships bound for ports in the United States, Canada, and Asia, in particular the South Korean port of Pusan, where vessel inspections are rare.

One of the most serious challenges to adequate fisheries enforcement is the rapid rise in so-called flag-of-convenience (FOC) fishing. Increasingly, commercial fishing companies register their ships in countries known to be lax enforcers of international fisheries laws or that are not members of major maritime agreements. By transferring their allegiance to these new "flags," the companies can easily enter the waters of their adopted countries or operate undercover on the high seas where only the country of registry (the flag state) can make an official arrest. Some vessels change their flags frequently in order to hide their origins or identities, making it difficult for enforcement officials to track them down.

An estimated 5 to 10 percent of the vessels in the world fishing fleet now fly flags of convenience. This includes more than 1,300 large industrial fishing vessels registered in such countries as Belize, Honduras, Panama, St. Vincent, and Equatorial Guinea.

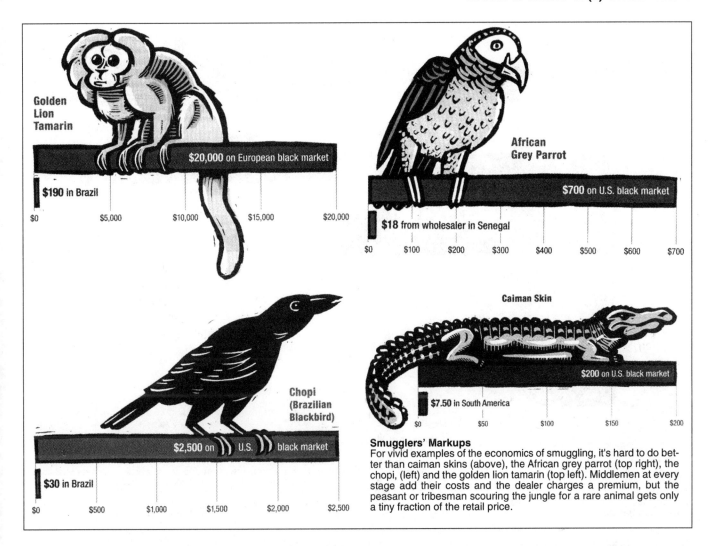

Golden Lion Tamarin
$20,000 on European black market
$190 in Brazil

African Grey Parrot
$700 on U.S. black market
$18 from wholesaler in Senegal

Chopi (Brazilian Blackbird)
$2,500 on U.S. black market
$30 in Brazil

Caiman Skin
$200 on U.S. black market
$7.50 in South America

Smugglers' Markups
For vivid examples of the economics of smuggling, it's hard to do better than caiman skins (above), the African grey parrot (top right), the chopi, (left) and the golden lion tamarin (top left). Middlemen at every stage add their costs and the dealer charges a premium, but the peasant or tribesman scouring the jungle for a rare animal gets only a tiny fraction of the retail price.

The countries and companies that own the reflagged vessels are also partially to blame for the problem. Fishing companies from Europe (primarily Spain) and Taiwan own the highest numbers of FOC vessels. Many of these firms receive special government subsidies to register their ships abroad.

FOC vessels also pose a serious problem in the cruise and shipping industries. Freight companies may re-flag their vessels in order to avoid higher taxes, labor, and operating costs at home, or to take advantage of the lower safety standards and weaker pollution laws in countries like Liberia, Panama, and Malta. The International Transport Workers' Federation reports that FOC ships accounted for nearly a quarter of all large freight vessels in 2000, and carried 53 percent of the world's gross tonnage. The re-flagged ships account for a disproportionate share of pollution and accidents at sea, as well as detentions in ports for violations of maritime laws.

Some of the most serious offenses have been breaches of MARPOL, the international treaty regulating the disposal of garbage and other pollutants at sea. Vessel owners may decide to dump their waste overboard illegally either because they lack the storage or incineration capacity onboard or because they wish to avoid the high costs of eventual disposal on land. Often this at-sea discharge occurs in the waters of countries where monitoring and enforcement are minimal and the activity can go

undetected. In 1999, however, U.S. officials fined Royal Caribbean Cruises Ltd. a record $18 million for releasing oil waste from its ships into U.S. waters, among other violations. In April 2002, rival Carnival Corporation was also fined $18 million for similar offenses.

As the scope of maritime violations becomes increasingly evident, many countries are redoubling their efforts to combat illegal fishing, including by exchanging information about illicit activity, making vessel registries more transparent, increasing inspections at sea and in ports, denying landing and trans-shipment rights for illegally caught fish, and improving monitoring and surveillance of fleets. These steps appear to be working, at least in some places: in the North Pacific, Canadian officials attribute the recent drop in the number of vessels using illegal driftnets to stepped-up air patrols by Canada and its partners (the United States, Russia, and China) in regional enforcement.

Significant effort has also been made at the global level to crack down on illegal fishing. In March of last year, 114 countries agreed to a non-binding plan to combat illicit activity, including that by flag-of-convenience vessels. Developed under the auspices of the UN Food and Agriculture Organization, the plan calls for improved oversight of vessels and coastal waters by flag states, better coordination and sharing of information

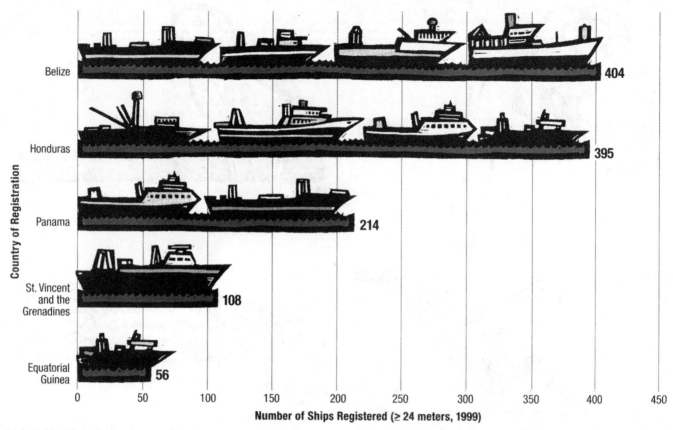

Country of Registration (y-axis)

- Belize — **404**
- Honduras — **395**
- Panama — **214**
- St. Vincent and the Grenadines — **108**
- Equatorial Guinea — **56**

Number of Ships Registered (≥ 24 meters, 1999)

False Colors
Treaties to control overfishing and other marine crimes generally require the nation where a ship is registered to do the policing. Commercial fishers often register their ships with countries unable, or deliberately reluctant, to enforce the treaties. Such flag-of-convenience (FOC) vessels can operate with little risk of prosecution, which leads to overfishing and depletion of fish stocks. As much as one-tenth of the world's fishing vessels fly flags of convenience. The top five FOC registration countries are shown here, and the opposite page shows the top five countries where firms owning the vessels are based.

among countries, and stronger efforts to ratify, implement, and enforce existing fisheries accords. The plan is expected to work in conjunction with other international efforts to protect global fisheries, including a 1993 pact known as the Compliance Agreement and the 1995 UN Agreement Relating to the Conservation and Management of Straddling Fish Stocks and Highly Migratory Fish Stocks.

Dumping on Land

A Japanese district court last March sentenced Hiromi Ito, the president of a waste disposal company, to four years in prison and slapped him with a 5 million yen ($40,000) fine. Ito and his accomplices had masterminded an elaborate scheme to illegally dump some 2,700 tons of industrial and medical waste in the Philippines, in containers marked 'paper for recycling.' When the Filipino importer opened the 122 shipping containers, each 40 feet long, he found not just paper but also hazardous materials, including contaminated hypodermic needles and bandages, used plastic sheeting, and old electronics equipment.

With Ito's conviction, the Japanese government closed the books on an embarrassing international incident. But the episode points to a much larger global challenge: effectively regulating the 300 to 500 million tons of hazardous waste generated

worldwide each year. This vast waste mountain includes everything from used batteries, electronic wastes, old ships, and toxic incinerator ash to industrial sludge and contaminated medical and military equipment.

Roughly 10 percent of this waste is shipped legally across international borders, under the terms of the 1989 Basel Convention on the Control of Transboundary Movements of Hazardous Wastes and Their Disposal. The agreement requires member countries to obtain prior consent from the importing country for waste shipments and uses a system of permits to track the pathway to disposal. Most of the waste originates in and moves among industrial countries, but some also travels to and within the developing world.

As waste disposal problems mount in countries like Japan, so does the likelihood of a lucrative illegal trade in hazardous wastes. Asia is thought to be one of the biggest destinations for the illicit waste. Greenpeace India reports that more than 100,000 tons of unauthorized wastes entered India in 1998 and 1999, including toxic zinc ash and residues, lead waste, used batteries, and scraps of chromium, cadmium, thallium, and other heavy metals. The illicit imports, originating in places like Australia, Belgium, Germany, Norway, and the United States, violated both the Basel treaty's notification rules and a 1997 Indian government ban on waste imports.

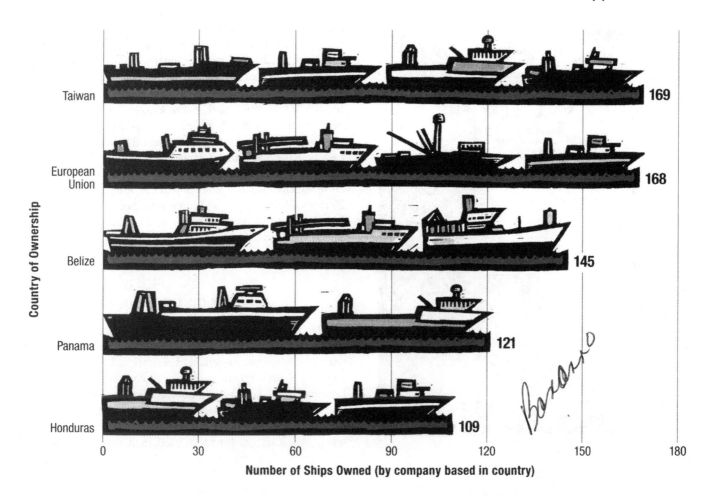

Number of Ships Owned (by company based in country)

China receives a massive flow of illegally imported hazardous electronic waste each year. A recent study by the Basel Action Network (BAN), a watchdog group that monitors implementation of the Basel treaty, and the California-based Silicon Valley Toxics Coalition reported that workers in Chinese recycling factories risk serious exposure to heavy metals and other poisonous chemicals as they salvage components from old computer circuitboards, monitors, batteries, and other equipment. This toxic trade continues despite a Chinese import ban on the material, and hence violates Basel rules that forbid the export of wastes to countries that have banned their import.

Roughly half of this "e-waste" originates in the United States, where an estimated 20 million computers become obsolete each year. Yet the U.S. government doesn't consider the high-tech shipments to be illegal because the waste isn't technically classified as hazardous. The United States is also the only industrial country that hasn't ratified the Basel Convention.

Developing countries are particularly vulnerable to the health and environmental effects associated with the illegal waste trade. Many governments lack the infrastructure or equipment to dispose of or recycle waste safely, prevent exposure to workers and communities, clean up dumpsites, or monitor waste movements. For this reason, a bloc of developing countries secured passage in 1995 of a far-reaching amendment to the Basel Convention, known as the Basel Ban, which would outlaw all transfers of hazardous wastes from industrial countries to the developing world. Though many countries already observe its terms voluntarily and the European Union has already implemented it, the Ban is not yet in strict legal force and still faces serious opposition from a few industrialized countries, including the United States.

Even so, the Basel Ban has helped slow the flow of hazardous wastes from industrial countries to the developing world, says Jim Puckett, coordinator of BAN. And once the amendment enters into legal force (it needs 32 more ratifications), it should be even harder for waste traders to operate, as violators would face strict criminal penalties. If the ban fails to enter into force, however, or if it is weakened through continued opposition, the waste flood could resume.

But the ban alone would likely fail to wipe out the illegal waste trade. Smugglers rely on a wide range of tactics, including false permits, bribes, and mislabeling of wastes as raw materials, less dangerous substances, or other products, to evade the laws. Moreover, no port in the world can check all sea-going containers, let alone developing-country ports.

One growing trend is the export of hazardous wastes under the pretext of "recycling." Like Hiromi Ito, illicit waste dealers increasingly pass off waste as recyclable material, which in many cases frees them from strict government oversight. Greenpeace estimates that as much as 90 percent of waste shipped to the developing world—particularly plastics and

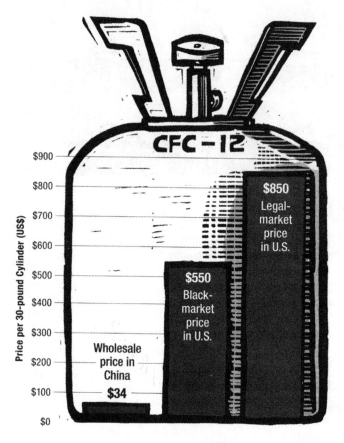

The Price of Cool
The Montreal Protocol ended CFC manufacture in the industrialized world, but developing nations have until 2010 to phase it out. Some of their output feeds illicit demand in the United States and Europe.

heavy metals—is now labeled as destined for recycling. Much of this waste, however, is never recycled—or, as with e-waste in China, is recycled in highly polluting operations that are little better than dumping. Typically, the end result is the same: the export of a serious pollution problem from a rich country to a poor one.

In many cases, organized crime is thought to be behind large-scale waste trading, which can be closely linked with money laundering, the illegal arms trade, and other criminal activities. The Italian mafia is reportedly a key player in the robust trade in radioactive metal waste from Eastern Europe and the former Soviet Union, which it re-sells to smelters as "safe" scrap metal.

Efforts to combat the illegal waste trade face serious challenges. A recent survey by the Basel secretariat revealed that many countries lack adequate—or any—legislation for preventing and punishing illegal waste traffic. On a global scale, the absence of uniform definitions of hazardous waste and of coordinated enforcement efforts among customs officers and port authorities has contributed to the spread of illegal waste trading. Even for countries that do have the resources, the lack of hard data on the extent or geographic flow of this trade makes it hard for officials to know how to allocate the resources properly.

The member countries of the Basel Convention have taken some steps to give the treaty sharper teeth. In 1994, the Basel secretariat strengthened information sharing among member countries by agreeing to set up a centralized system for reporting suspect activity. The treaty also requires members to pass laws to curb and punish illegal waste traffic and outlines how to handle illegally traded waste once it is discovered.

A new liability protocol to the convention, negotiated in December 1999, could further discourage illegal activity—if it ever comes into force. It makes exporters and disposers of hazardous waste liable for any harm that might occur during transport, both legal and illegal. It also requires dealers to be insured against the damage and to provide financial compensation to those affected. But the protocol will not be legally binding until 20 ratifications are received (none have been registered so far). Moreover, it still only covers harm that occurs in transit, not after disposal, and only applies to damage suffered in the jurisdiction of a treaty member.

CFCs on the Loose

The landmark Montreal Protocol on Substances That Deplete the Ozone Layer, adopted in 1987, mandated far-reaching restrictions in the use of certain chemicals that damage the thin, vital veil of stratospheric ozone that protects the earth and its inhabitants from excessive ultraviolet radiation. The Protocol and its later amendments set target dates for the phaseout of 96 different ozone-depleting substances, most notably CFCs and halons, chemicals once widely used in a range of industrial applications. Industrialized countries were required to halt production and import of CFCs in 1996, while developing countries have until 2010 to complete the phaseout.

Considered one of the world's most successful environmental treaties, the Protocol has resulted in a dramatic decline in the overall use of ozone-depleting substances. But starting in the mid-1990, the different phaseout schedules for industrial and developing countries helped stoke a flourishing illegal trade in the banned chemicals. CFCs that were still legally produced in the developing world began to make their way to lucrative black markets in the United States and Europe, where demand for substances like Freon (used in older-model auto air conditioners) remained high.

Government and industry reports suggest that in the mid-1990s, as much as 15 percent of global annual production of the chemicals, or 38,000 tons, was smuggled into industrial countries. Early shipments originated in Russia, but today the bulk of the smuggled CFC supply is thought to originate in China and India, which together account for more than half of the world's remaining CFC production.

CFC importers resort to fraud and other evasive tactics to smuggle the banned chemicals. For instance, traders abuse existing loopholes in the Montreal Protocol and domestic laws to pass off shipments of new CFCs as recycled material or as CFC replacements, neither of which are restricted under the treaty. The chemicals are typically colorless and odorless, making them easy to disguise and virtually impossible to differentiate without chemical analysis.

In the United States, the illegal CFC trade is believed to have peaked in the mid-1990s, just after the initial phaseout. At that

LOGGING ILLOGIC

Money doesn't grow on trees, but some trees might as well be pure gold. The world's voracious (and growing) appetite for wood, paper, and other forest products is driving a stampede to mow down forests.

Much of this logging is illegal. Illegally cut wood accounted for up to 65 percent of world supply in 2000, according to the World Resources Institute. Estimates of illegal logging as a share of the total range from 35 percent in Malaysia, to 50 percent in Cameroon, 50 percent in eastern Russia, 70 percent in Gabon, 73 percent in Indonesia, and up to 80 percent in parts of Brazil. About 40 percent of the wood processed in the pulp and paper industry in Indonesia is of questionable origin, and up to 46 percent of the domestic demand in the Philippines is supplied from illegal sources. Precise global data are unavailable, but in terms of commodity value, illegal logging may be the most serious transnational environmental crime.

Illegal logging activities include logging in national parks and protected areas; harvesting protected timber species; overcutting and underreporting of timber volume, grade, and species; logging for illegal commercial charcoal and fuelwood production; logging without permits; smuggling; and violating forest laws and restrictions. Like money, timber can be "laundered;" in May the BBC reported on illegally logged Indonesian timber that was imported openly into Malaysia, processed into garden furniture, and exported as of Malaysian origin.

National governments have taken a range of measures against these crimes—overhaul of forest legislation, reforming permit processes, adjusting tax codes and royalty systems, various bans—with mixed success. Indonesia, for instance, banned log exports last year and in May declared a temporary moratorium on logging concessions. Logging persists, however, to supply pulp and paper mills and (via bulldozing and burning) to clear forests for palm oil plantations. Logging bans often increase illegal logging in neighboring countries; a recent Chinese logging ban has boosted demand for timber from Cambodia, Vietnam and far eastern Russia. Thailand's logging ban in the 1990s encouraged the extraction of timber in Laos and Myanmar, especially teak.

On the plus side, in northern Tanzania, pilot village forest management programs were so successful in controlling illegal forest activities, including logging, that Tanzania's forest policy now promotes community involvement in forest management nationwide. Ecuador recently launched Vigilancia Verde, a coalition of the Environment Ministry, the armed forces, police, and environmental NGOs, to collaborate in monitoring the transport of timber to markets and mills. In its first year it seized five times the volume of timber confiscated by the government the year before.

Controlling illegal logging is dangerous. After Brazil banned mahogany logging, one of the promoters of the ban was gunned down at his house. In Guerrero, Mexico, the army shot one farmer and jailed and tortured two others who protested logging abuses. A campaign by the Philippine Department of Environment and Natural Resources against illegal logging and its own corruption has made progress—in 2001, over 12 thousand cubic meters of illegal timber confiscated, 76 criminal cases filed, and 14 DENR personnel suspended—but at the cost of nearly 80 DENR staff killed in the line of duty over the years.

The risk is commensurate with the profits, which are high because of strong demand for timber products, especially form the developed world. According to the nonprofit Environmental Investigation Agency, the European Union imported about $1.5 billion worth of illegal timber in 1999; in 2000 the United States imported an estimated $330 million worth of illegal timber from Indonesia alone. The G8 nations (Britain, Canada, France, Germany, Italy, Japan, Russia, and the United States) account for about three-quarters of global timber and wood products imports, yet to date no G8 nation has laws requiring routine seizure of illegal timber imports.

—Christine Haugen

Ms. Haugen (chaugen976@aol.com) is a consultant on resource and energy issues. She is the author of the Directory of Tropical Forestry Journals and Newsletters *(FAO-Bangkok, 1997) and co-author of* Keeping it Green; Tropical Forestry Opportunities for Mitigating Climate Change *(World Resources Institute, 1995).*

time, as much as 10,000 tons of the chemicals entered the country each year. By 1995, CFCs were considered the most valuable contraband entering Miami, after cocaine. Following a crackdown on large consignments through East Coast ports, much of the illegal trade shifted to the Mexican border.

The U.S. Department of Justice estimates that in total, some 10,000–20,000 tons of CFCs have been smuggled into the United States since 1992. Despite stronger enforcement efforts, officials have recovered only a fraction of this contraband. So far, activities under the North American CFC-Anti-Smuggling Initiative, an interagency task force established in 1994, have

led to 114 convictions and the seizure of some 1,125 tons of smuggled CFCs.

Today, a second, smaller, spike in black market activity is occurring as the remaining U.S. stockpiles of legal CFC-12 are depleted. Traders can once again earn a high profit on the contraband supplies: a 30-pound cylinder of CFC-12, bought in China for as little as $30 to $60, can be resold on the U.S. black market for as much as $600. For auto repair shops and other end users, obtaining this illegal product can still be cheaper than buying the legal supplies, which now cost as much as $1,000 per cylinder.

Europe has been another significant market for illegal CFCs. In the mid 1990s, researchers with the London-based Environmental Investigation Agency (EIA) uncovered a thriving regional trade amounting to between 6,000 and 20,000 tons of the chemicals annually. Well after the phaseout deadline, supplies were still abundant and prices disproportionately low, suggesting that the market was being swamped with illegal imports. Meanwhile, regional sales of CFC replacements were slower than expected.

The European black market has thrived in part because regional refrigerant management programs have been poorly organized, and because consumers perceive alternatives to CFCs to be too costly and less efficient. In Central and Eastern Europe, where the illegal CFC trade is thought to be increasing, a major problem is that border officials are typically untrained in identifying the chemicals and have difficulty deciphering their often vague customs codes.

European efforts to control CFC smuggling have generally lagged behind those of the United States, but there are signs that European enforcement is improving. In 1997, authorities in Belgium, Germany, the Netherlands, and the United Kingdom jointly nabbed a multimillion-dollar crime ring that had illegally imported more than 1,000 tons of Chinese-made CFCs for redistribution in Europe and the United States. And in an unprecedented move, in September 2000 the European Union adopted a regional ban on CFC sales and use.

Demand for contraband CFCs has also been high in Japan, especially as retail prices for the chemicals have skyrocketed. Although the Japanese government banned the use of Freon in new cars in 1994, 15 to 20 million vehicles in the country still use the refrigerant. The *Japan Times* reports that auto repair shops in the country circulated more than 100,000 canisters of illicit CFC-12 in 2001, most of them originating in China and other developing countries.

As the CFC phaseout begins to take hold in the developing world (countries were required to freeze consumption in July 1999), black markets are beginning to emerge in places like Asia, where there are still millions of users of CFC-based equipment, including old cars and refrigerators that have been exported from the industrial world. In October 2001, EIA reported on a growing multi-million dollar market for illegal CFCs in India, Pakistan, Bangladesh, Malaysia, the Philippines, and Vietnam. Between early 1999 and March 2000, smugglers slipped some 880 tons of ozone-depleting substances into India, representing an estimated 12 percent of national consumption.

To improve monitoring of the CFC trade and head off future black markets, parties to the Montreal Protocol recently adopted a new licensing amendment that entered into force in 1999. It requires member countries to issue licenses or permits for the import and export of all new, used, and recycled ozone-depleting substances and to exchange information regularly about these activities. By identifying who is and is not licensed to trade, the system should make it easier for police and customs officials to track the movement of the chemicals worldwide.

From Words to Action

Four years ago, the environment ministers of the leading economic powers expressed "grave concern about the ever-growing evidence of violations of international environmental agreements," and called for a range of cooperative actions aimed at stepped-up enforcement. This initiative was followed by the adoption last February of international guidelines to promote compliance with multilateral environmental agreements and prevent cross-border environmental crime. This August's World Summit on Sustainable Development will focus renewed international attention on the importance of adequately implementing and enforcing international environmental treaties and other agreements.

In other promising developments, the World Customs Organization is working with governments to harmonize classification systems for waste and other environmental contraband. The international police organization INTERPOL is training national enforcement officers and customs agents to identify illicitly traded goods more easily, and is also working with national police forces to bring international environmental criminals to justice. Both institutions have established close working relations with UNEP and with the CITES and Basel Convention secretariats.

NGOs are playing a strong role as well. Brazil's RENCTAS is cooperating with the police and the federal environment ministry to train officers in wildlife inspection and is investigating anonymous tips about wildlife smuggling left on its Web site. At the international level, TRAFFIC is tracking national customs enforcement efforts, documenting areas of unsustainable wildlife trade, identifying trade routes for wildlife commodities, and investigating smuggling allegations.

New technologies are also being deployed against international environmental crime. Remote sensing, for instance, was used by the U.S. Coast Guard in the late 1990s to gather the evidence of illegal dumping of oil in international waters that helped to bring Royal Carribean to justice. Satellite-linked Vessel Monitoring Systems are also increasingly being used to monitor the movement of fishing boats in order to detect illegal harvesting. DNA tracing is being used to monitor both fishing and wildlife trade by enabling researchers to link seafood items and wildlife parts back to the species or even the animal of origin.

Targeted industries and countries have bowed to the combined might of NGOs and consumers in several cases. In 1999, pressure from the World Wide Fund for Nature and increased public awareness of the threats that traditional medicine usage poses to wildlife caused leading practitioners and retailers in China to pledge not to prescribe or promote medicines containing parts from tigers, rhinos, bears, and other endangered species. Late last year, Belize, which Greenpeace calls "the world's most fish pirate-friendly country," struck five notorious pirate vessels from its shipping register after Greenpeace showcased the plight of the Patagonian toothfish.

FOR FURTHER INFORMATION

Wildlife Trade
CITES: www.cites.org
TRAFFIC: www.traffic.org
RENCTAS: www.renctas.org.br

Illegal Fishing
Take a Pass on Chilean Sea Bass campaign: http://environet.policy.net/marine/csb

International Transport Workers' Federation FOC campaign: www.itf.org.uk/seafarers/foc/foc.htm
FAO Fisheries Department: www.fao.org/fi

Hazardous Waste Trade
Basel Convention: www.basel.int
Basel Action Network: www.ban.org

CFC Smuggling
Montreal Protocol: www.unep.org/ozone

Miscellany
Environmental Investigation Agency: www.eia-international.org
UN Environment Programme: www.unep.org
World Summit on Sustainable Development: www.johannesburgsummit.org
INTERPOL: www.interpol.int
World Customs Organization: www.wcoomd.org

Although so far they are exceptions rather than the rule, these examples demonstrate that international environmental crime can be controlled through the combined efforts of governments, international institutions, businesses, NGOs, and ordinary citizens. That's good news for the integrity of the Earth's protective ozone layer, numerous threatened species, and the health of communities poisoned by hazardous wastes.

Lisa Mastny *is a research associate at Worldwatch Institute*. Hilary French *is director of the Institute's Global Governance Project.*

From *World Watch*, September/October 2002, pp. 12-23. © 2002 by Worldwatch Institute, www.worldwatch.org.

UNIT 2

The World's People: Population and Economy

Unit Selections

Key Points to Consider

• Assess the criticisms that some detractors have leveled against the organization of Planned Parenthood and its founder. Are these criticisms valid, and is the message of Planned Parenthood therefore discredited? Defend your answer.

• Why should policy makers in the more developed countries of the world become more aware of the true dimensions of the world's food problem? How can increased awareness of food scarcity and misallocation lead to solutions for both food production and environmental protection?

• How are present economic theories insufficient in addressing environmental issues? How might an "eco-economy" take on different characteristics than the traditional economies that exist in the world's developing countries?

• Explain some of the links between environmental degradation in both urban and rural areas and the increasing globalization of the world's economic system. Are there ways in which the advantages of globalized economies can be more equitably distributed? Explain.

 **Links: www.dushkin.com/online/**
These sites are annotated in the World Wide Web pages.

The Hunger Project
  http://www.thp.org
Poverty Mapping
  http://www.povertymap.net
World Health Organization
  http://www.who.int
World Population and Demographic Data
  http://geography.about.com/cs/worldpopulation/
WWW Virtual Library: Demography & Population Studies
  http://demography.anu.edu.au/VirtualLibrary/

One of the greatest setbacks on the road to the development of more stable and sensible population policies came about as a result of inaccurate population growth projections made in the late 1960s and early 1970s. The world was in for a population explosion, the experts told us back then. But shortly after the publication of the heralded works *The Population Bomb* (Paul Ehrlich, 1975) and *Limits to Growth* (D. H. Meadows et al., 1974), the growth rate of the world's population began to decline slightly. There was no cause and effect relationship at work here. The decline in growth was simply demographic transition at work, a process in which declining population growth tends to accompany increasing levels of economic development. Since the alarming predictions did not come to pass, the world began to relax a little. However, two facts still remain: population growth in biological systems must be limited by available resources, and the availability of Earth's resources is finite.

That population growth cannot continue indefinitely is a mathematical certainty. But it is also a certainty that contemporary notions of a continually expanding economy must give way before the realities of a finite resource base. Consider the following: In developing countries, high and growing rural population densities have forced the use of increasingly marginal farmland once considered to be too steep, too dry, too wet, too sterile, or too far from markets for efficient agricultural use. Farming this land damages soil and watershed systems, creates deforestation problems, and adds relatively little to total food production. In the more developed world, farmers also have been driven—usually by market forces—to farm more marginal lands and to rely more on environmentally harmful farming methods utilizing high levels of agricultural chemicals (such as pesticides and artificial fertilizers). These chemicals create hazards for all life and rob the soil of its natural ability to renew itself. The increased demand for economic expansion has also created an increase in the use of precious groundwater reserves for irrigation purposes, depleting those reserves beyond their natural capacity to recharge and creating the potential for once-fertile farmland and grazing land to be transformed into desert. The continued demand for higher production levels also contributes to a soil erosion problem that has reached alarming proportions in all agricultural areas of the world, whether high or low on the scale of economic development. The need to increase the food supply and its consequent effects on the agricultural environment are not the only results of continued population growth. For industrialists, the larger market creates an almost irresistible temptation to accelerate production, requiring the use of more marginal resources and resulting in the destruction of more fragile ecological systems, particularly in the tropics. For consumers, the increased demand for products means increased competition for scarce resources, driving up the cost of those resources until only the wealthiest can afford what our grandfathers would have viewed as an adequate standard of living.

The articles selected for this second unit all relate, in one way or another, to the theory and reality of population growth and its relationship to economic growth. In the first selection, "Population Control Today—and Tomorrow?" economist Jacqueline R. Kasun takes a sharply critical look at past population control ef-forts, with a particularly pointed criticism of Planned Parenthood and its founder, Margaret Sanger. Kasun claims that while population control has achieved success in reducing growth rates significantly, the funding of programs for further population control are still increasing. More important, she notes, money spent on population control in developing nations is often money that is not spent on basic medicines.

Next, in "Population and Consumption: What We Know and What We Need to Know," geographer and MacArthur Fellow Robert Kates argues that the present set of environmental problems are tied to both the expanding human population in strict numerical terms and the tendency of that growing population to demand more per capita shares of the world's dwindling resources.

In the following two articles in this unit—"An Economy for Earth" and "The Eco-Economic Revolution: Getting the Market in Sync With Nature"—Lester Brown takes first a theoretical approach and then a practical approach to the need to adjust the global economy to the demands of a sustainable environment. Economic theory, Brown contends, does not explain why so many environmental problems exist, whereas ecological theory does. What is required is a merger of the two sets of theories. A similar approach to understanding the links between population, resources, and economy is taken in the next article in the section. In "Poverty and Environmental Degradation: Challenges Within the Global Economy," Nigerian geographer Akin Mabogunje argues that the process of globalization—the expansion of information and technology that has expanded the reach of capitalist production systems throughout the world—has increased rather than decreased poverty. While enormous wealth exists in some parts of the world, over a third of the world's population exists on less than U.S. $2 per capita per day. As these deprived populations attempt to catch up with the wealthy, they pay little attention to the problems of environmental degradation. The last article in the unit also addresses the question of poverty, particularly that existing in the world's urban slums—another manifestation of the globalization of economic systems. Worldwatch Institute researcher Molly O'Meara Sheehan, in "Where the Sidewalk Ends," describes the plight of the urban poor. As more agricultural, mineral, and forest production is demanded by economic growth, the traditional patterns of subsistence and village life are broken down and enormous numbers of unskilled and destitute rural people flee the countryside for the cities. Here they exist without benefit of the urban infrastructure of sewers, utilities, and transportation that makes urban life livable for millions of the world's people. These urban poor and the slums they inhabit pose an enormous economic, social, and political challenge for developing regions.

All the authors of the selections in this unit make it clear that the global environment is being stressed by population growth and that more people means more pressure and more poverty. But while it should be evident that we can no longer afford to permit the unplanned and unchecked growth of the planet's dominant species, it should also be apparent that the unchecked growth of economic systems without some kind of environmental accounting systems is just as dangerous.

ANALYSIS

Population Control Today—and Tomorrow?

by Jacqueline R. Kasun

The success of the population control movement over the past four decades has been nothing less than astonishing. Places like Bangladesh and Kenya are awash in condoms (even though basic medicines are scarce), and population is actually falling in some countries and heading in that direction in many others.

Yet the movement is astonishing in another way, too: Despite its success, it is expanding at a breakneck pace in terms of both funding and programs.

One of the least-remarked events of the year 2000 was the announcement by the UN Population Division that "in the next 50 years, the populations of most developed countries are projected to become smaller and older as a result of low fertility and increased longevity.... Population decline is inevitable in the absence of replacement migration."

The division reported that 44 percent of the world's population lives in countries where birthrates are too low to prevent population decline. If present trends continue, there will be 100 million fewer people in Europe and 21 million fewer in Japan 50 years from now. The birthrate in the United States has fallen from 24.3 per thousand population in 1950–55 to 14.6 in 1998, a trend that is likely to continue for some time because the female population of reproductive age will decline by several million during the next decade (unless offset by immigration). Also, the U.S. death rate has been rising slightly but perceptibly, because the population is aging. (The death rate for a

fixed group of people is still 100 percent sooner or later, despite rising life expectancy, and older populations have higher death rates, other factors being equal.)

The UN Population Division predicts that world fertility will continue to decline from its present average of less than three children per woman (the one-child family is now typical in Europe and Japan) while the death rate rises. Thus, the proportion of people over 60 will rise to exceed the proportion of people under 15, for the first time in history.

Nevertheless, groups supporting population control continue to press for more funding for their programs both at home and abroad. Population Action International, for example, reported in June 2000 that "the Clinton administration intends to fight for additional funds" and that "Hollywood celebrities mingled with top policymakers and international family planning advocates on... World Health Day... to show the... administration's support for population assistance."

ROOTS OF THE MOVEMENT

The quest of those in power to control population is at least as old as the Exodus story of Pharaoh killing Hebrew baby boys. In our time, the movement has received stimuli from both eugenics and environmental worries.

Eugenics was a rather popular cause in the first half of the twentieth century in the United States and England. "More

children from the fit, less from the unfit," Margaret Sanger, the founder of Planned Parenthood, wrote in her popular magazine, *Birth Control Review*, in 1919. Thirty-one states passed compulsory sterilization laws in the first half of the century.

Early this year, the Virginia State legislature expressed its "regrets" to Raymond Ludlow for forcibly sterilizing him at the age of 16 in 1941 for repeatedly running away from home. Ludlow, one of thousands sterilized by force across the country, subsequently served as a radioman in the Army, earning a Bronze Star, a Purple Heart, and a Prisoner of War Medal.

To reduce the U.S. birthrate, Planned Parenthood proposed ideas like putting "fertility control agents" in the water supply and encouraging homosexuality.

At the close of World War II, Guy Irving Burch, the founder of the Population Reference Bureau, submitted his plan to solve all world problems through compulsory sterilization of "all persons who are inadequate, either biologically or socially," as he wrote in *Population*

Roads to Peace or War. Although Congress did not endorse Burch's plan, his bureau subsequently received millions of dollars in government grants and contracts for "population education" and other activities.

THE 'EXPLOSION'

New concerns emerged in the postwar years. A sudden spurt in population growth occurred in the 1960s as antibiotics and improvements in sanitation sharply reduced death rates. (Birthrates had been declining throughout the century as women joined the workforce and curtailed childbearing.) As death rates plunged below the falling birthrates, world population grew at an unprecedented pace. The response was intense.

Population Negation

- Birthrates have been declining precipitously around the world.

- Forty-four percent of the world's people live in nations whose population has shrunk or at least stalled in its growth.

- If present trends continue, Europe's population will fall by 100 million and Japan's by 21 million in the next 50 years.

- Yet funding and programs for population control are increasing.

- The population-control market is saturated, with a surfeit of contraceptives in many developing countries that otherwise lack basic medicines.

- Many Third World countries are suffering from the cultural seeds planted by the family planning movement, especially promiscuity, which spreads sexually transmitted diseases.

Congress held hearings. President Johnson recommended legislation, which Congress passed in 1965 and '67,

providing for the world's largest program of publicly financed birth control, targeted both at home and abroad.

In 1970, President Nixon appointed the Commission on Population and the American Future, under the chairmanship of John D. Rockefeller III, founder of the Population Council. That same year, Planned Parenthood published a list of "proposed measures to reduce U.S. fertility," among them putting "fertility control agents" in the water supply, encouraging homosexuality, imposing a "substantial" marriage tax, discouraging home ownership, requiring permits for couples to have children, making abortion compulsory, and mandating sterilization of all women who had borne two children. The United Nations proclaimed a World Population Year in 1974.

In a document that remained classified from 1974 to 1980, the U.S. State Department warned that "mandatory population control measures" might be necessary to bring about a "two-child family on the average" throughout the world by the year 2000. By 1975, the U.S. Agency for International Development (AID) was the world's chief player in world population control, spending more money on it than did all other countries combined.

In 1978, AID officials initiated and Congress enacted Section 104(d) of the new foreign aid legislation, which stipulated that "all… activities proposed for financing… shall be designed to build motivation for smaller families." The World Bank also began to impose population control requirements on its lending, as did other international institutions and countries. Henceforth, developing countries seeking international aid would be required to give evidence of their "commitment" to the "control of population growth."

PROMISED CALAMITIES

The justifications were a long, varied list of calamities that would ensue in the absence of swift, stern action. Starvation was looming, according to experts such as biologist Paul Ehrlich of Stanford University. The Sierra Club published his book *The Population Bomb*, which

became required reading in many high schools and colleges.

The House Select Committee on Population announced in 1978 that the "major biological systems that humanity depends upon… are being strained by rapid population growth… [and] in some cases, they are… losing productive capacity." Created by the Smithsonian Institution at about the same time, a traveling exhibit for schoolchildren called "Population: The Problem Is Us" featured a picture of a dead rat on a dinner plate as an example of "future food sources."

The Carter administration's Council on Environmental Quality and State Department together warned that "the staggering growth of human population… [was creating]… possibilities of… permanent damage to the planet's resource base." Robert McNamara, then director of the World Bank, warned in 1977 that continued population growth would cause "poverty, hunger, stress, crowding, and frustration" that would threaten social, economic, and military stability.

Sen. (later Vice President) Al Gore warned in his 1992 book *Earth in the Balance* of the approach of an "environmental holocaust without precedent," like a "black hole" caused by "expansion beyond the environment's carrying capacity." To stave off this catastrophe, he wrote, "the first strategic goal should be the stabilizing of world population."

Herman Daly, a World Bank economist, proposed in his 1990 book *For the Common Good* that, as a step toward the "sustainable society," births be limited by a government-operated licensing system. The UN Population Fund (UNFPA, not to be confused with the UN Population Division, a statistical agency) gave millions of dollars a year starting in 1979 to China's population control program, which featured forced abortion.

The funding increased along with the pressure. By 1994, federal and state governments were spending more than $2 billion a year directly on domestic and foreign population control. (Probably much more was being spent due to population-control requirements attached to

Population Control Pillar

Margaret Sanger, who founded the Planned parenthood Federation of America in 1942, is often viewed as the patron saint of the modern population control movement.

Her critics, citing numerous references in her writings, denounce her as a white supremacist, Nazi sympathizer, and an advocate of free sex.

Her supporters dismiss the criticism, saying that the references are confined to a small number of sources and are often taken out of context. Esther Katz, editor and director of the Margaret Sanger Papers Project, has said, "As a historian, I take issue with [such] gross misuse of historical sources to support those views."

One thing that critics often say about Sanger is that she viewed blacks as inferior and wanted to use birth control and abortion to reduce their numbers. They cite Sanger's quotation: "We don't want word to get out that we want to exterminate the Negro population."

Supporters say the full context of the quote proves Sanger did not want to eliminate blacks. In a letter to philanthropist Clarence Gamble in 1939 about her "Negro Project," she said: "The minister's work is also important and also he should be trained, perhaps by the [Birth Control] Federation [of America] as to our ideals and the goal that we hope to reach. We do not want word to go out that we want to exterminate the Negro population, and the minister is the man who can straighten out that idea if it ever occurs to any of their more rebellious members."

Alexander Sanger, president of Planned Parenthood of New York City and Sanger's grandson, says Sanger was committed to helping all women "regardless of race or nationality." He highlights her slogan "Let every child be a wanted child."

But Sanger's extensive written comments over several decades continue to make life difficult for population control advocates who would otherwise like to unreservedly embrace her. For example:

- In her 1922 *Pivot of Civilization*, she clearly called for the sterilization of "genetically inferior races," the elimination of "human weeds" and the cessation of charity.

- In the same book, she advocated the segration of "morons, misfits, and maladjusted."

- The *Birth Control Review*, founded by Sanger in 1917, sounded eugenics themes for decades and categorized blacks, southern Europeans, and other immigrants as mentally inferior, calling them a nuisance and a menace to society.

—**The Editor**

many other programs in the $12–15 billion a year U.S. foreign aid budget.)

Eventually, unmistakable signs of population-control market saturation became evident around the world. In 1994, at the International Conference on Population and Development in Cairo, Margaret Ogola, a Kenyan pediatrician, reported that clinics in her country had an abundance of every kind of contraceptive but lacked the "simplest medicines" to treat common childhood diseases. Similar reports came from other places. In Bangladesh, newspapers reported that unwanted birth control pills were piling up in warehouses.

In addition, many Third World countries have been put off by the cultural appurtenances of the family-planning movement. A sticking point in Cairo, for example, was the insistence by the United Nations that countries provide "sexual health care" for adolescents without their parents' supervision or knowledge, as is done in the United States. Gadul Haqq Ali Gadul Haqq, the grand imam sheikh of al-Azar University, one of many critics of this policy, said, "Islam can by no means agree to give young generations full freedom to do what they like." Other countries also objected, but International Planned Parenthood and other agencies funded by the United States have continued to promote sexual freedom.

CONDOM FAILURE

Stephen Karanja, a Kenyan gynecologist, visited the United States in 2000 to report on what he said were the devastating effects of U.S. population programs in his country. Under the pretext of preventing AIDS, he said, foreign-paid family-planning workers promote promiscuity by indiscriminately distributing condoms and are taking over the healthcare system to perform sterilizations.

"Over and over, we have seen it in Africa—condoms do not stop HIV/AIDS," he said. "In the last two years in Kenya, more than 100 million condoms have been used [while] the number of HIV/AIDS people doubled. Stopping HIV/AIDS is a behavior thing. It is a thing to do with not having sexual activity outside of marriage. We do not need the African family to be attacked."

As to the reputed economic benefits of lowering fertility, several countries said in Cairo that they had reduced or eliminated population growth without improving their economies. But those nations with free economies—even those as densely populated as South Korea—reported not only sturdy growth in wealth but no problems of overpopulation.

Shortly before the Cairo conference, economists at the IMF listed the causes of Africa's severe economic problems. They blamed excessive government spending, high taxes on farmers, inflation, restrictions on trade (a Zambian representative in Cairo said that "trade barriers by developed countries cost developing countries 10 times as much in lost trade as they receive in development assistance"), too much government ownership, overregulation of private economic activity, and government creation of "powerful vested interests." There was no mention of overpopulation—not surprisingly, since Africa has fewer than one-fourth as many people per square mile as prosperous Europe.

Nevertheless, the Sierra Club announced in Cairo its support for increased "international population assistance" and a "sustainable population level within the carrying capacity" of the United States—with its "local activists" being the ones determining that "carrying capacity."

MORE MONEY, MORE COMMITMENT

After the conference, the Clinton administration and the population-control network redoubled their efforts. By 1998, world flow of international aid for population programs amounted to $2.06 billion, with another $9 billion in local funds being reported by the targeted countries themselves.

In 1998, two of the largest U.S. recipients of federal family-planning funds, Planned Parenthood Federation of America and its affiliate, the Alan Guttmacher Institute, received $122 million from the federal government and additional amounts from the states. Medicaid alone spent $449 million for family-planning services in 1998, up over $100 million from 1994.

"More children from the fit, less from the unfit," wrote Margaret Sanger, the founder of Planned Parenthood.

AID, which cites stabilizing population growth as one of its foreign policy goals, asked for "total funding of $542 million from all grant-funded accounts" for population control and $569 million for "protecting human health" in its 2001 budget request. The agency also asked for $254 million for work against AIDS.

In the meantime, failing economies continue to fail, and, as the high-level negotiations regarding "sustainable development" and "reproductive health" proceed, the evidence of the programs' innate tendencies toward coercion mounts. Paid by the head for recruiting women and men for sterilization and other birth control procedures, local family-planning workers press forward to meet their targets in Asia, Africa, and Latin America.

Yet birthrates continue to fall throughout the world as more women work outside their homes. World food availability rose to unprecedented levels, according to the UN Food and Agriculture Organization (FAO). World forest acreage remained at the same levels as in the 1950s, according to FAO data. Some 19,000 scientists have signed a petition stating that there is "no convincing scientific evidence" that human release of greenhouse gases will cause "disruption of Earth's climate" (www.sitewave.net/pproject/s33p427.htm).

What the future will bring is anyone's guess. Perhaps the new Bush administration will exercise its conservative sinews and stanch the flow of federal funds to population control groups. Perhaps a cultural backlash in the Third World will gain strength and slow the population-control juggernaut to a crawl. But it may very well be that current programs, propelled by political inertia and sluiced by already open funding spigots, will continue and even grow.

Jacqueline R. Kasun is an economist and the author of **The War Against Population: The Economics and Ideology of World Population Control** *(Ignatius, 1999).*

Population and Consumption

What We Know, What We Need to Know

by Robert W. Kates

Thirty years ago, as Earth Day dawned, three wise men recognized three proximate causes of environmental degradation yet spent half a decade or more arguing their relative importance. In this classic environmentalist feud between Barry Commoner on one side and Paul Ehrlich and John Holdren on the other, all three recognized that growth in population, affluence, and technology were jointly responsible for environmental problems, but they strongly differed about their relative importance. Commoner asserted that technology and the economic system that produced it were primarily responsible.[1] Ehrlich and Holdren asserted the importance of all three drivers: population, affluence, and technology. But given Ehrlich's writings on population,[2] the differences were often, albeit incorrectly, described as an argument over whether population or technology was responsible for the environmental crisis.

Now, 30 years later, a general consensus among scientists posits that growth in population, affluence, and technology are jointly responsible for environmental problems. This has become enshrined in a useful, albeit overly simplified, identity known as IPAT, first published by Ehrlich and Holdren in *Environment* in 1972[3] in response to the more limited version by Commoner that had appeared earlier in *Environment* and in his famous book *The Closing Circle*.[4] In this identity, various forms of environmental or resource impacts (I) equals population (P) times affluence (A) (usually income per capita) times the impacts per unit of income as determined by technology (T) and the institutions that use it. Academic debate has now shifted from the greater or lesser importance of each of these driving forces of environmental degradation or resource depletion to debate about their interaction and the ultimate forces that drive them.

However, in the wider global realm, the debate about who or what is responsible for environmental degradation lives on. Today, many Earth Days later, international debates over such major concerns as biodiversity, climate change, or sustainable development address the population and the affluence terms of Holdrens' and Ehrlich's identity, specifically focusing on the character of consumption that affluence permits. The concern with technology is more complicated because it is now widely recognized that while technology can be a problem, it can be a

solution as well. The development and use of more environmentally benign and friendly technologies in industrialized countries have slowed the growth of many of the most pernicious forms of pollution that originally drew Commoner's attention and still dominate Earth Day concerns.

A recent report from the National Research Council captures one view of the current public debate, and it begins as follows:

For over two decades, the same frustrating exchange has been repeated countless times in international policy circles. A government official or scientist from a wealthy country would make the following argument: The world is threatened with environmental disaster because of the depletion of natural resources (or climate change or the loss of biodiversity), and it cannot continue for long to support its rapidly growing population. To preserve the environment for future generations, we need to move quickly to control global population growth, and we must concentrate the effort on the world's poorer countries, where the vast majority of population growth is occurring.

Government officials and scientists from low-income countries would typically respond:

If the world is facing environmental disaster, it is not the fault of the poor, who use few resources. The fault must lie with the world's wealthy countries, where people consume the great bulk of the world's natural resources and energy and cause the great bulk of its environmental degradation. We need to curtail overconsumption in the rich countries which use far more than their fair share, both to preserve the environment and to allow the poorest people on earth to achieve an acceptable standard of living.[5]

It would be helpful, as in all such classic disputes, to begin by laying out what is known about the relative responsibilities of both population and consumption for the environmental crisis, and what might need to be known to address them. However, there is a profound asymmetry that must fuel the frustration of the developing countries' politicians and scientists: namely, how much people know about population and how little they know about consumption. Thus, this article begins by examining these differences in knowledge and action and concludes with the alternative actions needed to go from more to enough in both population and consumption.[6]

Population

What population is and how it grows is well understood even if all the forces driving it are not. Population begins with people and their key events of birth, death, and location. At the margins, there is some debate over when life begins and ends or whether residence is temporary or permanent, but little debate in between. Thus, change in the world's population or any place is the simple arithmetic of adding births, subtracting deaths, adding immigrants, and subtracting outmigrants. While whole subfields of demography are devoted to the arcane details of these additions and subtractions, the error in estimates of population for almost all places is probably within 20 percent and for countries with modern statistical services, under 3 percent—better estimates than for any other living things and for most other environmental concerns.

Current world population is more than six billion people, growing at a rate of 1.3 percent per year. The peak annual growth rate in all history—about 2.1 percent—occurred in the early 1960s, and the peak population increase of around 87 million per year occurred in the late 1980s. About 80 percent or 4.8 billion people live in the less developed areas of the world, with 1.2 billion living in industrialized countries. Population is now projected by the United Nations (UN) to be 8.9 billion in 2050, according to its medium fertility assumption, the one usually considered most likely, or as high as 10.6 billion or as low as 7.3 billion.[7]

A general description of how birth rates and death rates are changing over time is a process called the demographic transition.[8] It was first studied in the context of Europe, where in the space of two centuries, societies went from a condition of high births and high deaths to the current situation of low births and low deaths. In such a transition, deaths decline more rapidly than births, and in that gap, population grows rapidly but eventually stabilizes as the birth decline matches or even exceeds the death decline. Although the general description of the transition is widely accepted, much is debated about its cause and details.

The world is now in the midst of a global transition that, unlike the European transition, is much more rapid. Both births and deaths have dropped faster than experts expected and history foreshadowed. It took 100 years for deaths to drop in Europe compared to the drop in 30 years in the developing world. Three is the current global average births per woman of reproductive age. This number is more than halfway between the average of five children born to each woman at the post World War II peak of population growth and the average of 2.1 births required to achieve eventual zero population growth.[9] The death transition is more advanced, with life expectancy currently at 64 years. This represents three-quarters of the transition between a life expectancy of 40 years to one of 75 years. The current rates of decline in births outpace the estimates of the demographers, the UN having reduced its latest medium ex-

pectation of global population in 2050 to 8.9 billion, a reduction of almost 10 percent from its projection in 1994.

Demographers debate the causes of this rapid birth decline. But even with such differences, it is possible to break down the projected growth of the next century and to identify policies that would reduce projected populations even further. John Bongaarts of the Population Council has decomposed the projected developing country growth into three parts and, with his colleague Judith Bruce, has envisioned policies that would encourage further and more rapid decline.[10] The first part is unwanted fertility, making available the methods and materials for contraception to the 120 million married women (and the many more unmarried women) in developing countries who in survey research say they either want fewer children or want to space them better. A basic strategy for doing so links voluntary family planning with other reproductive and child health services.

Yet in many parts of the world, the desired number of children is too high for a stabilized population. Bongaarts would reduce this desire for large families by changing the costs and benefits of childrearing so that more parents would recognize the value of smaller families while simultaneously increasing their investment in children. A basic strategy for doing so accelerates three trends that have been shown to lead to lower desired family size: the survival of children, their education, and improvement in the economic, social, and legal status for girls and women.

However, even if fertility could immediately be brought down to the replacement level of two surviving children per woman, population growth would continue for many years in most developing countries because so many more young people of reproductive age exist. So Bongaarts would slow this momentum of population growth by increasing the age of childbearing, primarily by improving secondary education opportunity for girls and by addressing such neglected issues as adolescent sexuality and reproductive behavior.

How much further could population be reduced? Bongaarts provides the outer limits. The population of the developing world (using older projections) was expected to reach 10.2 billion by 2100. In theory, Bongaarts found that meeting the unmet need for contraception could reduce this total by about 2 billion. Bringing down desired family size to replacement fertility would reduce the population a billion more, with the remaining growth—from 4.5 billion today to 7.3 billion in 2100—due to population momentum. In practice, however, a recent U.S. National Academy of Sciences report concluded that a 10 percent reduction is both realistic and attainable and could lead to a lessening in projected population numbers by 2050 of upwards of a billion fewer people.[11]

Consumption

In contrast to population, where people and their births and deaths are relatively well-defined biological events, there is no consensus as to what consumption includes. Paul Stern of the National Research Council has described the different ways physics, economics, ecology, and sociology view consumption.[12] For physicists, matter and energy cannot be consumed, so consumption is conceived as transformations of matter and energy with increased entropy. For economists, consumption is spending on consumer goods and services and thus distinguished from their production and distribution. For ecologists, consumption is obtaining energy and nutrients by eating something else, mostly green plants or other consumers of green plants. And for some sociologists, consumption is a status symbol—keeping up with the Joneses—when individuals and households use their incomes to increase their social status through certain kinds of purchases. These differences are summarized in the box below.

In 1977, the councils of the Royal Society of London and the U.S. National Academy of Sciences issued a joint statement on consumption, having previously done so on population. They chose a variant of the physicist's definition:

> *Consumption is the human transformation of materials and energy. Consumption is of concern to the extent that it makes the transformed materials or energy less available for future use, or negatively impacts biophysical systems in such a way as to threaten human health, welfare, or other things people value.*[13]

On the one hand, this society/academy view is more holistic and fundamental than the other definitions; on the other hand, it is more focused, turning attention to the environmentally damaging. This article uses it as a working definition with one modification, the addition of information to energy and matter, thus completing the triad of the biophysical and ecological basics that support life.

In contrast to population, only limited data and concepts on the transformation of energy, materials, and information exist.[14] There is relatively good global knowledge of energy transformations due in part to the common units of conversion between different technologies. Between 1950 and today, global energy production and use increased more than fourfold.[15] For material transformations, there are no aggregate data in common units on a global basis, only for some specific classes of materials including materials for energy production, construction, industrial minerals and metals, agricultural crops, and water.[16] Calculations of material use by volume, mass, or value lead to different trends.

Trend data for per capita use of physical structure materials (construction and industrial minerals, metals, and forestry products) in the United States are relatively complete. They show an inverted S shaped (logistic) growth pattern: modest doubling between 1900 and the depression of the 1930s (from two to four metric tons), followed by a steep quintupling with economic recovery until the early 1970s (from two to eleven tons), followed by a leveling off since then with fluctuations related to economic downturns (see Figure 1).[17] An aggregate analysis of all current material production and consumption in the United States averages more than 60 kilos per person per day (excluding water). Most of this material flow is split between energy and related products (38 percent) and minerals for construction (37 percent), with the remainder as industrial minerals (5 percent), metals (2 percent), products of fields (12 percent), and forest (5 percent).[18]

A massive effort is under way to catalog biological (genetic) information and to sequence the genomes of microbes, worms, plants, mice, and people. In contrast to the molecular detail, the number and diversity of organisms is unknown, but a conservative estimate places the number of species on the order of 10 million, of which only one-tenth have been described.[19] Although there is much interest and many anecdotes, neither concepts nor data are available on most cultural information. For example, the number of languages in the world continues to decline while the number of messages expands exponentially.

What Is Consumption?

Physicist: "What happens when you transform matter/energy"

Ecologist: "What big fish do to little fish"

Economist: "What consumers do with their money"

Sociologist: "What you do to keep up with the Joneses"

Trends and projections in agriculture, energy, and economy can serve as surrogates for more detailed data on energy and material transformation.[20] From 1950 to the early 1990s, world population more than doubled (2.2 times), food as measured by grain production almost tripled (2.7 times), energy more than quadrupled (4.4 times), and the economy quintupled (5.1 times). This 43-year record is similar to a current 55-year projection (1995–2050) that assumes the continuation of current trends or, as some would note, "business as usual." In this 55-year projection, growth in half again of population (1.6 times) finds almost a doubling of agriculture (1.8 times), more than twice as much energy used (2.4 times), and a quadrupling of the economy (4.3 times).[21]

Thus, both history and future scenarios predict growth rates of consumption well beyond population. An attractive similarity exists between a demographic transition that moves over time from high births and high deaths to low births and low deaths with an energy, materials, and information transition. In this transition, societies will use increasing amounts of energy and materials as consumption increases, but over time the energy and materials input per unit of consumption decrease and information substitutes for more material and energy inputs.

Some encouraging signs surface for such a transition in both energy and materials, and these have been variously labeled as decarbonization and dematerialization.[22] For more than a century, the amount of carbon per unit of energy produced has been decreasing. Over a shorter period, the amount of energy used to produce a unit of production has also steadily declined. There is also evidence for dematerialization, using fewer materials for a unit of production, but only for industrialized countries and for some specific materials. Overall, improvements in technology

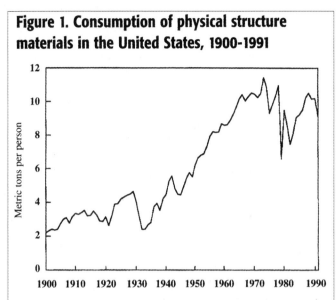

Figure 1. Consumption of physical structure materials in the United States, 1900-1991

SOURCE: I. Wernick, "Consuming Materials: The American Way," *Technological Forecasting and Social Change*, 53 (1996): 114.

and substitution of information for energy and materials will continue to increase energy efficiency (including decarbonization) and dematerialization per unit of product or service. Thus, over time, less energy and materials will be needed to make specific things. At the same time, the demand for products and services continues to increase, and the overall consumption of energy and most materials more than offsets these efficiency and productivity gains.

What to Do about Consumption

While quantitative analysis of consumption is just beginning, three questions suggest a direction for reducing environmentally damaging and resource-depleting consumption. The first asks: *When is more too much for the life-support systems of the natural world and the social infrastructure of human society?* Not all the projected growth in consumption may be resource-depleting—"less available for future use"—or environmentally damaging in a way that "negatively impacts biophysical systems to threaten human health, welfare, or other things people value."[23] Yet almost any human-induced transformations turn out to be either or both resource-depleting or damaging to some valued environmental component. For example, a few years ago, a series of eight energy controversies in Maine were related to coal, nuclear, natural gas, hydroelectric, biomass, and wind generating sources, as well as to various energy policies. In all the controversies, competing sides, often more than two, emphasized environmental benefits to support their choice and attributed environmental damage to the other alternatives.

Despite this complexity, it is possible to rank energy sources by the varied and multiple risks they pose and, for those concerned, to choose which risks they wish to minimize and which they are more willing to accept. There is now almost 30 years of experience with the theory and methods of risk assessment and 10 years of experience with the identification and setting of en-

vironmental priorities. While there is still no readily accepted methodology for separating resource-depleting or environmentally damaging consumption from general consumption or for identifying harmful transformations from those that are benign, one can separate consumption into more or less damaging and depleting classes and *shift* consumption to the less harmful class. It is possible to *substitute* less damaging and depleting energy and materials for more damaging ones. There is growing experience with encouraging substitution and its difficulties: renewables for nonrenewables, toxics with fewer toxics, ozone-depleting chemicals for more benign substitutes, natural gas for coal, and so forth.

The second question, *Can we do more with less?*, addresses the supply side of consumption. Beyond substitution, shrinking the energy and material transformations required per unit of consumption is probably the most effective current means for reducing environmentally damaging consumption. In the 1997 book, *Stuff: The Secret Lives of Everyday Things*, John Ryan and Alan Durning of Northwest Environment Watch trace the complex origins, materials, production, and transport of such everyday things as coffee, newspapers, cars, and computers and highlight the complexity of reengineering such products and reorganizing their production and distribution.[24]

Yet there is growing experience with the three Rs of consumption shrinkage: reduce, recycle, reuse. These have now been strengthened by a growing science, technology, and practice of industrial ecology that seeks to learn from nature's ecology to reuse everything. These efforts will only increase the existing favorable trends in the efficiency of energy and material usage. Such a potential led the Intergovernmental Panel on Climate Change to conclude that it was possible, using current best practice technology, to reduce energy use by 30 percent in the short run and 50–60 percent in the long run.[25] Perhaps most important in the long run, but possibly least studied, is the potential for and value of substituting information for energy and materials. Energy and materials per unit of consumption are going down, in part because more and more consumption consists of information.

The third question addresses the demand side of consumption—*When is more enough?*[26] Is it possible to reduce consumption by more satisfaction with what people already have, by *satiation*, no more needing more because there is enough, and by *sublimation*, having more satisfaction with less to achieve some greater good? This is the least explored area of consumption and the most difficult. There are, of course, many signs of *satiation* for some goods. For example, people in the industrialized world no longer buy additional refrigerators (except in newly formed households) but only replace them. Moreover, the quality of refrigerators has so improved that a 20-year or more life span is commonplace. The financial pages include frequent stories of the plight of this industry or corporation whose markets are saturated and whose products no longer show the annual growth equated with profits and progress. Such enterprises are frequently viewed as failures of marketing or entrepreneurship rather than successes in meeting human needs sufficiently and efficiently. Is it possible to reverse such views, to create a standard of satiation, a satisfaction in a need well met?

Can people have more satisfaction with what they already have by using it more intensely and having the time to do so? Economist Juliet Schor tells of some overworked Americans who would willingly exchange time for money, time to spend with family and using what they already have, but who are constrained by an uncooperative employment structure.[27] Proposed U.S. legislation would permit the trading of overtime for such compensatory time off, a step in this direction. *Sublimation*, according to the dictionary, is the diversion of energy from an immediate goal to a higher social, moral, or aesthetic purpose. Can people be more satisfied with less satisfaction derived from the diversion of immediate consumption for the satisfaction of a smaller ecological footprint?[28] An emergent research field grapples with how to encourage consumer behavior that will lead to change in environmentally damaging consumption.[29]

A small but growing "simplicity" movement tries to fashion new images of "living the good life."[30] Such movements may never much reduce the burdens of consumption, but they facilitate by example and experiment other less-demanding alternatives. Peter Menzel's remarkable photo essay of the material goods of some 30 households from around the world is powerful testimony to the great variety and inequality of possessions amidst the existence of alternative life styles.[31] Can a standard of "more is enough" be linked to an ethic of "enough for all"? One of the great discoveries of childhood is that eating lunch does not feed the starving children of some far-off place. But increasingly, in sharing the global commons, people flirt with mechanisms that hint at such—a rationing system for the remaining chlorofluorocarbons, trading systems for reducing emissions, rewards for preserving species, or allowances for using available resources.

A recent compilation of essays, *Consuming Desires: Consumption, Culture, and the Pursuit of Happiness*,[32] explores many of these essential issues. These elegant essays by 14 well-known writers and academics ask the fundamental question of why more never seems to be enough and why satiation and sublimation are so difficult in a culture of consumption. Indeed, how is the culture of consumption different for mainstream America, women, inner-city children, South Asian immigrants, or newly industrializing countries?

Why We Know and Don't Know

In an imagined dialog between rich and poor countries, with each side listening carefully to the other, they might ask themselves just what they actually know about population and consumption. Struck with the asymmetry described above, they might then ask: "Why do we know so much more about population than consumption?"

The answer would be that population is simpler, easier to study, and a consensus exists about terms, trends, even policies. Consumption is harder, with no consensus as to what it is, and with few studies except in the fields of marketing and advertising. But the consensus that exists about population comes from substantial research and study, much of it funded by governments and groups in rich countries, whose asymmetric concern readily identifies the troubling fertility behavior of others and only reluctantly considers their own consumption behavior. So while consumption is harder, it is surely studied less (see Table 1).

The asymmetry of concern is not very flattering to people in developing countries. Anglo-Saxon tradition has a long history of dominant thought holding the poor responsible for their con-

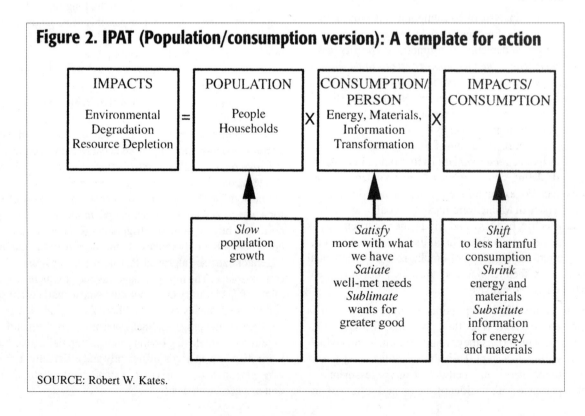

Figure 2. IPAT (Population/consumption version): A template for action

| IMPACTS

Environmental
Degradation
Resource Depletion | = | POPULATION

People
Households | X | CONSUMPTION/
PERSON
Energy, Materials,
Information
Transformation | X | IMPACTS/
CONSUMPTION |

| *Slow*
population
growth | *Satisfy*
more with what
we have
Satiate
well-met needs
Sublimate
wants for
greater good | *Shift*
to less harmful
consumption
Shrink
energy and
materials
Substitute
information
for energy
and materials |

SOURCE: Robert W. Kates.

dition—they have too many children—and an even longer tradition of urban civilization feeling besieged by the barbarians at their gates. But whatever the origins of the asymmetry, its persistence does no one a service. Indeed, the stylized debate of population versus consumption reflects neither popular understanding nor scientific insight. Yet lurking somewhere beneath the surface concerns lies a deeper fear.

Table 1. A comparison of population and consumption

Population	Consumption
Simpler, easier to study	More complex
Well-funded research	Unfunded, except marketing
Consensus terms, trends	Uncertain terms, trends
Consensus policies	Threatening policies

SOURCE: Robert W. Kates.

Consumption is more threatening, and despite the North–South rhetoric, it is threatening to all. In both rich and poor countries alike, making and selling things to each other, including unnecessary things, is the essence of the economic system. No longer challenged by socialism, global capitalism seems inherently based on growth—growth of both consumers and their consumption. To study consumption in this light is to risk concluding that a transition to sustainability might require profound changes in the making and selling of things and in the opportunities that this provides. To draw such conclusions, in the absence of convincing alternative visions, is fearful and to be avoided.

What We Need to Know and Do

In conclusion, returning to the 30-year-old IPAT identity—a variant of which might be called the Population/Consumption (PC) version—and restating that identity in terms of population and consumption, it would be: $I = P*C/P*I/C$, where I equals environmental degradation and/or resource depletion; P equals the number of people or households; and C equals the transformation of energy, materials, and information (see Figure 2).

With such an identity as a template, and with the goal of reducing environmentally degrading and resource-depleting influences, there are at least seven major directions for research and policy. To reduce the level of impacts per unit of consumption, it is necessary to separate out more damaging consumption and shift to less harmful forms, *shrink* the amounts of environmentally damaging energy and materials per unit of consumption, and *substitute* information for energy and materials. To reduce consumption per person or household, it is necessary to *satisfy* more with what is already had, *satiate* well-met consumption needs, and *sublimate* wants for a greater good. Finally, it is possible to *slow* population growth and then to *stabilize* population numbers as indicated above.

However, as with all versions of the IPAT identity, population and consumption in the PC version are only proximate driving forces, and the ultimate forces that drive consumption, the consuming desires, are poorly understood, as are many of the major interventions needed to reduce these proximate driving forces. People know most about slowing population growth, more about shrinking and substituting environmentally damaging consumption, much about shifting to less damaging consumption, and least about satisfaction, satiation, and sublimation. Thus the determinants of consumption and its alternative patterns have been identified as a key understudied topic for an emerging sustainability science by the recent U.S. National Academy of Science study.[33]

But people and society do not need to know more in order to act. They can readily begin to separate out the most serious problems of consumption, shrink its energy and material throughputs, substitute information for energy and materials, create a standard for satiation, sublimate the possession of things for that of the global commons, as well as slow and stabilize population. To go from more to enough is more than enough to do for 30 more Earth Days.

Robert W. Kates is an independent scholar in Trenton, Maine; a geographer; university professor emeritus at Brown University; and an executive editor of *Environment*. The research for "Population and Consumption: What We Know, What We Need to Know" was undertaken as a contribution to the recent National Academies/National Research Council report, *Our Common Journey: A Transition Toward Sustainability*. The author retains the copyright to this article. Kates can be reached at RR1, Box 169B, Trenton, ME 04605.

NOTES

1. B. Commoner, M. Corr, and P. Stamler, "The Causes of Pollution," *Environment*, April 1971, 2–19.

2. P. Ehrlich, *The Population Bomb* (New York: Ballantine, 1966).

3. P. Ehrlich and J. Holdren, "Review of The Closing Circle," *Environment*, April 1972, 24–39.

4. B. Commoner, *The Closing Circle* (New York: Knopf, 1971).

5. P. Stern, T. Dietz, V. Ruttan, R. H. Socolow, and J. L. Sweeney, eds., *Environmentally Significant Consumption: Research Direction* (Washington, D.C.: National Academy Press, 1997), 1.

6. This article draws in part upon a presentation for the 1997 De Lange-Woodlands Conference, an expanded version of which will appear as: R. W. Kates, "Population and Consumption: From More to Enough," in *In Sustainable Development: The Challenge of Transition*, J. Schmandt and C. H. Wards, eds. (Cambridge, U.K.: Cambridge University Press, forthcoming), 79–99.

7. United Nations, Population Division, *World Population Prospects: The 1998 Revision* (New York: United Nations, 1999).

8. K. Davis, "Population and Resources: Fact and Interpretation," K. Davis and M. S. Bernstam, eds., in *Resources, Environment and Population: Present Knowledge, Future Options*, supplement to *Population and Development Review*, 1990: 1–21.

9. Population Reference Bureau, *1997 World Population Data Sheet of the Population Reference Bureau* (Washington, D.C.: Population Reference Bureau, 1997).

10. J. Bongaarts, "Population Policy Options in the Developing World," *Science*, 263: (1994), 771–776; and J. Bongaarts and J. Bruce, "What Can Be Done to Address Population Growth?" (unpublished background paper for The Rockefeller Foundation, 1997).

11. National Research Council, Board on Sustainable Development, *Our Common Journey: A Transition Toward Sustainability* (Washington, D.C.: National Academy Press, 1999).

12. See Stern, et al., note 5 above.

13. Royal Society of London and the U.S. National Academy of Sciences, "Towards Sustainable Consumption," reprinted in *Population and Development Review*, 1977, 23 (3): 683–686.

14. For the available data and concepts, I have drawn heavily from J. H. Ausubel and H. D. Langford, eds., *Technological Trajectories and the Human Environment.* (Washington, D.C.: National Academy Press, 1997).

15. L. R. Brown, H. Kane, and D. M. Roodman, *Vital Signs 1994: The Trends That Are Shaping Our Future* (New York: W. W. Norton and Co., 1994).

16. World Resources Institute, United Nations Environment Programme, United Nations Development Programme, World Bank, *World Resources*, 1996–97 (New York: Oxford University Press, 1996); and A. Gruebler, *Technology and Global Change* (Cambridge, Mass.: Cambridge University Press, 1998).

17. I. Wernick, "Consuming Materials: The American Way," *Technological Forecasting and Social Change*, 53 (1996): 111–122.

18. I. Wernick and J. H. Ausubel, "National Materials Flow and the Environment," *Annual Review of Energy and Environment*, 20 (1995): 463–492.

19. S. Pimm, G. Russell, J. Gittelman, and T. Brooks, "The Future of Biodiversity," *Science*, 269 (1995): 347–350.

20. Historic data from L. R. Brown, H. Kane, and D. M. Roodman, note 15 above.

21. One of several projections from P. Raskin, G. Gallopin, P. Gutman, A. Hammond, and R. Swart, *Bending the Curve: Toward Global Sustainability*, a report of the Global Scenario Group, Polestar Series, report no. 8 (Boston: Stockholm Environmental Institute, 1995).

22. N. Nakicénovíc, "Freeing Energy from Carbon," in *Technological Trajectories and the Human Environment*, eds., J. H. Ausubel and H. D. Langford. (Washington, D.C.: National Academy Press, 1997); I. Wernick, R. Herman, S. Govind, and J. H. Ausubel, "Materialization and Dematerialization: Measures and Trends," in J. H. Ausubel and H. D.

Langford, eds., *Technological Trajectories and the Human Environment* (Washington, D.C.: National Academy Press, 1997), 135–156; and see A. Gruebler, note 16 above.

23. Royal Society of London and the U.S. National Academy of Science, note 13 above.

24. J. Ryan and A. Durning, *Stuff: The Secret Lives of Everyday Things* (Seattle, Wash.: Northwest Environment Watch, 1997).

25. R. T. Watson, M. C. Zinyowera, and R. H. Moss, eds., *Climate Change 1995: Impacts, Adaptations, and Mitigation of Climate Change—Scientific-Technical Analyses* (Cambridge, U.K.: Cambridge University Press, 1996).

26. A sampling of similar queries includes: A. Durning, *How Much Is Enough?* (New York: W. W. Norton and Co., 1992); Center for a New American Dream, *Enough!: A Quarterly Report on Consumption, Quality of Life and the Environment* (Burlington, Vt.: The Center for a New American Dream, 1997); and N. Myers, "Consumption in Relation to Population, Environment, and Development," *The Environmentalist*, 17 (1997): 33–44.

27. J. Schor, *The Overworked American* (New York: Basic Books, 1991).

28. A. Durning, *How Much Is Enough?: The Consumer Society and the Future of the Earth* (New York: W. W. Norton and Co., 1992); Center for a New American Dream, note 26 above; and M. Wackernagel and W. Ress, *Our Ecological Footprint: Reducing Human Impact on the Earth* (Philadelphia. Pa.: New Society Publishers, 1996).

29. W. Jager, M. van Asselt, J. Rotmans, C. Vlek, and P. Costerman Boodt, *Consumer Behavior: A Modeling Perspective in the Contest of Integrated Assessment of Global Change*, RIVM report no. 461502017 (Bilthoven, the Netherlands: National Institute for Public Health and the Environment, 1997); and P. Vellinga, S. de Bryn, R. Heintz, and P. Molder, eds., *Industrial Transformation: An Inventory of Research*. IHDP-IT no. 8 (Amsterdam, the Netherlands: Institute for Environmental Studies, 1997).

30. H. Nearing and S. Nearing. *The Good Life: Helen and Scott Nearing's Sixty Years of Self-Sufficient Living* (New York: Schocken, 1990); and D. Elgin, *Voluntary Simplicity: Toward a Way of Life That Is Outwardly Simple, Inwardly Rich* (New York: William Morrow, 1993).

31. P. Menzel, *Material World: A Global Family Portrait* (San Francisco: Sierra Club Books, 1994).

32. R. Rosenblatt, ed., *Consuming Desires: Consumption, Culture, and the Pursuit of Happiness* (Washington, D.C.: Island Press, 1999).

33. National Research Council, Board on Sustainable Development, *Our Common Journey: A Transition Toward Sustainability* (Washington, D.C.: National Academy Press, 1999).

From *Environment*, April 2000, pp. 10–19. Reprinted with permission of the Helen Dwight Reid Educational Foundation. Published by Heldref Publications, 1319 Eighteenth St., NW, Washington, DC 20036-1802. © 2000.

AN
Economy
FOR THE
Earth

by LESTER R. BROWN

In 1543, Polish astronomer Nicolaus Copernicus published "On the Revolutions of the Celestial Spheres," in which he challenged the view that the sun revolves around the Earth, arguing instead that the Earth revolves around the sun. With his new model of the solar system, he began a wide-ranging debate among scientists, theologians, and others. His alternative to the earlier Ptolemaic model, which had the Earth at the center of the universe, led to a revolution in thinking, to a new worldview.

Today we need a similar shift in our worldview in how we think about the relationship between the Earth and the economy. The issue now is not which celestial sphere revolves around the other but whether the environment is part of the economy or the economy is part of the environment. Economists see the environment as a subset of the economy. Ecologists, on the other hand, see the economy as a subset of the environment.

Like Ptolemy's view of the solar system, the economists' view is confusing efforts to understand our modern world. This has resulted in an economy that is out of sync with the ecosystem on which it depends.

Economic theory and economic indicators don't explain how the economy is disrupting and destroying the Earth's natural systems. Economic theory doesn't explain why Arctic Sea ice is melting, why grasslands are turning into desert in northwestern China, why coral reefs are dying in the South Pacific, or why the Newfoundland cod fishery collapsed. Nor does it explain why we are in the early stages of the greatest extinction of plants and animals since the dinosaurs disappeared 65 million years ago. Yet economics is essential to measuring the cost to society of these excesses.

Evidence that the economy is in conflict with the Earth's natural systems can be seen in the daily news reports of shrinking forests, eroding soils, deteriorating rangelands, expanding deserts, rising carbon dioxide levels, falling water tables, rising temperatures, more destructive storms, melting glaciers, rising sea level, dying coral reefs, collapsing fisheries, and disappearing species. These trends, which mark an increasingly stressed relationship between the economy and the ecosystem, are taking a growing economic toll. At some point this could overwhelm the worldwide forces of progress, leading to economic decline. The challenge for our generation is to reverse these trends before environmental deterioration leads to long-term economic decline—as it did for so many earlier civilizations.

These increasingly visible trends indicate that, if the operation of the subsystem—the economy—is not compatible with the behavior of the larger system—the ecosystem—both will eventually suffer. The larger the economy becomes relative to the ecosystem, and the more it presses against the Earth's natural limits, the more destructive this incompatibility will be.

An environmentally sustainable economy—an *eco-economy*—requires that the principles of ecology establish the framework for the formulation of economic policy and that economists and ecologists work together to fashion the new economy. Ecologists understand that all economic activity, indeed all life, depends on the Earth's ecosystem—the complex of individual species living together, interacting with each other and their physical habitat. These millions of species exist in an intricate balance, woven together by food chains, nutrient cycles, the hydrological cycle, and the climate system. Economists know how to translate goals into policy. Ecologists and economists working together can design and build an eco-economy that can sustain progress.

Just as recognition that the Earth is not the center of the solar system set the stage for advances in astronomy, physics, and related sciences, so will recognition that the economy isn't the center of our world create the conditions to sustain economic progress and improve the human condition.

Converting the world economy into an eco-economy, however, is a monumental undertaking, as the current gap between economists and ecologists in their perception of the world could not be wider. There is no precedent for transforming an economy shaped largely by market forces into one shaped by the principles of ecology.

The scale of projected economic growth outlines the dimensions of the challenge. The growth in world output of goods and services from $6 trillion in 1950 to $43 trillion in 2000 has caused environmental devastation on a scale that we could not easily have imagined a half-century ago. If the world economy continues to expand at 3 percent annually, the output of goods and services will increase fourfold over the next half-century, reaching $172 trillion.

Building an eco-economy in the time available requires rapid systemic change. We won't succeed with a project here and a project there. We are winning occasional battles, but we are losing the war because we don't have a strategy for the systemic economic change that will put the world on a development path that is environmentally sustainable.

Although the concept of environmentally sustainable development evolved a quarter-century ago, not one country has a strategy to build an eco-economy—to restore balances, to stabilize population and water tables, and to conserve forests, soils, and diversity of plant and animal life. We can find individual countries that are succeeding with one or more elements of restructuring but not one that is progressing satisfactorily on all fronts.

Nevertheless, glimpses of the eco-economy are clearly visible in some countries. For example, thirty-one nations in Europe, as well as Japan, have stabilized their population size, satisfying one of the most basic conditions of an eco-economy. Europe has stabilized its population within its food-producing capacity, leaving it with an exportable surplus of grain to help meet the deficits in developing countries. Furthermore, China—the world's most populous country—now has lower fertility than the United States and is moving toward population stability.

Currently Denmark is the eco-economy leader. It has stabilized its population, banned the construction of coal-fired power plants, banned the use of nonrefillable beverage containers, and is now getting 15 percent of its electricity from wind. In addition, it has restructured its urban transport network; now 32 percent of all trips in Copenhagen are on bicycle. Denmark is still not close to balancing carbon emissions and fixation, but it is moving in that direction.

Other countries have also achieved specific goals. A reforestation program in South Korea, begun more than a generation ago, has blanketed that country's hills and mountains with trees. Costa Rica has a plan to shift entirely to renewable energy by 2025. Iceland plans to be the world's first hydrogen-powered economy.

Building an eco-economy will affect every facet of our lives. It will alter how we light our homes, what we eat, where we live, how we use our leisure time, and how many children we have. It will give us a world in which we are **a part of** *nature instead of estranged from it.*

According to Seth Dunn in the November/December 2000 issue of *World Watch* magazine, a consortium of corporations led by Shell Hydrogen and DaimlerChrysler reached an agreement in 1999 with the government of Iceland to establish this new economy. Shell is interested because it wants to begin developing its hydrogen production and distribution capacity, and DaimlerChrysler expects to have the first fuel cell-powered automobile on the market. Shell plans to open its first chain of hydrogen stations in Iceland.

So we can see pieces of the eco-economy emerging, but systemic change requires a fundamental shift in market signals—signals that respect the principles of ecological sustainability. Unless we are prepared to shift taxes from income to environmentally destructive activities—such as carbon emissions and the wasteful use of water—we won't succeed in building an eco-economy.

It is a huge undertaking to restore the balances of nature. For energy, this will depend on shifting from a carbon-based economy to a hydrogen-based one. Even the most progressive oil companies, such as British Petroleum and Royal Dutch Shell, which are all talking extensively about building a solar/hydrogen energy economy, are still investing overwhelmingly in oil, with funds going into climate-benign sources accounting for a minute share of their investment.

Reducing soil erosion to the level of new soil formation will require changes in farming practices. In some situations, it will mean shifting from intense tillage to minimum tillage or no tillage. Agroforestry will loom large in an eco-economy.

Restoring forests that recycle rainfall inland and control flooding is itself a huge undertaking. It means reversing decades of tree cutting and land clearing with forest restoration—an activity that will require millions of people planting billions of trees.

Building an eco-economy will affect every facet of our lives. It will alter how we light our homes, what we eat, where we live, how we use our leisure time, and how many children we have. It will give us a world in which we are *a part of* nature instead of estranged from it.

In May 2001, the Bush White House released with great fanfare a twenty-year plan for the United States' energy economy. It disappointed many people because it largely overlooked the enormous potential for raising energy efficiency. It also overlooked the huge potential of wind power, which is likely to add more to U.S. generating capacity over the next twenty years than coal does. The plan was indicative of the problems some governments are having in fashioning an energy economy that is compatible with the Earth's ecosystem.

Prepared under the direction of Vice-President Dick Cheney, the administration's plan centers on expanding production of fossil fuels—something more appropriate for the early twentieth century than the early twenty-first. It emphasizes the role of coal, but its architects were apparently unaware that world coal use peaked in 1996 and has declined some 7 percent since then as other countries have turned away from this fuel. Even China, which rivals the United States as a coal-burning country, has reduced its coal use by an estimated 14 percent since 1996.

Bush's energy plan notes that the 2 percent of U.S. electricity generation that today comes from renewable sources, excluding hydropower, would increase to 2.8 percent in 2020. But months before the Bush plan was released, the American Wind Energy Association was projecting a staggering 60 percent growth in U.S. wind-generating capacity in 2001. Worldwide, use of wind power alone has multiplied nearly fourfold over the last five years—a growth rate matched only by the computer industry.

The solar cell is a relatively new source of alternative energy and, after wind power, is already the second fastest growing source. In 1952, three scientists at Bell Labs in Princeton, New Jersey, discovered that sunlight striking a silicon-based material produces electricity. The discovery of this photovoltaic, or solar cell, opened up a vast new potential for power generation. Initially very costly, solar cells were used mostly for high-value purposes such as providing the electricity to operate satellites. As the solar cell became economical, it opened the potential for providing electricity to remote sites not yet linked to an electrical grid. It is already more economical in remote areas to install solar cells than to build a power plant and connect villages by grid. By the end of 2000, about a million homes worldwide were getting their electricity from solar cell installations. An estimated 700,000 of these were in villages in developing countries.

Today, as the cost of solar cells continues to decline, this technology is becoming competitive with large, centralized power sources. For many of the two billion people in the world who don't have access to conventional electricity sources, small solar cell arrays provide an affordable shortcut. In the developing world, in some communities not serviced by a centralized power system, local entrepreneurs are investing in solar cell generating facilities and selling the energy to village families.

Perhaps the most exciting technological advance has been the development of a photovoltaic roofing material in Japan. A joint effort involving the construction industry, the solar cell manufacturing industry, and the Japanese government plans to have 4,600 megawatts of electrical generating capacity in place by 2010—enough to satisfy all of the electricity needs of a country like Estonia. With photovoltaic roofing material, the roof of a building becomes the power plant. In some countries, including Germany and Japan, buildings now have a two-way meter—selling electricity to the local utility when they have an excess and buying it when they don't have enough.

In contrast to these other sources of renewable energy, geothermal energy comes from within the Earth itself. Produced radioactively and by the pressures of gravity, it is a vast resource, most of which is deep within the planet. Geothermal energy can be economically tapped when it is relatively close to the surface, as evidenced by hot springs, geysers, and volcanic activity. It is used directly both to supply heat and generate electricity.

Geothermal energy is much more abundant in some parts of the world than in others. The richest region is the vast Pacific Rim: along the western coastal regions of Latin America, Central America, and North America; and widely distributed in eastern Russia, Japan, the Korean Peninsula, China, and island nations like the Philippines, Indonesia, New Guinea, Australia, and New Zealand.

This energy source is essentially inexhaustible. Hot baths, for example, have been used for millennia. It is possible to extract heat faster than it is generated at any local site, but this is a matter of adjusting the extraction of heat to the amount generated. In contrast to oil fields, which are eventually depleted, properly managed geothermal fields keep producing indefinitely. In a time of mounting concern about climate change, many governments are beginning to exploit the geothermal potential—as in Iceland, where it heats some 85 percent of buildings; as in Japan, for hot baths when springs bring geothermal energy to the surface; as in the United States for generating electricity. In fact, the U.S. Department of Energy announced in 2000 that it was launching a program to develop the rich geothermal energy resources in the western United States. The goal is to have 10 percent of the electricity in the West coming from geothermal energy by 2020.

Although the Bush energy plan doesn't reflect it, the world energy economy is on the verge of a major transformation. Historically, the twentieth century was the century of fossil fuels: first coal, then oil, and finally natural gas were the workhorses of the world economy. But with the advent of the twenty-first century, the sun is setting on the fossil fuel era. The last several decades have shown a steady shift from the most polluting and climate-disrupting fuels toward clean, climate-benign energy sources.

Even the oil companies are now beginning to recognize that the time has come for an energy transition. After years of denying any link between fossil fuel burning and climate change, John Browne, the chief executive officer of BP, announced his new position in a historic speech at Stanford University in May 1997:

> My colleagues and I now take the threat of global warming seriously. The time to consider the policy dimensions of climate change is not when the link between greenhouse gases and climate change is conclusively proven but when the possibility cannot be discounted and is taken seriously by the society of which we are a part. We in BP have reached that point.

At an energy conference in Houston, Texas, in February 1999, Michael Bowlin, CEO of ARCO, said that the beginning of the end of the age of oil was in sight. He went on to discuss the need to shift from a carbon-based to a hydrogen-based energy economy.

The signs of restructuring the global energy economy are unmistakable. Events are moving far faster than would have been expected even a few years ago, driven in part by the mounting evidence that the Earth is indeed warming up and that the burning of fossil fuels is responsible. But can we do what needs to be done fast enough?

We know that social change often takes time. In Eastern Europe, it was fully four decades from the imposition of communism until its demise. Thirty-four years passed between the first U.S. Surgeon General's report on smoking and health and the landmark agreement between the tobacco industry and state governments. Thirty-eight years have passed since biologist Rachel Carson published *Silent Spring,* the wakeup call that gave rise to the modern environmental movement.

Sometimes things move much faster, especially when the magnitude of the threat is understood and the nature of the response is obvious, such as the U.S. response to the attack on Pearl Harbor. Within one year, the U.S. economy had largely been reconstructed. In less than four years, the war was over.

Accelerating the transition to a sustainable future means overcoming the inertia of both individuals and institutions. In some ways, inertia is our worst enemy. As individuals we often resist change. When we are gathered into large organizations, we resist it even more.

At the institutional level, we are looking for massive changes in industry, especially in energy. We are looking for changes in the material economy, shifting from a throwaway mentality to a closed loop/recycle mindset. If future food needs are to be satisfied adequately, we need a worldwide effort to reforest the land, conserve soil, and raise water productivity. Stabilizing population growth means quite literally a revolution in human reproductive behavior—one that recognizes a sustainable future is possible only if we average two children per couple. This isn't a debatable point. It is a mathematical reality.

The big remaining challenge is on the educational front: how can we help literally billions of people in the world understand not only the need for change but how that change can bring a life far better than they have today?

In this connection, I am frequently asked if it is too late. My response is: "Too late for what?" Is it too late to save the Aral Sea? Yes, the Aral Sea is dead; its fish have died, and its fisheries have collapsed. Is it too late to save the glaciers in Glacier National Park in the United States? Most likely. They are already half gone, and it would be virtually impossible now to reverse the rise in temperature in time to save them. Is it too late to avoid a rise in temperature from the buildup of greenhouse gases? Yes. A greenhouse gas-induced rise in temperature is apparently already underway. But is it too late to avoid runaway climate change? Perhaps not, if we quickly restructure the energy economy.

For many specifics, the answer is, yes, it is too late. But there is a broader, more fundamental question: is it too late to reverse the trends that will eventually lead to economic decline? Here I think the answer is no—not if we act quickly.

Perhaps the biggest challenge we face is shifting from a carbon-based to a hydrogen-based energy economy—basically moving from fossil fuels to renewable sources of energy, such as solar, wind, and geothermal. How fast can we make this change? Can it be done before we trigger irreversible damage, such as a disastrous rise in sea level? As I indicated, we know from the United States' response to the attack on Pearl Harbor that economic restructuring can occur at an incredible pace if a society is convinced of the need for it.

We study the archaeological sites of civilizations that moved onto economic paths that were environmentally destructive and could not make the needed course corrections in time. We face the same risk.

There is no middle path. Do we join together to build an economy that is sustainable, or do we stay with our environmentally unsustainable economy until it declines? It isn't a goal that can be compromised. One way or another, the choice will be made by our generation. But what we choose will affect life on Earth for all generations to come.

Lester R. Brown is president and founder of the Earth Policy Institute; the founder and former president of the Worldwatch Institute; a MacArthur Fellow; and the recipient of twenty-two honorary degrees and many awards, including the 1987 UN Environment Prize, the 1989 World Wide Fund for Nature Gold Medal, the 1991 Humanist of the Year Award, and the 1994 Blue Planet Prize for his "exceptional contributions to solving global environmental problems." He has authored or coauthored forty-seven books, nineteen monographs, and countless articles. This article is adapted from Eco-Economy: Building an Economy for the Earth, *which is available online at www.earthpolicy.org.*

Originally appeared in *The Humanist,* May/June 2002, pp. 32-34. Excerpted from Chapters 1 & 5 in Lester R. Brown's, *Eco-Economy: Building an Economy for the Earth,* W. W. Norton & Company, NY: 2001. © 2002 by Earth Policy Institute. Reprinted by permission.

the eco-economic revolution

Getting the Market in Sync with Nature

If we want economic progress to continue, we must systematically restructure the global economy to make it environmentally sustainable. Here's a description of a future eco-economy, along with tips on future industries and job possibilities.

By Lester R. Brown

Today's global economy has been shaped by market forces, not by the principles of ecology. Unfortunately, by failing to reflect the full costs of goods and services, the market provides misleading information to economic decision makers at all levels. This has created a distorted economy that is out of sync with the earth's ecosystem—an economy that is destroying its natural support systems.

An economy is sustainable only if it respects the principles of ecology. These principles are as real as those of aerodynamics. If an aircraft is to fly, it has to satisfy certain principles of thrust and lift. So, too, if an economy is to sustain progress, it must satisfy the basic principles of ecology. If it does not, it will decline and eventually collapse. There is no middle ground. An economy is either sustainable or it is not.

The market does not recognize basic ecological concepts of sustainable yield, nor does it respect the balances of nature. For example, it pays no attention to the growing imbalance between carbon emissions and nature's capacity to "fix" carbon, much less to the role of burning fossil fuels in creating the imbalance. For most economists, a rise in carbon dioxide levels is of little concern. For an ecologist, such a rise—driven by the use of fossil fuels—is a signal to shift to other energy sources in order to avoid rising temperatures, melting ice, and rising sea level.

An eco-economy would be one that satisfies our needs without jeopardizing the prospects of future generations to meet their needs. Creating such an economy is not a trivial undertaking; it is nothing less than an Environmental Revolution.

Ecology Over Economics

Ecologists understand the processes that support life on Earth. They understand the fundamental role of pho-

tosynthesis, the concept of sustainable yield, the role of nutrient cycles, the hydrological cycle, the sensitive role of climate, and the intricate relationship between the plant and animal kingdoms. They know that the earth's ecosystems supply services as well as goods and that the former are often more valuable than the latter.

A sustainable economy respects the sustainable yield of the ecosystems on which it depends: fisheries, forests, rangelands, and croplands. A particular fishery can sustain a catch of a certain size, but if the demands on the fishery exceed the sustainable yield by even 2% a year, the fish stocks will begin to shrink and will eventually disappear. As long as the harvest does not exceed the sustainable yield, it can be sustained in perpetuity. The same is true for forests and rangelands.

Nature also relies on balances. These include balances between soil erosion and new soil formation, between carbon emissions and carbon fixation, and between trees dying and trees regenerating.

Nature depends on cycles to maintain life. In nature, there are no linear flow-throughs, no situations where raw materials go in one end and garbage comes out the other. In nature, one organism's waste is another's sustenance, and nutrients are continuously cycled. This system works. Our challenge is to emulate it in the design of the economy.

Ecologists appreciate the role of photosynthesis, the process by which plants convert solar energy into the biochemical energy that supports life on the earth. Anything that reduces the photosynthetic product, such as desertification, the paving of productive land, or the acidification of lakes by acid rain, reduces the productivity of the earth in the most fundamental sense.

Despite this long-standing body of ecological knowledge, national governments have expanded economic activity with little regard for sustainable yields or the fragile balances in nature. Over the last half century, the sevenfold expansion of the global economy has pushed the demand on local ecosystems beyond the sustainable yield in country after country. The fivefold growth in the world fish catch since 1950 has pushed the demand of most oceanic fisheries past their ability to produce fish sustainably. The sixfold growth in the worldwide demand for paper is shrinking the world's forests. The doubling of the world's herds of cattle and flocks of sheep and goats since 1950 is damaging rangelands, converting them to desert.

The clear-cutting of a forest may be profitable for a logging firm, but it is economically costly to society.

An ecologist not only recognizes that the services provided by ecosystems may sometimes be worth more than the goods, but that the value of services needs to be calculated and incorporated into market signals if they are to be protected. Although calculating services is not a simple matter, any reasonable estimate is far better than assuming that the costs are zero, as is now the case. For example, a forest in the upper reaches of a watershed may provide services—such as flood control and the recycling of rainfall inland—that are several times more valuable than its timber yield. Unfortunately, market signals do not reflect this, because the loggers who are cutting the trees do not bear the costs of the reduction in services. National economic policies and corporate strategies are based largely on market signals. The clear-cutting of a forest may be profitable for a logging firm, but it is economically costly to society.

Another major failure of the market to provide reliable information comes when governments subsidize the depletion of resources or environmentally destructive activities. For example, over several decades the U.S. Forest Service used taxpayer money to build roads into national forests so that logging companies could clear-cut forests. This subsidy only artificially lowered the costs of lumber and paper, and it led to flooding, soil erosion, and the silting of streams and rivers. In the Pacific Northwest, it destroyed highly productive salmon fisheries. And all this destruction was underwritten by taxpayers.

In a world where the demands of the economy are pressing against the limits of natural systems, relying on distorted market signals to guide investment decisions is a recipe for disaster. Historically, for example, when the supply of fish was inadequate, the price would rise, encouraging investment in additional fishing trawlers. When there were more fish in the sea than we could ever hope to catch, the market worked well. Today, with the fish catch often exceeding the sustainable yield, investing in more trawlers in response to higher prices will simply accelerate the collapse of these fisheries.

A similar situation exists with other natural systems, such as aquifers, forests, and rangelands. Once the climbing demand for water surpasses the sustainable yield of aquifers, the water tables begin to fall and wells go dry. The market says drill deeper wells. Farmers engage in a competitive orgy of well drilling, chasing the water table downward. On the North China Plain, where 25% of the country's grain is produced, this process is under way. In Hebei Province, data for 1999 show 36,000 wells, mostly shallower ones, being abandoned during the year as 55,000 new, much deeper wells were drilled. In Shandong Province, 31,000 were abandoned and 68,000 new wells were drilled.

In an eco-economy, drilling additional wells would be banned once a water table showed signs of falling. Instead of spending money to dig deeper wells, investments would be channeled into measures to boost water efficiency and to stabilize population in order to bring water use into balance with the sustainable supply.

Evidence is accumulating that our global economy is slowly undermining itself on several fronts. If we want economic progress to continue, we have little choice but to systematically restructure the global economy in order to make it environmentally sustainable.

Imagining the Scope of Change

Converting our economy into an eco-economy is a monumental undertaking. There is no precedent for transforming an economy shaped largely by market forces into one shaped by the principles of ecology.

The scale of projected economic growth outlines the dimensions of the challenge. The growth in world output of goods and services from $6 trillion in 1950 to $43 trillion in 2000 has caused environmental devastation on a scale that we could not easily have imagined a half century ago. If the world economy continued to expand at 3% annually, the output of goods and services would increase fourfold over the next half century, reaching $172 trillion.

Building an eco-economy in the time available requires rapid systemic change. We will not succeed with a project here and a project there. We are winning occasional battles now, but we are losing the war because we do not have a strategy for the systemic economic change that will put the world on an environmentally sustainable development path.

Although the concept of environmentally sustainable development evolved a quarter century ago, not one country has a strategy to build an eco-economy—to restore carbon balances, to stabilize population and water tables, and to conserve its forests, soils, and diversity of plant and animal life. We can find individual countries that are succeeding with one or more elements of the restructuring, but not one that is progressing satisfactorily on all fronts.

Today's Economy vs. Tomorrow's Eco-Economy

Today's Economy	Eco-Economy
Shaped by market forces.	Respects principles of ecology.
Unsustainable: Maximizes profit regardless of consequences to the ecosystem.	Sustainable: Respects carrying capacity of systems; e.g., does not exceed sustainable oceanic fish catch.
Disregards nature's services.	Recognizes ecosystems' natural services.
Consumes dwindling supply of fossil fuels.	Relies on renewable resources such as wind, solar, and geothermal energy.
Pollutes the environment and destabilizes climate.	Minimal pollution, climate-neutral.
Carbon-based auto industry dependent on oil from politically unstable Middle East.	Hydrogen-based fuel-cell auto industry not reliant on specific country source.
Contributes to noisy, congested, and polluted cities.	Will create rail-centered, bicycle-friendly cities that offer less stress and pollution, more exercise.
Likely to decline in not-too-distant future as natural supplies deteriorate.	Will create major new industries; e.g., the wind industry will bring income and jobs from manufacturing, installation, and maintenance.

Source: *Eco-Economy*

An eco-economy would be one that satisfies our needs without jeopardizing the prospects of future generations to meet their needs.

Nevertheless, glimpses of the eco-economy are clearly visible in some countries. For example, 31 countries in Europe, plus Japan, have stabilized their population size, satisfying one of the most basic conditions of an eco-economy. Europe has stabilized its population within its food-producing capacity, leaving it with an exportable surplus of grain to help fill the deficits in developing countries. China—the world's most populous country—now has lower fertility than the United States and is moving toward population stability.

Denmark is the eco-economy leader. It has stabilized its population, banned the construction of coal-fired power plants, banned the use of non-refillable beverage containers, and is now getting 15% of its electricity from wind. In addition, it has restructured its urban transport network; now 32% of all trips in Copenhagen are on bicycle. Denmark is still not close to balancing carbon emissions and fixation, but it is moving in that direction.

Other countries have also achieved specific goals. A reforestation program in South Korea, begun more than a generation ago, has blanketed the country's hills and mountains with trees. Costa Rica has a plan to shift entirely to renewable energy by 2025. Iceland, working with a consortium of corporations led by Shell and Daimler-Chrysler, plans to be the world's first hydrogen-powered economy.

So we can see pieces of the eco-economy emerging, but systemic change requires a fundamental shift in market signals—signals that respect the principles of ecological sustainability. Unless we are prepared to shift taxes from income to environmentally destructive activities, such as carbon emissions and the wasteful use of water, we will not succeed in building an eco-economy.

Restoring the balances of nature in energy production depends on shifting from a carbon-based economy to a hydrogen-based one. Even the most progressive oil companies, such as BP and Royal Dutch/Shell, that are talking extensively about building a solar/hydrogen energy economy are still investing overwhelmingly in oil, with funds going into climate-benign sources accounting for a minute share of their investment.

Reducing soil erosion to the level of new soil formation will require changes in farming practices. In some situations, it will mean shifting from intense tillage to minimum tillage or no tillage. Agroforestry will loom large in an eco-economy. Restoring forests that recycle rainfall inland and control flooding means reversing decades of tree cutting and land clearing with forest restoration, an activity that will require millions of people planting billions of trees.

Building an eco-economy will affect every facet of our lives. It will alter how we light our homes, what we eat, where we live, how we use our leisure time, and how many children we have. It will give us a world where we are a part of nature, instead of estranged from it.

Restructuring the Economy

We can now see what an eco-economy looks like. Instead of running on fossil fuels, it will be powered by renewable sources of energy, such as wind and sunlight, and by geothermal energy from within the earth. It will be hydrogen-based instead of carbon-based. Cars and buses will run on fuel-cell engines powered by electricity produced with an electrochemical process using hydrogen as the fuel instead of internal combustion engines. With fuel cells powered by hydrogen, there is no climate-disrupt-

Declining Industries in an Eco-Economy

Industry	Description
Coal mining	The 7% decline in world coal burning since it peaked in 1996 will continue in the years ahead.
Oil pumping	Projections based on shrinking oil reserves indicate production will peak and start declining in the next 5-20 years. Concern about global warming could bring the decline closer.
Nuclear power	Although public concern focuses on safety issues, it is the high cost that is ensuring the industry's decline.
Clear-cut logging	The rapid spread in eco-labeling of forest products will likely force logging firms to change to sustainable harvesting or be driven out of business.
Manufacture of throwaway products	As efforts to close the materials cycle intensify, many throwaway products will either be banned or taxed out of existence.
Automobile manufacturing	As world population urbanizes, the conflict between the automobile and the city will intensify, reducing dependence of automobiles.

Source: *Eco-Economy*

ing carbon dioxide or noxious health-damaging pollutants; only water is emitted.

In the new economy, atmospheric carbon dioxide levels will be stable. In contrast to today's energy economy, where the world's reserves of oil and coal are concentrated in a handful of countries, energy sources in the eco-economy will be widely dispersed—as widely distributed as sunlight and wind. The world's heavy dependence on the Middle East for much of its energy will likely decline as the new climate-benign energy sources and fuel-cell engines take over.

The energy economy will be essentially a solar/hydrogen economy with various energy sources deriving from the sun used either directly for heating and cooling or indirectly to produce electricity. Wind-generated electricity, which is likely to be the lowest-cost source of energy, will be used to electrolyze water, producing hydrogen. This provides a means of both storing and transporting wind energy. Initially, existing natural gas pipelines will be used to distribute hydrogen. But over the longer term, both natural gas and oil pipeline networks can be adapted to carry hydrogen as the world shifts from a carbon-based to a hydrogen-based economy.

The transport systems of cities have already begun to change. Instead of the noisy, congested, polluting, auto-centered transport systems of today, cities will have rail-centered transport systems, and they will be bicycle- and pedestrian-friendly, offering more mobility, more exercise, cleaner air, and less frustration.

Urban transport systems will have the same components as they do today: automobile, rail, bus, and bicycle. The difference will be in the mix. As more city planners recognize the inherent conflict between the automobile and the city, cleaner and more efficient transport systems will develop. Urban personal mobility will increase as automobile use and traffic congestion decline.

The materials sector of the eco-economy will look far different, too, as it shifts from the linear economic model, where materials go from the mine or forest to the landfill, to the reuse/recycle model, yielding no waste and nothing for the landfills.

One of the keys to reversing the deforestation of the earth is paper recycling; the potential here has been only partly realized. A second key is developing alternative energy sources that will reduce the amount of wood used as fuel. In addition, boosting the efficiency of wood burning can measurably lighten the load on forests.

Another promising option is the use of carefully designed, ecologically managed, and highly productive tree plantations. A small area devoted to plantations may be essential to protecting forests at the global level. Plantations can yield several times as much wood per hectare as can a natural forest.

Examples of Expanding Industries in an Eco-Economy

Industry	Description
Fish farming	Although growth will slow from the double-digit rate of the last decade, rapid expansion is likely to continue.
Bicycle manufacturing	Because bicycles are affordable, nonpolluting, quiet, require little parking space, and provide much-needed exercise in exercise-deprived societies, they will become increasingly common.
Wind-farm construction	Wind-electricity generation, including off-shore wind farms, will grow rapidly over the next few decades, until wind is supplying most of the world's electricity.
Wind-turbine manufacturing	Today the number of utility-scale wind turbines is measured in the thousands, but soon it will be measured in the millions, creating an enormous manufacturing opportunity.
Hydrogen generation	As the transition from a carbon-based to a hydrogen-based energy economy progresses, hydrogen generation will become a key industry.
Fuel-cell manufacturing	As fuel cells replace internal-combustion engines in automobiles and begin generating power in buildings, a huge market will evolve.
Solar-cell manufacturing	For many of the 2 billion people living in rural Third World communities who lack electricity, solar cells will be the best bet for electrification.
Light-rail construction	As people tire of the traffic congestion and pollution associated with the automobile, cities in industrial and developing countries alike will be turning to light rail to provide mobility.
Tree planting	As efforts to reforest the earth gain momentum and as tree plantations expand, tree planting will emerge as a leading economic activity.

Source: *Eco-Economy*

In the economy of the future, the use of water will be in balance with supply. Water tables will be stable, not falling. The economic restructuring will be designed to raise water productivity in every facet of economic activity.

In this environmentally sustainable economy, harvests from oceanic fisheries, a major source of animal protein in the human diet, will be reduced to the sustainable yield. Additional demand will be satisfied by fish farming. This is, in effect, an aquatic version of the same shift that occurred during the transition from hunting and gathering to farming. The freshwater, herbivorous carp polyculture on which the Chinese rely heavily for their vast production of farmed fish offers an ecological model for the rest of the world.

A somewhat similar situation exists for rangelands. One of the keys to alleviating the excessive pressure on rangelands is to feed livestock the crop residues that are otherwise being burned for fuel or for disposal. This trend, already well under way in India and China, may hold the key to stabilizing the world's rangelands.

And finally, the new economy will have a stable population. Over the longer term, the only sustainable society is one in which couples have an average of two children.

Creating New Industries

Describing the eco-economy is obviously a somewhat speculative undertaking. In the end, however, it is not as open ended as it might seem, because the eco-economy's broad outlines are defined by the principles of ecology.

What is not so clear is how ecological principles will translate into economic design. For example, each country has a unique combination of renewable energy sources that will power its economy. Some countries may draw broadly on all their renewable energy sources, while others may concentrate heavily on one that is particularly abundant, such as wind or solar energy. A country with a wealth of geothermal energy may choose to structure its energy economy around this subterranean energy source.

Building a new economy involves phasing out old industries, restructuring existing ones, and creating new ones. World coal use is already being phased out, dropping 7% since peaking in 1996. It is being replaced by efficiency gains in some countries, by natural gas in others (such as the United Kingdom and China), and by wind power in others (such as Denmark).

The automobile industry faces a major restructuring as it changes power sources, shifting from the gasoline-powered internal combustion engine to the hydrogen-powered fuel-cell engine. This shift from the explosive energy that derives from igniting gasoline vapor to a chemical reaction that generates electricity will require both a retooling of engine plants and the retraining of automotive engineers and automobile mechanics.

The new economy will also bring major new industries, ones that either do not yet exist or that are just beginning. Wind electricity generation is one such industry. Now in its embryonic stage, it promises to become the foundation of the new energy economy. Millions of turbines soon will be converting wind into electricity, becoming part of the global landscape. In many countries, wind will supply both electricity and, through the electrolysis of water, hydrogen. Together, electricity and hydrogen can meet all the energy needs of a modern society.

In effect, there will be three new subsidiary industries associated with wind power: turbine manufacturing, installation, and maintenance. Manufacturing facilities will be found in scores of countries, industrial and developing. Installation, which is basically a construction industry, will be more local in nature. Maintenance, since it is a day-to-day activity, will be a source of ongoing local employment.

The robustness of the wind turbine industry was evident in 2000 and 2001 when high-tech stocks were in a free fall worldwide. While high-tech firms as a group were performing poorly, sales of wind turbines were climbing, pushing the earnings of turbine manufacturers to the top of the charts. Continuing growth of this sector is expected for the next few decades.

As wind power emerges as a low-cost source of electricity and a mainstream energy source, it will spawn another industry: hydrogen production. Once wind turbines are in wide use, there will be a large, unused capacity during the night when the demand for electricity drops. With this essentially free electricity, turbine owners can turn on the hydrogen generators and convert the wind power into hydrogen, ideal for fuel-cell engines. Hydrogen generators will start to replace oil refineries. The wind turbine will replace both the coal mine and the oil well. Both wind turbines and hydrogen generators will be widely dispersed as countries take advantage of local wind resources.

Changes in the world food economy will also be substantial. Some of these, such as the shift to fish farming, are already under way. The fastest-growing subsector of he world food economy during the 1990s was aquaculture, expanding at more than 11% a year. Fish farming is likely to continue to expand simply because of its efficiency in converting grain into animal protein.

Even allowing for slower future growth in aquaculture, fish farm output will likely overtake beef production before 2010. Perhaps more surprising, fish farming could eventually exceed the oceanic fish catch. Indeed, for China—the world's leading consumer of seafood—fish farming already supplies two-thirds of the seafood, while the oceanic catch accounts for the other third. With this development, new jobs will be created: aquatic ecologist, fish nutritionist, and marine veterinarian.

Another growth industry of the future is bicycle manufacturing and servicing. Because the bicycle is nonpolluting, frugal in its use of land, and provides the exercise much needed in sedentary societies, future reliance on it is expected to grow. As recently as 1965, the production of cars and bikes was essentially the same, but today more than twice as many bikes as cars are manufactured each year. Among industrial countries, the urban transport model being pioneered in the Netherlands and Denmark, where bikes are featured prominently, gives a sense of the bicycle's future role worldwide.

As bicycle use expands, interest in electrically assisted bikes is also growing. These bikes are similar to existing bicycles, except for a tiny battery-powered electric motor that can either power the bicycle entirely or assist elderly riders or those living in hilly terrain, and their soaring sales are expected to continue climbing in the years ahead.

Just as the last half century has been devoted to raising land productivity, the next half century will be focused on another growth industry: raising water productivity. Virtually all societies will be turning to the management of water at the watershed level in order to manage available supply most efficiently. Irrigation technologies will become more efficient. Urban wastewater recycling will become common. At present, water tends to flow into and out of cities, carrying waste with it. In the future, water will be used over and over, never discharged. Since water does not wear out, there is no limit to how long it can be used, as long as it is purified before reuse.

Another industry that will play a prominent role in the new economy, one that will reduce energy use, is teleconferencing. Increasingly for environmental reasons and to save time, individuals will be "attending" conferences electronically with both audio and visual connections. This industry involves developing the electronic global infrastructure, as well as the services, to make teleconferencing possible. One day there may be thousands of firms organizing electronic conferences.

New Jobs in the Eco-Economy

Restructuring the global economy will create not only new industries, but also new jobs—indeed, whole new professions and new specialties within professions. For example, as wind becomes an increasingly prominent energy source, thousands of wind meteorologists will be needed to analyze potential wind sites, monitor wind speeds, and select the best sites for wind farms. The better the data on wind resources, the more efficient the industry will become.

Wind engineers will be hired to design customized wind turbines. The appropriate turbine size and design can vary widely according to site. It will be the job of wind engineers to tailor designs to specific wind regimes in order to maximize electricity generation.

Environmental architecture is another fast-growing profession. Among the signposts of an environmentally sustainable economy are buildings that are in harmony with the environment. Environmental architects design buildings that are energy- and materials-efficient and that maximize natural heating, cooling, and lighting. In a future of water scarcity, watershed hydrologists will be in demand. It will be their responsibility to understand the hydrological cycle, including the movement of underground water, and to know the depth of aquifers and determine their sustainable yield. They will be at the center of watershed management regimes.

As the world shifts from a throwaway economy, engineers will be needed to design products that can be recycled—from cars to computers. Once products are designed to be disassembled quickly and easily into component parts and materials, comprehensive recycling is relatively easy.

Technologies used in recycling are sometimes quite different from those used in producing from virgin raw materials. Within the U.S. steel industry, for example, where nearly 60% of all steel is produced from scrap, the technologies used differ depending on the feedstock. Steel manufactured in electric arc furnaces from scrap uses far less energy than traditional open-hearth furnaces using pig iron. Recycling engineers will be responsible for closing the materials loop, converting the linear flow-through economy into a comprehensive recycling economy.

In countries with a wealth of geothermal energy, it will be up to geothermal geologists to locate the best sites either for supplying power plants or for tapping directly to heat buildings. Retraining petroleum geologists to master geothermal technologies is one way of satisfying the likely surge in demand for geothermal geologists.

If the world is to stabilize population sooner rather than later, it will need far more family-planning midwives in Third World communities. This growth sector will be concentrated largely in developing countries, where millions of women lack access to family planning. The same family-planning counselors who advise on reproductive health and contraceptive use can also play a central role in controlling the spread of HIV.

In scale, the Environmental Revolution is comparable to the Agricultural and Industrial Revolutions that preceded it.

Another pressing need, particularly in developing countries, is for sanitation-system engineers who can design sewage systems not dependent on water, a trend that is already under way in some water-scarce countries. As it becomes clear that using water to wash waste away is a reckless use of a scarce resource, a new breed of sanitation engineers will be in wide demand. Washing waste away is even less acceptable today as marine ecosystems are overwhelmed by nutrient flows. Apart from the ecological disruption of a water-based disposal method, there

are also much higher priorities in the use of water, such as drinking, bathing, and irrigation.

Yet another new specialty that is likely to expand rapidly in agriculture as productive farmland becomes scarce is that of the agronomist who specializes in multiple cropping and intercropping. This position requires expertise both in the selection of crops that can fit together well in a tight rotation in various locales and in agricultural practices that facilitate multiple cropping.

Investing in the Environmental Revolution

Restructuring the global economy so that economic progress can be sustained represents the greatest investment opportunity in history. The conceptual shift is comparable to that of the Copernican Revolution in the sixteenth century. In scale, the Environmental Revolution is comparable to the Agricultural and Industrial Revolutions that preceded it.

The Agricultural Revolution involved restructuring the food economy, shifting from a nomadic lifestyle based on hunting and gathering to a settled lifestyle based on tilling the soil. Although agriculture started as a supplement to hunting and gathering, it eventually replaced these practices almost entirely. The Agricultural Revolution entailed clearing one-tenth of the earth's land surface of either grass or trees so it could be plowed. Unlike the hunter-gatherer culture that had little effect on the earth, this new farming culture literally transformed the surface of the earth.

The Industrial Revolution has been under way for two centuries, although in some countries it is still in its early stages. At its foundation was a shift in sources of energy from wood to fossil fuels, a shift that set the stage for a massive expansion in economic activity. Indeed, its distinguishing feature is the harnessing of vast amounts of fossil energy for economic purposes. While the Agricultural Revolution transformed the earth's surface, the Industrial Revolution is transforming the earth's atmosphere.

The additional productivity that the Industrial Revolution made possible unleashed enormous creative energies. It also gave birth to new lifestyles and to the most environmentally destructive era in human history, setting the world firmly on a course of eventual economic decline. The Environmental Revolution resembles the Industrial Revolution in that each is dependent on the shift to a new energy source. And like both earlier revolutions, the Environmental Revolution will affect the entire world.

There are differences in scale, timing, and origin among the three revolutions. Unlike the other two, the Environmental Revolution must be compressed into a matter of decades. And while the other revolutions were driven by new discoveries and advances in technology, this revolution is being driven more by our instinct for survival.

Expanding Professions in an Eco-Economy

Profession	Description
Wind meteorologists	Wind meteorologists will play a role in the new energy economy comparable to that of petroleum geologists in the old one.
Family-planning midwives	If world population is to stabilize soon, literally millions of family-planning midwives will be needed.
Foresters	Reforesting the earth will require professional guidance on what tree species to plant where and in what combination.
Hydrologists	As water scarcity spreads, the demand for hydrologists to advise on watershed management, water sources, and water efficiency will increase.
Recycling engineers	Designing consumer applications so they can be easily disassembled and completely recycled will become an engineering specialty.
Aquacultural veterinarians	Until now, veterinarians have typically specialized in either large animals or small animals, but with fish farming likely to overtake beef production by 2010, marine veterinarians will be in demand.
Ecological economists	As it becomes clear that the basic principles of ecology must be incorporated into economic planning and policy making, the demand for economists able to think like ecologists will grow.
Geothermal geologists	With the likelihood that large areas of the world will turn to geothermal energy both for electricity and for heating, the demands for geothermal geologists will climb.
Environmental architects	Architects are learning the principles of ecology so they can incorporate them into the buildings where we will live and work.
Bicycle mechanics	As the world turns to the bicycle for transportation and exercise, bicycle mechanics will be needed to keep the fleet running.
Wind-turbine engineers	With millions of wind turbines likely to be installed in the decades ahead, there will be strong worldwide demand for wind-turbine engineers.

Source: *Eco-Economy*

There has not been an investment situation like this before. The amount that the world spends now each year on oil, the leading source of energy, provides some insight into how much it could spend on energy in the eco-economy. In 2000, the world used nearly 28 billion barrels of oil, some 76 million barrels per day. At $27 a barrel, the total comes to $756 billion per year. How many wind turbines, solar rooftops, and geothermal wells will it take to produce this much energy?

One big difference between the investments in fossil fuels and those in wind power, solar cells, and geothermal energy is that the latter will supply energy in perpetuity. These "wells" will not run dry. If the money spent on oil in one year were invested in wind turbines, the electricity generated would be enough to meet one-fifth of the world's needs.

Investments in the infrastructure for the new energy economy, which would eventually have to be made as fossil fuels are depleted, will obviously be huge. These include the transmission lines that connect wind farms with electricity consumers and the pipelines that link hydrogen supply sources with end users. Much of the infrastructure for the existing energy economy—the transmission lines for electricity and the pipelines for natural gas—can be used in the new energy economy as well. The local pipeline distribution network in various cities for natural gas can easily be converted to hydrogen.

For developing countries, the new energy sources promise to reduce dependence on imported oil, freeing up capital for investment in domestic energy sources. Although few countries have their own oil fields, all have wind and solar energy. In terms of economic expansion and job generation, these new energy technologies are a godsend.

Investments in energy efficiency are also likely to grow rapidly simply because they are so profitable. In virtually all countries, industrial and developing, saved energy is the cheapest source of new energy. Replacing inefficient incandescent light bulbs with highly efficient compact fluorescent lamps offers a rate of return that stock markets are unlikely to match.

There are also abundant investment opportunities in the food economy. It is likely that the world demand for seafood, for example, will increase at least by half over the next 50 years, and perhaps much more. If so, fish-farming output—now 31 million tons a year—will roughly need to triple, as will investments in fish farming. Although aquaculture's growth is likely to slow from the 11% a year of the last decade, it is nonetheless likely to be robust, presenting a promising opportunity for future investment.

A similar situation exists for tree plantations. At present, tree plantations cover some 113 million hectares (280 million acres). An expansion of these by at least half, along with a continuing rise in productivity, is likely to be needed both to satisfy future demand and to eliminate one of the pressures that are shrinking forests. This, too, presents a huge opportunity for investment. No sector of the global economy will be untouched by the Environmental Revolution. In this new economy, some companies will be winners and some will be losers. Those who anticipate the emerging eco-economy and plan for it will be the winners. Those who cling to the past risk becoming part of it.

About the Author
Lester R. Brown is board chairman of the Worldwatch Institute and president of the Earth Policy Institute, 1350 Connecticut Avenue, N.W., Washington, D.C. 20036. Telephone 1-202-496-9290; email epi@earth-policy.org; Web site www.earth-policy.org.

This article draws from his book, *Eco-Economy: Building an Economy for the Earth* (W.W. Norton, 2001, paperback), which is available from the Futurist Bookstore for $15.95 ($14.50 for Society members), cat. no. B-2382.

Originally published in the March/April 2002 issue of *The Futurist,* pp. 23-32. Used with permission from the World Future Society, 7910 Woodmont Avenue, Suite 450, Bethesda, Maryland 20814. Telephone: 310/656-8274; Fax: 301/951-0394; http://www.wfs.org.

POVERTY
AND ENVIRONMENTAL DEGRADATION:
CHALLENGES WITHIN THE GLOBAL ECONOMY

by Akin L. Mabogunje

THE LINK BETWEEN DEEPENING POVERTY AND ENVIRONMENTAL degradation confronts anyone living in a developing country on a daily basis. Perhaps most striking is the increased visibility and extent of both of these phenomena since the end of the Second World War, despite the organized efforts of the United Nations and related international agencies to promote global development in a series of well-publicized social and economic development efforts. Although some countries have made significant progress in this respect, and some individual groups and social classes have escaped poverty, millions remain mired in desperation.

According to the *World Development Report 2000/2001*, 1.2 billion of the world's 6 billion people live on less than $1 per day, and 2.8 billion people, or almost half of the of the world's population, live on less than $2 per day. In 1998, at least 40 percent of the population in South Asia and more than 46 percent in sub-Saharan Africa was living on less than $1 per day.[1]

However, poverty can no longer be adequately defined in terms of income alone—it must be recognized as a multifaceted phenomenon. In an attempt to represent the complexity of poverty, the United Nations Development Programme (UNDP) distinguishes between income poverty and human poverty.[2] According to UNDP, income poverty oc-

curs when the income level of an individual falls below a nationally defined poverty line. Income-based measures of poverty attempt to express the failure of economic resources to meet basic minimum needs—especially food; they also facilitate comparative assessments of countries' progress in poverty reduction. For example, the World Bank has established an international poverty line at $1 per day per person for the purpose of international comparison.

UNDP defines human poverty as the denial or deprivation of opportunities and choices that would enable an individual "to lead a long, healthy, creative life and to enjoy a decent standard of living, freedom, dignity, self-respect and the respect of others."[3] To measure human poverty, UNDP proposes three indices: The first relates to an individual's vulnerability to death at a relatively early age and is measured by the percentage of the population expected to die before the age of 40; the second relates to an individual's exclusion from the world of reading and communication and is measured by the percentage of adults who are illiterate; the third index relates to the standard of living and is measured by the percentage of people with access to health services and safe water and the percentage of malnourished children less than 5 years old.

The failure of these definitions to relate poverty to the environment reflects a

shortcoming in the approach to solving these problems. In an address to the high session of the Economic and Social Council of the United Nations in June 1993, Boutros Boutros-Ghali, then UN secretary-general, demonstrated an effort to change this approach when he emphasized that poverty is only one aspect of the generally dehumanizing phenomenon of deprivation:

Deprivation is a multidimensional concept. In the sphere of economics, deprivation manifests itself as poverty; in politics, as marginalization; in social relations, as discrimination; in culture, as rootlessness; in ecology, as vulnerability. The different forms of deprivation reinforce one another Often the same household, the same region, the same country is the victim of all these forms of deprivation. We must attack deprivation in all its forms. None of the other dimensions of deprivation, however, can be tackled unless we address the problem of poverty and unemployment.[4]

This conceptualization places poverty in a broader web of deprivation. Because the poor are often cut off from the decisionmaking process of their communities, discriminated against by other stations in society, cut off from abiding roots in the community, and relegated to occupy environmentally unsafe areas of societal space, the solutions to their dilemma require a multifaceted approach based on an understanding of broader deprivation.

Globalization, Poverty, and the Environment

Perhaps the single most important development in the world today is what is generally referred to as "globalization." Globalization is partly a result of the tremendous advances in information technology that have, in effect, shrunk the world and linked distant parts of the Earth, creating global relationships. Globalization is also a result of the expanding reach of the capitalist mode of production. Changes in technology and manufacturing organization have fostered the emergence of transnational corporations that have been able to amass wealth within and beyond individual nation-states such that their roles in the economy of their countries and of other countries rival or exceed those of nation-states. Communication technologies have enabled enormous financial resources to be moved from one end of the world to another in a matter of minutes. The instantaneous transfer of vast economic resources has the potential to make or break the economic fortunes of countries and affect the lives and employment opportunities of large numbers of people. Therefore, nation-states are forced to vigorously compete for foreign investment to enhance the rate of growth of their economies. To attract these investments, nation-states must achieve minimum levels of infrastructure development and, more importantly, maintain a certain degree of economic, social, and political stability.

Manufacturing operations have evolved from the classic model (epitomized by the vehicle production operations of Henry Ford in the early twentieth century), in which a huge factory produces all the components as well as the end-product, to an increasingly flexible method of production whereby components are produced in different countries and then assembled at another location close to the market site.

Therefore, although globalization exacerbates poverty in some places and among some groups, it also has a democratizing potential that may be essential to breaking out of poverty. Developing economies cannot get out of poverty without attracting transnational corporations, but at the same time, they cannot attract these corporations unless they have achieved a certain level of development. Because many developing-country economies are in the early stage of the transformation to a free-market, capitalist mode, the conditions necessary to attract international investment are difficult to fulfill, particularly for South Asia and sub-Saharan Africa.

Integrating Developing Economies into the Global Market

In the early stage of capitalism, it is crucial to treat the different factors of production—land, labor, capital, and Entrepreneurship—as commodities, so they can be brought to the free, self-regulating market of supply and demand. To enter the global economy, the poor of the developing world must transform themselves from self-employed peasants to wage-earning labor. However, the price for their labor must be determined by supply and demand. Although labor is treated as a discrete commodity, in reality it is an attribute of human beings. Other commodities can be shoved about, used indiscriminately, and even left unused, but labor cannot be treated in this manner without severe human consequences. For this reason, when the capitalist mode of production emerged in Europe, masses of people were thrown into abject and humiliating poverty. Developing economies are likely to face similar consequences as they are integrated into the global capitalist market.

Commenting on the capitalist system in England in the first half of the nineteenth century, Karl Polanyi noted,

The system, in disposing of a man's labour power, would incidentally dispose of the physical, psychological and moral entity, "man," attached to that tag. Robbed of the protective covering of cultural institutions, human beings would per-

ish from the effects of social exposure. They would die as victims of acute social dislocation through vice, perversion, crime and starvation. Nature would be reduced to its elements, neighbourhoods and landscapes defiled, rivers polluted, military safety jeopardized, the power to produce food and raw materials destroyed.... Undoubtedly, labour, land and money markets are essential to a market economy. But no society could stand the effects of such a system of crude fictions even for the shortest stretch of time, unless its human natural substance as well as its business organization was protected against the ravages of this satanic mill.[5]

In the face of such social consequences, it is no wonder that societies in Western Europe and later in North America, from the middle of the twentieth century until today, were forced to protect individuals from full exposure to the ravages of the free-market economy. The protections came in the form of trade unions and centralized representative governments committed to restraining the socially disruptive potency of capitalism through regulations such as tariff laws, factory laws, social security and pension laws, labor codes, and a host of other social welfare legislation.

Globalization has extended the reach of capitalism and has thus forced countries in the initial phase of capitalism to confront the poverty among their populations. Such populations are caught in producing raw materials for a global market that is becoming increasingly sophisticated and has the option to use industrially produced substitutes, such as the replacement of copper wire with fiber optics. Developing countries with nascent industrial production have to compete with cheaper and better substitutes from the developed countries. Developing countries' capacity to compete in the emerging areas of technological innovation is constrained by weaknesses in their educational system and their institutional capabilities. Thus, prices for their primary production show a decline in real value.

This situation has not been helped by developed countries' tendency to protect the value of their own agricultural products with tariffs, quotas, and export subsidies. This practice cuts into the international trade of developing countries. Although agricultural exports are important for the foreign-exchange earnings of many developing countries, where more than two-thirds of poor people live in rural areas, world trade in agricultural products grew between 1985 and 1994 at only 1.8 percent per annum compared with the expansion of 5.8 percent per annum for manufactured products during the same period.[6]

In addition to globalization, political instability and regional conflicts have been major factors in deepening poverty in many developing countries. Between 1987 and 1997, more than 85 percent of the world's armed conflicts were civil wars fought within the borders of individual countries.[7] Fourteen such conflicts were in African countries, including Sudan, Somalia, Angola, Rwanda, Burundi, Liberia, and Sierra Leone. Asia recorded 14 conflicts in Cambodia, Vietnam, Sri Lanka, and Indonesia, while Europe contended with the breakup of the former Yugoslavia. Although the populations caught in these conflicts come from different socioeconomic classes, the casualties occur predominantly among the poor. A significant number of those whose assets (social and material) and sources of livelihood were destroyed and who have been displaced as a result of armed conflict are often added to the ranks of the poor. Population displacement creates masses of refugees, disrupts markets and other forms of economic and social institutions, and diverts human resources and public expenditure away from productive activities. In 1998, it was estimated that there were 12.4 million international refugees and 18 million internally displaced people, almost half of them in Africa.[8]

Poverty and Environmental Degradation

Despite the constraints globalization places on economic growth and the insecurity that arises from regional armed conflicts, advances in health sciences—especially in epidemiology—have led to a human population explosion. Between 1960 and 2000, the world's population grew from less than 3 billion to some 6 billion. World population reached 6.1 billion in mid-2000 and is currently growing at an annual rate of 1.2 percent (about 77 million people). The United Nations estimates that by 2050, world population will reach between 7.9 billion and 10.9 billion people.[9] Population in developed countries is expected to change little during the next 50 years and is even expected to decrease in some countries. However, in the developing world, population is expected to increase by 3.3 million between 2000 and 2050.[10]

A remarkable shift of global population from rural to urban areas has occurred in developed and developing regions of the world. In fact, by 2030, the urban population is expected to be twice the size of the rural population globally.[11] The shift in population distribution from rural to urban areas has been accompanied by a shift in the concentration of the poor. Poverty in urban centers has been increasing more rapidly than in rural areas. According to a United Nations estimate, 600 million people in urban areas in developing countries (almost 28 percent of the developing world's urban population) cannot meet their basic needs for shelter, water, and health. In fact, about half the urban population in poor countries is living below official poverty levels.[12] This number is expected to rise phenomenally over the next few decades. Rapid population growth and urbanization, coupled with the need to produce for export, has negatively affected the environment in at least seven ways:

Deforestation. The agricultural practices in developing countries are still relatively primitive, depending largely on bush-fallow cultivation, with the repeated clearing and burning of shrub and forest to make room for food crops. Equally significant is the deforestation arising from the need for firewood. For instance, it is estimated, that firewood and brush provide about 52 percent of the domestic energy supply in sub-Saharan Africa; charcoal, another forest product, is also major source of domestic energy.[13]

Desertification. Overcultivation and overgrazing on marginal lands are the major causes of desertification. Although desertification results from many factors and occurs in a variety of environments, rangelands are particularly at risk because they are often found in arid and semiarid regions. In the tropical grassland regions that often border deserts, overgrazing is a potent cause of desertification because feeding the livestock population (which is rapidly expanding to meet increased demand) requires frontier expansion. Desertification also arises from the removal of wood for fuel and the salinization of croplands caused by poorly managed irrigation.

Biodiversity loss. The wide range of ecosystems on which the poor eke out a living has been degraded, and the ecosystems' diverse communities of plants and animals have been put at risk in the process. According to the World Resources Institute, most scientists agree that between 5 and 10 percent of closed tropical forest species will become extinct each decade at current rates of forest loss and disturbance. This loss amounts to about 100 species a day.[14] Indeed, about one-third of the forests that existed in 1950 have been cleared, primarily for agriculture, grazing, or firewood collection.[15] The U.S. National Academy of Sciences estimates that more than 50 percent of all the Earth's species live in tropical rain-forests: A typical four-square-mile patch of rain-forest contains as many as 1,500 species of flowering plants, 750 tree species, 125 mammal species, 400 bird species, 100 reptile species, 60 amphibian species, and 150 butterfly species.[16]

Erosion. Population pressure has led to a reduction in the fallow period and to overcultivation of cropland, particularly in developing countries, where poverty ensures that bush-fallow farming predominates. As a result of overcultivation and forest clearing, soil erosion has become widespread. For instance, in Ethiopia, annual topsoil losses of up to 296 metric tons per hectare have been recorded on relatively steep slopes. Even in countries with somewhat moderate slopes, erosion can proceed rapidly where such areas are unprotected by vegetation. In West Africa, losses of 30 to 55 metric tons per hectare have been noted on slopes of only 1 to 2 percent.[17] In regions with unstable sedimentary rock formations, such as in southeastern Nigeria, gully erosion devastates a considerable area of land. Wind erosion is also significant in drier, marginal lands close to the deserts.

Urban pollution. Urban pollution represents an increasing feature of cities and metropolitan areas in developing countries. It begins with the difficult shelter conditions of squatter settlements, which consist of makeshift huts on land to which the poor have no ownership rights and which usually lack adequate water supply and sanitation facilities. Air pollution becomes a serious concern in such areas. The dependence of the poor on biomass fuels for cooking and other domestic uses increases the concentration of suspended particulates, which often reach levels that exceed World Health Organization (WHO) standards in areas where the poor are concentrated. The need of the poor for cheap means of transport within urban areas has encouraged the proliferation of highly polluting transportation modes such as single-stroke engine motorcycles. Poorly maintained secondhand vehicles heighten the level of air pollution in most cities of developing countries.[18]

Water pollution. Contaminated drinking water transmits diseases such as diarrhea, typhoid, and cholera. In developing countries, diarrheal diseases are believed to have killed about 3 million children annually in the early 1990s and 1 million adults and children older than 5 years annually in the mid-1980s.[19] The lack of solid waste management in squatter settlements is also visibly disturbing. These areas generally receive minimal garbage collection service or none at all. For example, in 1993, in Dhaka, Bangladesh, 90 percent of the slum areas did not have regular garbage collection services.[20] The problems resulting from such conditions are obvious—odors, disease vectors, pests that are attracted to garbage (including rats, mosquitoes, and flies), and the overflowing drainage channels clogged with garbage. Leachate from decomposing and putrefying garbage con-

taminates water sources. Because the poorest areas of cities are generally those that receive the fewest sanitation services, the uncollected solid wastes usually include a significant proportion of fecal matter.[21]

Water pollution is also a serious problem in areas where farmers have been given fertilizers and pesticides to increase agricultural productivity. In India, pollution caused by the leaching of nitrogen fertilizers has been detected in the ground water in many areas. In parts of India's Haryana State, for example, well water with nitrate concentrations ranging from 114 milligrams per liter (mg/L) to 1,800 mg/L (far above the 45 mg/L national standard) have been reported.[22] Pesticides that governments have supplied to peasant farmers contaminate sources of ground water, endanger local water supplies, and pollute aquatic systems. Indeed, according to a 1990 estimate published by WHO, occupational pesticide poisoning may affect as many as 25 million people, or 3 percent of the agricultural workforce each year in developing countries. In Africa alone, where some 80 percent of the populace is involved in agriculture, as many as 11 million cases of acute pesticide exposures occur each year.[23]

Climate change. The melting snows of Kilimanjaro provide dramatic evidence that climate change is already affecting the Earth.[24] The Third Assessment Report of the Intergovernmental Panel on Climate Change projects that African countries are especially vulnerable to climate change because much of its agriculture is rain-fed, and it experiences a high frequency of droughts and floods. In particular, grain yields are projected to decline. Coastal settlements in such regions as Egypt, southeastern Africa, the Gulf of Guinea, Senegal, and Gambia will be affected by rising sea levels and coastal erosion. All over the world, the range of infectious disease will likely increase, and significant extinctions of plant and animal species are projected. Desertification is expected to worsen, and most importantly, the numbers and impact of extreme droughts and floods are expected to grow.[25] All of these projections are made worse by the limited

ability of the poor to adapt to climate change.[26]

Poverty, Environmental Hazards, and Natural Disasters

The relationship between poverty and the environment is an obvious feature of life in the developing world. The poor degrade the environment in various ways, while the environment—in the form of environmental hazards and natural disasters—takes a particularly devastating toll on the poor. A review of 30 case studies in developing countries (14 from in Africa) found three major spirals of impoverishment and environmental decline driven by two forces external to the case study locales— development/ commercialization and natural hazard events—and two internal to the communities studied—population growth and existing poverty.[27] In the first spiral, poor people were displaced from their resources by richer claimants or by competition for existing land or employment. They were driven by development activities, commercialization, and population growth. For these displaced people, division of the remaining resources followed, or else forced migration to other areas that are usually more marginal. In the second, driven by population growth and poverty, meager resources were further divided to meet the needs of generations or the exigencies of poverty. Remaining resources were then degraded by excessive use of divided lands or inappropriate use of environments unable to sustain the requisite resource use. And in the third, driven by poverty and natural hazard events, poor families were unable to maintain protective works against natural hazards of disease, drought, flood, soil erosion, landslides, and pests or to restore resources damaged by these hazards.[28]

Thus, although the poor in urban and rural areas tend to have a negative impact on the environment, they are also the most vulnerable to environmental hazards and natural disasters. Environmental hazards represent ever-present dangers of life-threatening proportions, and natural disasters tend to be episodic and of varying duration. In many developing countries, inadequate attention is given to environmental management in areas occupied by the poor, and therefore, they are exposed to numerous environmental hazards. For example, carbon-monoxide poisoning, one of the most serious environmental hazards, arises from the use of biomass fuels in poorly ventilated dwellings.[29]

The environmental vulnerability of the poor is most pronounced when natural disaster strikes. Earthquakes, droughts, floods, landslides, volcanic eruptions, windstorms, and forest fires devastate the poor and greatly diminish their chances of escaping poverty. Developing countries, especially in densely populated regions, are known to suffer the brunt of natural disasters. According to the World Bank, between 1990 and 1998, 94 percent of the world's 568 major disasters and more than 97 percent of all natural disaster-related deaths occurred in developing countries.[30] In Bangladesh alone, three storms, four floods, one tsunami, and two cyclones killed more than 400,000 people and affected another 42 million. In southern Africa in 1991–92, Malawi, South Africa, Zambia, and Zimbabwe experienced severe droughts. Between 1995 and 1999, in Latin America and the Caribbean, major natural disasters associated with El Niño, Hurricane Mitch, Hurricane Georges, and the Quindio earthquake in Colombia claimed thousands of lives and caused billions of dollars of damage.

Because the poor often live in insubstantial, makeshift, overcrowded conditions in disaster-prone areas, they are the primary victims of natural disasters. Given that labor is the only economic asset for the majority of the poor, any injury, disability, or loss of life resulting from such vulnerability results in a major loss of economic power. For example, Thomas Reardon, professor of Population Resources, Economic Development, and Applied Microeconomics at the University of California at Davis, and Edward J. Taylor, professor of International Agricultural Development and Agribusiness/Marketing at Michigan State University, observed that the 1984 drought in Burkina Faso resulted in a 50-percent decline in the income of the poorest third

of the rural population residing in the difficult agroclimatic zone of the Sahelian Savanna; those living in the Sudan Savanna (an easier agroclimatic zone) suffered only a 7-percent decline in income.[31] Similarly, Rob Vos, Margarita Velasco, and Edgar de Labastida estimated that in Ecuador, El Niño may have increased the incidence of poverty in affected areas by more than 10 percent.[32]

Women and children are especially vulnerable to such hazards. Indeed, female heads of households tend to suffer more during natural disasters than male heads of households, not only because of their much smaller asset base but also because of customary usages and social relations in many of the communities in developing countries. John Hoddinott, a research fellow at the International Food Policy Research Institute, and Bill Kinsey, professor in the Institute of Development Studies at the University of Zimbabwe, noted, for instance, that although the 1994–95 drought in Zimbabwe had no impact on the health of the men, women and young children were adversely affected.[33] Although the effect of the drought on womens' health was temporary and largely in terms of their loss of body mass (which they regained with the good rains in the following year), the consequence for children, especially those under 2 years of age, was more permanent—an average reduction of 1.5 to 2.0 centimeters of linear growth.

Other hazards derive from disease vectors that are fostered by unsanitary surroundings and settlement on polluted or poorly drained sites. For instance, the sandfly that transmits leishmaniases (a group of parasitic diseases) breeds in piles of refuse or in pit latrines, and the mosquito *culex quinquefasciatus*, a vector for bancroftian filariasis (or elephantiasis), breeds in open or cracked septic tanks, flooded pit latrines, and drains. D. Sapir noted that leptospirosis (a bacterial disease) outbreaks have been associated with an increase in the level of water in the poorly drained areas occupied by the poor in Sao Paulo and Rio de Janeiro. The disease passes to human beings through water contaminated with the urine of infected rats or certain domestic animals.[34] Indeed, many of the diseases

passed on by insect vectors that were once predominantly rural are now threatening urban areas. The poor are also exposed to toxic and radioactive waste, when, for instance, their economic desperation forces them to scavenge through landfill sites.

HIV/AIDS is another factor that increases the vulnerability of the poor to environmental threats. The Joint United Nations Programme on HIV/AIDS (UNAIDS) estimates that more than 40 million people worldwide are infected with HIV. About one-third of those infected are between ages 15 and 24. Almost three-quarters of the people currently infected with HIV live in sub-Saharan Africa.[35] Although HIV/AIDS is primarily discussed as a health threat, its destructive effects are also clear in economic, social, and environmental sectors, particularly because the disease makes it difficult for infected people to escape poverty. People with the disease demonstrate decreased productivity, and high mortality rates often result in single-parent families or in orphans, who have little capacity to support themselves. The increased desperation that results from these scenarios makes it unlikely that infected people will benefit from education about improved planning and development that will help improve environmental practices. Thus, HIV/AIDS stresses the environment in much the same way it stresses other sectors.

The Challenge of Poverty Reduction and Environmental Improvement

The spiral of degradation of the environment and vulnerability to environmental hazards and natural disasters characteristic of the poor in the developing world cannot be allowed to continue. Strategic policies and programs are critically needed to address these problems. One factor that underscores the exigent nature of the problem relates to the globalization process. The information technology that fueled the initial poverty-inducing consequences of globalization can also serve as a tool to change the relationship between states and civil society in virtually all countries of the developing world: The instantaneous transmission of live events across the globe can empower societies to demand rights from their governments. Already, the expanded availability of information has spawned the growth of worldwide, voluntary non-governmental organizations (NGOs) that catalyze, mobilize, and give direction to civil society. Availability of information also forces government to be more responsive to the needs and interests of its population. NGOs make it difficult for states to ignore problems of income distribution, environmental conditions, poverty, employment, and housing for the vast majority of their population.

Good governance is central to issues of poverty reduction as well as improvement in environmental quality. Good governance requires three basic conditions: decentralization (the authority structures must be decentralized and devolved); inclusiveness (decisionmaking processes must be participatory and all-inclusive); and accountability (government strategies and activities must be transparent and accountable to the populace).

Decentralization begins with the relation of local governments to the higher levels of government and the lower systems of administrative delegation within the administrative unit, and down to the neighborhood levels, especially those populated largely by the poor. This means a review of the responsibilities delegated to local governments and their implications for the welfare of citizens at the ward and the neighborhood levels. Such delegation must be accompanied by commensurate powers and resources so that local governments can effectively carry out the responsibilities assigned to them. It also entails creating transparent systems, which ensures that local taxation and other statutory transfers of resources are used in a manner to create necessary opportunities for poverty reduction. This involves providing necessary social services (e.g., schools and health services), physical infrastructure (e.g., roads, water, and electricity), environmental improvement (e.g., sanitation, potable water, solid waste disposal, and air pollution reduction) and creating jobs, credit, and market opportunities.

Good governance also emphasizes a participatory process of decisionmaking.

It encourages the involvement of all citizens, especially disadvantaged groups—such as women, the urban poor, and the disabled—in the affairs of the community. An emerging strategy for ensuring the efficacy of this process is to engage all stakeholders, including the poor, in the preparation of the budget of local governments both in urban and rural areas. Participatory budgeting ensures that priorities are determined on the basis of reconciling the divergent interests of all sections of the populace for the common good. It facilitates the collective commitment of all, not only to the long-term, strategic vision of sustainable human development but also to the more immediate goals of poverty reduction. Thus, the community as a whole becomes engaged in the design, implementation, and monitoring of local priorities. Such a participatory process of decisionmaking not only facilitates partnership with private-sector organizations but also results in at least three major beneficial consequences: efficiency, equity, and sustainability in the planning and management of community affairs.

Efficiency in community effort comes first, because participation—especially by the poor—is a first and important factor in their empowerment. When all segments of a population are involved in decisionmaking, they will be more likely to fulfill their civic responsibilities. Where the poor contribute to local development of needed services, infrastructure, and environmental improvements, greater efficiency can be achieved in the overall delivery of such services. This entails not only making their own tax contribution to the local revenue base but also ensuring that they have the resources to do so. Experiments in which banks provide credit to the poor have shown that, rather than subsidized credit, the poor require access to noncollaterized credit, entailing minimal transaction costs to lenders and borrowers, while allowing for charging market-based interest rates. This is especially true when accompanied by fair and predictable legal and regulatory frameworks, which recognize and legitimize the predominantly informal characteristics of their economic activities. The many ways in which the poor make, sell, or trade things hardly register in conven-

tional economic measures, but in aggregate, it constitutes a significant sector of the economy.

The equity effects of involving the poor in decisionmaking are no less substantial. Participation promotes the development of human resources and ensures the provision of facilities for primary education, primary health care, nutrition, family planning, employment and livelihood, shelter, safe drinking water, sanitation, and other basic services. This is crucial not only for improving the labor productivity and income of the poor but also for giving their children opportunities to escape from the trap of poverty. In particular, participation enhances the status of poor women and encourages their upward mobility by giving them a voice. It promotes actions that reduce the vulnerability of the poor to natural and human-made disasters and hazards in the environment.

Perhaps the most important consequence of an all-inclusive and participatory decisionmaking process is its potential to foster sustainable development (i.e., meeting the needs of the present without compromising the ability of future generations to meet their needs).[36] Because of the emphasis on consultation and the reconciliation of divergent interests of various groups, it is possible to reach agreement on issues such as environmental planning and management of cities. Participation promotes strong local democracies and fosters the ethic of civic responsibility among all segments of the community, including the poor. The heightened civic engagement that results encourages the development of a consensual vision for the community's future. Realizing such a vision engenders concern with the quality of the environment and the well-being of present and future generations.

Accountability is the third major feature of good governance. A fundamental tenet of good governance is that local authorities must be accountable to their citizens. Access of all citizens to all relevant information is critical for citizens to understand what is happening in the local government and who is benefiting from the decisions and actions of the government. Therefore, there is a need for some institutional framework to ensure that local governments are directly accountable to their electorates. These frameworks can be fostered through regularly organized and open consultations of all citizens and public feedback mechanisms such as an ombudsman, hotlines, complaint offices and procedures, local government report cards, and procedures for public petitioning and/or public interest litigation.

In any country, strategies for alleviating and reducing poverty have greater chances of succeeding under conditions of national stability and sustained economic growth. Consequently, poverty-reduction measures must relate to how well the macroeconomic and sectoral policies of national governments are oriented to promoting increased productivity and output in the economy. Such policies must not only stimulate economic growth but must also create more employment opportunities. Moreover, in order to ensure the enhanced participation of the poor in economic growth, special attention must be given to reforming laws and regulations that impede access to land, credit, and public infrastructure and services. Improving the access of the poor to land through secure tenure is one way of enhancing their sense of stewardship of the portion of the Earth's surface they occupy.

Conclusion

Any discussion of poverty and the environment must be posited within a political economy framework that focuses on the globalization process and its emphasis on the social relations of production and reproduction. Globalization compels different countries to approach problems of their environment and their poor as resources to be developed. Thus, the 1992 Earth Summit in Rio, in pursuing its Agenda 21 for sustainable development, identified the role local authorities could play in support of Agenda.[37] This was a recognition that local authorities, as the level of governance closest to the people, are able to play the pivotal role of educating, mobilizing, and responding to the public to promote sustainable development.[38]

Despite this recognition, the record shows that not many local governments

in developing countries are aware of their role in meeting the expectations of Agenda 21. A few, especially in Latin America and Asia, have developed their own Agenda 21. In the last few years, the International Council for Local Environment Initiatives (ICLEI) has been furthering these initiatives in various countries. Particularly with respect to developing countries, ICLEI noted that lack of resources and technical capacity within most local authorities has been a major constraint in carrying out this agenda. In many countries, this problem is further compounded by local governments, which are often restricted by central governments in raising finances and in other activities.

Clearly, the issue of local governance has to be given greater emphasis in any program to reduce poverty and enhance the quality of the environment. Equally important in this regard is the need to strengthen the role of networks of NGOs committed to mobilizing, educating, and defending the interests of the poor. Several such networks are gaining global visibility. In 1997, one network representing the alliance of grassroots organizations, researchers, and international organizations formed Women in Informal Employment: Globalizing and Organizing, whose mission is to promote better statistics, research, and policy in support of poor women.[39] Another network, known as HomeNet, was created in the mid-1990s by unions, grassroots organizations, and NGOs working with home-based workers and street vendors in developing and developed countries. HomeNet is concerned with the adverse impact of globalization on the livelihoods of poor women in the informal economy.

In the decades ahead, much work remains to be done. Despite the increasing proactive interventionism from different tiers of governments and the network of NGOs, it is important to recognize our limited understanding of the complex and dynamic interaction between poverty and the environment. Because the collective impact of the poor on the environment is nonlinear, complex, and often does not show up immediately, it is necessary to begin to reconceptualize research proce-

dures and methodologies to account for the role of various social actors.

It is important to develop effective systems and mechanisms that can better provide needed scientific and technical knowledge to support local efforts to address the needs of both the environment and the poor. The emerging paradigm of sustainability science encourages processes of coproduction of knowledge in which scholars and stakeholders, including the poor, interact to define important questions, relevant evidence, and convincing arguments that are scientifically sound and rooted in social understanding.

Sustainability science emphasizes that when the stakeholders are involved in the production of such requisite knowledge, they will become active agents of sustainable and equitable development.[40] Only in this way is it possible to deal effectively with the twin problems of poverty and environmental degradation in developing countries and to begin the arduous task of attempting to eliminate poverty in the context of sustainable development.

NOTES

1. World Bank. *World Development Report 2000/2001: Attacking Poverty* (New York: Oxford University Press, 2000), 3.
2. *UNDP* (United Nations Development Programme) *Human Development Report 1997* (New York: Oxford University Press, 1997), 3.
3. Ibid., page 15.
4. United Nations. 1993.
5. K. Polanyi, *The Great Transformation: The Political and Economic Origins of Our Time* (New York: Holt, Rinehart and Winston, 1944), 75.
6. World Bank. note I above, 180.
7. D. A. Pottebaum, *Economic and Social Implications of War and Conflict* (Ithaca, N.Y.: Cornell University. Agricultural Economics Department, 1999).
8. World Bank, note 1 above. page 127.
9. United Nations Population Division, *World Population Prospects: The 2000 Revision*, draft, 28 February 2001.
10. Ibid.
11. United Nations Centre for Human Settlements (UNCHS), *An Urbanizing World: Global Report on Human Settlements 1996* (Oxford: Oxford University Press, 1996).
12. A. Marshall, ed., *The State of the World Population 1996: Changing Places: Population, Development and the Urban*

Future (United Nations Population Fund. 1996). Available at http://www.un-fpa.org/swp/1996/SWP96MN.HTM.

13. World Resources Institute (WRI), *World Resources 1994–95: A Guide to the Global Environment* (New York: Oxford University Press, 1994), 10.

14. WRI, available at http://www.wri.org/biodivtropical/html.

15. P. H. Raven and T. Williams, eds., *Human Nature and Society: The Quest for a Sustainable World*, Proceedings of the 1997 Forum on Biodiversity, Board on Biology, National Research Council (National Academy Press. 2000).

16. Rainforest Web.Org. Available at http://www.rainforestweb.org_information/biodiversity/?state=more.

17. WRI, *World Resources 1987: An Assessment of the Resource Base that Supports the Global Economy* (New York: Basic nooks Inc., 1987), 3.

18. United Nations Environment Programme (UNEP) and World Health Organization (WHO), "Air Pollution in the World's Megacities," *Environment*, March 1994, 4–13, 25–37; and UNEP and WHO, "Monitoring the Global Environment: An Assessment of Urban Air Quality." *Environment*, October 1998, 6–13, 26–37.

19. UNCHS, *An Urbanizing World: Global Report on Human Settlements 1996* (Oxford: Oxford University Press, 1996), 171.

20. Economic and Social Commission for Asia and the Pacific (ESCAP), *State of Urbanization in Asia and the Pacific*, ST/ESCAP/1300 (New York: United Nations, 1992).

21. UNCHS, note 19 above, page 270.

22. India, *Status of Ground Water Pollution in India, Central Ground Water Board, Ministry of Water Resources* (Lucknow, Uttar Pradesh: Ground Water Pollution Directorate, 1991), 32.

23. J. Jeyaratnam, "Acute Pesticide Poisoning: A Major Global Health Problem," *World Health Statistics Quarterly* 43 (1990): 139–43.

24. J. McCarthy and M. McKenna, "How Earth's Ice Is Changing," *Environment*, December 2000, 8–18.

25. Intergovernmental Panel on Climate Change, *Climate Change 2000: Impacts, Adaptation, and Vulnerability* (New York: Cambridge University Press, 2001), 14.

26. R. W. Kates, "Cautionary Tales: Adaptation and the Global Poor." *Climatic Change 2000* 45, no. 1 (2000): 5–17.

27. R. W. Kates and V. Haarmann, "Where the Poor Live: Are the Assumptions Correct?" *Environment*, May 1992, 4–11, 25–28.

28. Ibid.

29. G. McGranahan and J. Songsore, "Wealth, Health, and the Urban Household: Weighing Environmental Burdens in Accra, Jakarta, and São Paulo," *Environment*, July/August 1994, 4–11, 40–45.

30. World Bank, note 1 above, page 182.

31. T. Reardon and J. E. Taylor, "Agroclimatic Shock, Income Inequality and Poverty: Evidence from Burkina Faso," *World Development* 24, no. 5 (1996): 901–14.

32. R. Vos, M. Velasco, and E. de Labastida, *Economic and Social Effects of El Niño in Ecuador, 1997–1998* (Washington, D.C.: Inter-America Development Bank, 1999).

33. J. Hoddinott and B. Kinsey, *Child Growth in the Time of Drought* (Washington, D.C.: International Food Policy Research Institute, 1998).

34. D. Sapir, *Infectious Disease Epidemics and Urbanization: A Critical Review of the Issues*, Paper prepared for the WHO Commission on Health and Environment (Geneva: WHO, Division of Environmental Health, 1990).

35. Joint United Nations Progmmme on HIV/AIDS and WHO, "AIDS Epidemic Update—December 2001." December 2001. Available at http://www.unaids.org/epidemic_update/report_dec01/index*full.

36. World Commission on Environment and Development, *Our Common Future* (Oxford: Oxford University Press, 1987).

37. For more on Agenda 21, see P. Haas, M. A. Levy, and E. A. Parson. "Appraising the Earth Summit: How Should We Judge UNCED's Success?" *Environment* October 1992, 6–11, 26–33.

38. N. A. Robinson, ed., *Agenda 21: Earth's Action Plan* (New York: Oceania, 1996), 63.

39. World Bank, note 1 above, page 187.

40. R. W. Kates et al., "Sustainability Science," *Science*, 27 April 2001, 641–42. Also available at http://www.sustainabilityscience.org.

Akin L. Mabogunje is a geographer whose research and policy efforts center on urban and regional development. He is chairman of the Development Policy Centre, Ibadan, Nigeria, and coconvener of the international initiative on science and technology for sustainability. A foreign associate of the U.S. National Academy of Sciences, he was recently honored with Nigeria's highest distinction: Commander of the Order of the Niger. He is also a contributing editor of *Environment*.

This article is taken from a paper presented at the 2001 Open Meeting of the Human Dimensions of Global Environmental Change Research Community in Rio de Janeiro, Brazil, on 6–8 October 2001 under the auspices of the Brazilian Academy of Sciences.

WHERE THE SIDEWALKS END

One out of every seven people now lives in a slum—or at least that's the UN's best estimate. More and more slum residents are organizing to improve their lot, as their numbers swell in cities all over the world.

by Molly O'Meara Sheehan

SQUINTING IN THE SUNLIGHT, George Ng'ang'a leads me up a mound of dirt and rubbish on the edge of his Nairobi neighborhood to take in the view. To the south unfolds a safari scene of grassy plains dotted with acacia bushes as far as I can see. To the north stands a dense gathering of gangly shacks cobbled together with cloth, mud, tin, rocks, and sheets of plastic. There are about 800 homes in all crowded onto some 5 to 6 hectares, says Ng'ang'a.

On city maps, the location of this settlement—called "Mtumba" by the 6,000 people who live there—shows up as prime habitat for rhino and giraffe. That's because this unsanctioned community lies on the edge of Nairobi National Park. Mtumba is only one of the many slums around Nairobi. In fact, more than half of the residents of Kenya's capital city cannot afford to live in "formal" housing, and have been forced to find shelter in slums like this one.

Ng'ang'a turns to me and tells me to call him "Castro," which, he says, is his nickname. He has the physique of a bear and is clean shaven, but he insists he was thin and bearded in his youth. I'm not sure if he's joking about the physical resemblance, but it's clear that he's passionate and politically active. For several years in a row the people of Mtumba have chosen Castro to be the leader of the community's governing council in informal elections—informal because the city government does not serve slums, so the people of Mtumba have found their own ways to organize and police themselves.

"We can't depend on the government for anything," says Castro as we walk through the settlement. One of his neighbors, a solemn man named Tom Werunga, joins in our stroll. Werunga, who carries a Bible, tells me that he's a pastor. He points out a water tap—one of two small spigots that supply water for the entire settlement. But no city water is piped here. Instead, these taps are fed by pri-

vate companies that truck in tanks. And they sell their water at a premium. As of yet, no company has seen fit to establish any sort of business setting up toilets or sewers. Instead the 6,000 people who live here share three flimsy pit latrines. "Flying toilets," I learn, are baggies of human excrement that are flung atop roofs or into rubbish piles.

NAIROBI

"When you wake up in the morning, the important thing to do first is to find out where are your shoes. Why shoes are useful: when you walk without them your legs can get injured by anything dangerous like bones, thorns and many others. So I will suggest that shoes are the most useful object in our home."—Serah Waithera, a 15-year old girl living in Nairobi's Mathare slum. Reprinted from: *Shootback: Photos by Kids from the Nairobi Slums* (London: Booth-Clibborn Editions, 1999).

I am scribbling notes, trying to pay attention to the latrines Castro is showing me, but my eyes are stinging in the acrid air. Cinders and fumes from untended piles of burning trash mingle with ash and smoke from charcoal cooking fires where women prepare meals. At night, kerosene fumes from lanterns join the stew. More than 80 percent of Nairobi's households use charcoal for cooking, but the air is worst in neighborhoods such as this, which lack both electricity and trash removal.

Everything in Mtumba, it seems, is insecure and informal. There is no land ownership. There is no public infrastructure. And there is no protection provided by the law. Mtumba's families have moved together twice before, says Castro. They landed in this location in 1992. Since then Nairobi officials have threatened to evict the community several times. And on one occasion, he says, officials sent in bulldozers to completely demolish the settlement. Some families have seen their homes destroyed as many as 10 times. "Every day we are waiting for the demolition squad," says Castro. "We are refugees in our own country."

IT IS NEIGHBORHOODS LIKE MTUMBA—not Greenwich Village in Manhattan or the Rive Gauche in Paris—that are setting the trends for modern urban living. The UN estimates that somewhere between 835 million and 2 billion people now live in some type of slum, whether in a *kampung* in Indonesia, a *favela* in Brazil, a *gecekondu* in Turkey, or a *katchi abadi* in Pakistan. The population of slum dwellers in some of the world's largest cities—Bombay, Bogotá, and Cairo, for example—now outnumbers the population of people living in formal housing.

NAIROBI

Families living in slums often have no access to state-sponsored education. Lacking government support, the Mtumba slum in Nairobi pooled resources for a schoolhouse. Community members built the structure, and now support four teachers.

In many cities—particularly in sub-Saharan Africa and South Asia—explosive urban growth is combining with the world's worst poverty to fuel the proliferation of slums. The world's population increased by 2.4 billion in the past 30 years, and half of that growth was in cities. Over the next three decades, global population is expected to increase by another 2 billion. Demographers expect that nearly *all* of that population increase will end up in developing-country cities, due to urban migration and high birth rates (see graph, next page).

While most poor people still live in rural areas, poverty is rapidly urbanizing. As of 1998, more than 1.2 billion people were living in extreme poverty (on less than the equivalent of about $1 a day), unable to meet even basic food needs. Martin Ravallion of the World Bank estimates that the urban share of the world's extreme poverty is currently 25 percent. He projects that it is likely to reach 50 percent by 2035.

A number of factors are driving the growth of cities worldwide. Rural economies in many regions have been hard hit by environmental degradation, military or ethnic conflicts, and the mechanization of agriculture, which has curbed the number of rural jobs. The prospect of better-paying jobs has drawn many people to cities.

Latin America is by far the most urbanized region of the developing world. About 75 percent of people in

Latin America live in cities—along with 75 percent of the poor. While only 37 and 38 percent of Asians and Africans live in cities respectively, a number of nations in these regions are beginning to see poverty shift to urban centers. For instance, the proportion of people living below the poverty line in rural Kenya between 1992 and 1996 increased from 48 to 53 percent, while the share of people living below the poverty line in Nairobi doubled from 25 to 50 percent.

Castro tells me that his family's land was taken by the colonial Kenyan government in 1952 to build a golf course. "My father was a businessman," he says, "so we went to different places, like nomads." Castro continued the itinerant lifestyle as a young man, but then he got married and began looking for a better life for his family. Eventually, he says, "we came to the Nairobi slums, even though I have an education."

IN GENERAL, THE "OFF-THE-BOOKS" NATURE OF MTUMBA and other informal communities confers certain advantages. Rents are lower than in formal housing. There are no property taxes. Residents can skirt cumbersome zoning laws that separate housing from businesses, and set up shop inside their homes or just outside. Mtumba's commercial strip boasts rows of brightly painted storefronts, each about 1 meter wide. There are produce stands, coffee shops, a "movie house" showing videos, a barber shop, and an outfit that collects old newspapers. But the short-term benefits of living and working outside the formal economy rarely outweigh the long-term costs to residents—and to the cities that have failed to address their needs.

Slums are often located in a city's least-desirable locations—situated on steep hillsides, in floodplains, or downstream from industrial polluters—leaving residents vulnerable to disease and natural disasters. Another long-term cost is the premium residents pay for basic services. The African Population and Health Research Center recently released a report showing that Nairobi's slum dwellers pay more than residents of wealthy housing estates for water—and, as a result, use less than is adequate to meet health needs. "A family needs 100 liters per day for drinking and cleaning," says Mtumba's Tom Werunga. As that much water costs 25 Kenyan shillings (30 cents), it could easily eat up half the income of people who, on average, make about 50 to 60 shillings (60 to 75 cents) per day.

Landlords operating in slums can easily gouge their tenants without fear of legal recourse. And the proportion of renters in slums is higher than commonly thought, as vacant land close to employment opportunities tends to be quickly developed by enterprising landlords. In fact, four out of five slum residents in Nairobi are renters, according to a study done by the Kenyan government and UN-HABITAT, the UN agency for human settlements, which happens to be headquartered in Nairobi. The

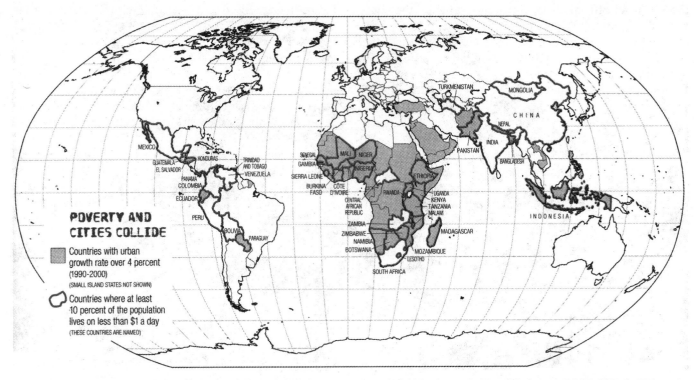

POVERTY AND CITIES COLLIDE

- ▨ Countries with urban growth rate over 4 percent (1990-2000) (SMALL ISLAND STATES NOT SHOWN)
- ◯ Countries where at least 10 percent of the population lives on less than $1 a day (THESE COUNTRIES ARE NAMED)

Sources: *World Urbanization Prospects: The 1999 Revision* (New York: United Nations, 2001); and *World Development Indicators* (Washington, DC: World Bank, 2001)

shacks are lucrative investments, finds the survey, yielding a return in less than two years (compared to 10 to 15 years in the formal property market). Yet landlords do not typically reinvest their profits in the shacks by repairing them or hooking them up to electricity or water, and tenants have no way to hold landlords accountable.

Lacking adequate access to water, toilets, and trash removal, crowded slums also breed diseases that threaten the public health of entire cities. More than half of Nairobi's 3 million people live in slums, squeezed into just 5 percent of the city's land area. In urban centers throughout the developing world, the AIDS virus is facilitating outbreaks of tuberculosis—and both diseases are spreading rapidly. In the Nairobi slums, the mortality rate of children under five years of age is 151 per 1,000 births, far higher than the average of 61 per 1,000 for the city as a whole.

Economic inequalities may significantly hamper public health, according to several new studies. *The Society and Population Health Reader* has brought together journal articles showing that economic inequality in the United States and parts of Europe correlates with reduced public health. In Nairobi, where slums occasionally abut posh, gated enclaves, the economic disparities are as glaring as the public health nightmare.

The growth of slums in an era of unprecedented economic prosperity may also contribute to tensions that threaten local, national, and even global security. "Poor urban settlements are breeding grounds for disease, crime, and terrorism," warned Anna Tibaijuka, the head of UN-HABITAT, in April 2002. While desperate situa-

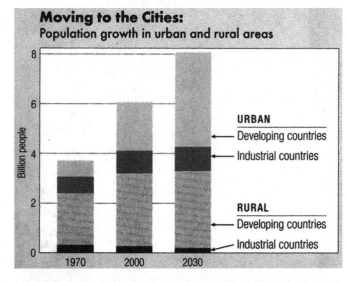

Moving to the Cities:
Population growth in urban and rural areas

URBAN
- Developing countries
- Industrial countries

RURAL
- Developing countries
- Industrial countries

URBAN POVERTY By 2007 more than half of the world's population will reside in cities. In the next 30 years the global population is expected to grow by 2 billion, and demographers project that developing-country cities—many of which are already faced with dire poverty—will be the locations of the most growth.

tions may foster problems, it is the poor who are disproportionately the victims of crime. Some slums are crime ridden and other are nearly crime free, but those that lack municipal or community policing are usually more dangerous.

Following the September 11, 2001 attacks on the United States, *New York Times* columnist Tom Friedman wrote that in an increasingly interconnected world, it will be impossible to ignore the problems of people living in desperate conditions at home or abroad: "if you don't visit a bad neighborhood, a bad neighborhood will visit you."

WALKING AGAIN WITH CASTRO, I AM BEING PURSUED by a friendly, giggling swarm of small children, none taller than my waist. They want to hold my hands. My tour guide is talking about the three vehicles owned by various people in Mtumba—one old car and two bicycles—but my attention is drawn to the children. Many of them have no shoes, yet are following us over sharp rocks, human and animal waste, and all sorts of garbage.

I looked at the kids' feet and I cringed. It is impossible to watch bright-eyed children play in toxic trash and human waste, and listen to their articulate parents describe their efforts and their hopes to build a better life, and not feel obliged to help somehow. This well-intentioned impulse to help slum dwellers into better housing, however, has been carried out with rather disastrous consequences throughout history.

In the United States, for example, the 1949 Housing Act paved the way for cities to raze blighted neighborhoods and build giant public housing projects to house the newly displaced inhabitants. Brazil, Colombia, Egypt, and South Korea were among the developing nations that launched huge public housing campaigns in the 1960s. These costly efforts destroyed the networks of family and friends that poor people had used to survive. Communities often had to move from inner-city locations to outlying areas with fewer job prospects. Added transportation costs meant less could be spent on food. In many cases, the people whose homes were destroyed could not afford the new public projects, which ended up housing wealthier residents. "Urban renewal" projects often had the perverse effect of worsening living conditions for the people they were intended to help.

A major shift began to occur in the 1970s, as city planners were faced with the fact that poor people had been improving their neighborhoods more effectively and with less money than many government projects. Drawing on his experiences working in the slums of Lima, Peru in the 1960s, British architect John F. C. Turner challenged the prevailing orthodoxy with his influential 1972 book, *Freedom to Build*, warning that officials should stop doing more harm than good.

Lacking city services, some communities have managed to close the gap themselves. One of the trailblazers was Akhter Hameed Khan, who in 1980 began mobilizing the community of Orangi, the largest squatter settlement in Karachi, Pakistan. He started a research institute called the Orangi Pilot Project to help residents organize and build a sewer system. Each block collected money and began construction of their own sewers, which served some 90 percent of Orangi's residents by the late 1990s. Between 1982 and 1991, infant mortality rates in the settlement dropped from 130 per thousand to 37 per thousand.

IN THE SLUMS OF NAIROBI, COMMUNITIES LONG NEGLECTED by the government are just beginning to gain some level of political effectiveness. In Mtumba, for instance, residents have begun to organize. "On our own," says Tom Werunga, "we have built a school." Four teachers juggle morning and afternoon shifts to teach more than 400 children in three classrooms. The classroom I saw boasted a small chalkboard, and about 30 to 40 small children, who jumped up smiling from their desks as we passed.

RIO DE JANEIRO

One out of every six of Rio's 5.9 million residents lives in a favela, or slum, according to government estimates, but citizen's groups say the share is as high as one-third of the metro region's 12 million people. Entire communities have been wiped out by mudslides, as erosion has claimed land under makeshift houses on Rio's hillsides. "It raises doubts in people's minds if you give an address here—you have to give a false address to be treated fairly," said a woman in the Nova Brasilia section of Rio to U.S. scholar Janice Perlman. A city-wide program, Favela-Bairro, aims to integrate these neighborhoods into the city with pavement, water, sewers, and electricity.

With the help of a local nongovernmental organization, the Pamoja Trust, Mtumba has started a savings scheme and opened a bank account to pool funds. They hope to save up enough to purchase land at a better location. So far, they have saved about 300,000 Kenyan shillings ($3,800) altogether. According to Pamoja Trust's Jack Makau, his organization would like to match the savings accrued by the Mtumba families, shilling for shilling, and help them invest it, to speed the time necessary to reach the 5 million or so shillings that will be needed.

The residents of Mtumba and Nairobi's other slums are starting to flex some political muscle, bolstered by a city-wide federation, *Muungano wa Wanavijiji*. "Unity is strength," says Jane Weru, the head of Pamoja Trust, which is supporting the federation in 40 of Nairobi's more than 100 slums. *Muungano* members are setting up savings groups, which help build trust and can be turned into revolving loan funds. They are also collecting data on their neighborhoods and sharing experiences to help build coalitions that will help sway government policies in their favor.

Slum residents in Nairobi are also learning from their counterparts around world, loosely organized by Slum/Shack Dwellers International (SDI). The Group was founded in 1996 when the Asian Coalition for Housing Rights joined forces with the South African Homeless People's Federation. Today, the group boasts members from Argentina, Cambodia, Colombia, India, Kenya, Madagascar, Namibia, Nepal, the Philippines, South Af-

TOILET POWER

Jockin Arputham is President of the National Slum Dwellers Federation of India and leader of Slum Dwellers International, a network of grassroots organizations in Cambodia, India, Kenya, South Africa and Thailand, among other countries. He has led the struggle for housing rights in Mumbai, India, since the 1970s, where he started the National Slum Dwellers Federation, which now spans 34 Indian cities. Rasna Warah, Editor of Habitat Debate, *the journal of UN-HABITAT, interviewed him during the World Urban Forum in Nairobi in April 2002. (Printed with permission.)*

Rasna Warah: Tell me a bit about your background.

Jockin Arputham: I am a slum dweller who has lived in Mumbai's Mankhurd Janata Colony for the last 35 years. I was not always a slum dweller. I belonged to an upper middle class family, which lost everything in 1963. That is when I left my home in Kolargold Field for Bangalore and then for Mumbai in 1964. I had never lived in a slum before that, so it was a new culture and new concept for me.

My first experience with activism was when I brought together children from the slum and organized them to carry uncollected garbage in the slum and dump it in front of the Mumbai Municipality's offices. Then in 1974, I invited officers from the Municipality to visit our slum. The idea was to lock one of the officers in the public toilet for a whole day, which we did. He was there for eight hours. In the end, the police had to take him out. But we had made our point. That incident changed people's attitudes towards cleanliness in public toilets.

RW: One of the main issues your organization agitates around is that of toilets. Why is this such an important issue for you?

JA: In India, a public toilet is not simply a toilet. Public toilets are community centers where people meet to exchange news about what is happening in the community. When you go to a public toilet, you get all the news about the settlement. In fact, today in India, if you want to mobilize people, you first go to the public toilets.

In India, 90 per cent of slum dwellers use public toilets maintained by the municipality. But I call these toilets "monuments" because the minute they become the responsibility of the municipality, the service becomes defunct. The National Slum Dwellers Federation has been urging the government to allow slum communities to plan, design, construct, and maintain public toilets. As a result, in the city of Mumbai, we are now constructing toilet blocks—10,000 toilet seats in total.

The Prime Minister has also given a grant of 10 billion rupees to build toilets all over India. These toilets will be planned, designed and maintained by the communities themselves.

RW: What is Slum Dwellers International?

JA: SDI is not a political movement or a social service organization. It is a platform through which urban poor communities take responsibility for improving their lives. It is a voice of the urban poor. When the poor don't take responsibility for their lives, everyone from NGOs to the World Bank will take responsibility away from them. They will tell you how to live, how to eat, how to dress. They will even tell you how to use a toilet. What nonsense! I am an adult, why should anyone control my life? However, instead of merely protesting against these institutions, we are going a step forward by saying, "involve us."

Institutions such as the World Bank have been unsuccessful in many cases because their projects lacked one key ingredient—the involvement of the community. For instance, the World Bank once took seven years to build one toilet in a slum in Mumbai. Seven years! Why? Because they spent much of the time "studying" Indian culture, Indian values, even Indian ways of shitting. Tell me, how can a consultant from London know these things?

But since the World Bank started to work with the National Slum Dwellers Federation, we have managed to build 100 toilet blocks, or 2,200 toilets, within one year. At this rate, for the first time in the history of independent India, no one in Mumbai will have to squat on the streets because there will be one public toilet for every 50 people.

RW: But don't governments have a responsibility towards the urban poor? Shouldn't they be providing these services?

JA: Certainly governments have this responsibility. But they are not doing it. We used to have the handout approach. But now we are saying that we need to split responsibility. In the case of toilets, we are saying to governments, "You pay the capital costs, but we will plan, design and maintain the toilets."

RW: You still live in a slum, even though your economic conditions have improved. Why?

JA: I think I continue to live there because my whole life's work began there. Everything I learned, all the knowledge I gained, was from the slum. Besides, both my children are married and it is just me and my wife. Why would we need a bigger place? I've lived in my current home for 15 years, why not another 25? If I move out, I'll feel as if I exploited the very people who made me who I am. I am very proud to be a slum dweller. I don't see my work as a job; it is my life.

rica, Swaziland, Thailand, Zambia, and Zimbabwe. "A lot of what we do in Nairobi," says Pamoja Trust's Jack Makau, "has been tried out in other cities by the SID network."

IN RECENT YEARS, THESE NEW COALITIONS HAVE ARTICULATED ground-breaking strategies for urban development, where governments engage slum dwellers as equal partners in efforts to improve communities. "We are not coming here to beg," declared Jockin Arputham, the head of Slum Dwellers International, at UN-HABITAT's World Urban Forum in Nairobi in May 2002. "We can sit together with you—national governments, city authorities, and bilateral aid agencies—to plan the city" (see "Toilet Power").

Where local and national governments have been willing to seriously engage those living in urban slums, the

MUMBAI (BOMBAY)

Some 40 percent of Mumbai's 18 million people live in slums or some type of degraded housing, and 5 to 10 percent of the city lives on the pavement, with no housing at all. They are crowded onto "8 percent of the land area of a city smaller than the two New York boroughs of Manhattan and Queens," writes Arjun Appadurai of the University of Chicago in *Environment & Urbanization.*

partnership has often produced significant results. But for the most part, governments still have a long way to go to help address the problems faced by people living in slums. In general, slum leaders like Arputham have identified three key obstacles that governments must surmount in order to become more effective partners:

1) HOME SECURITY

"Land is the key to implement any project for development," says a Mtumba woman who is involved in the community's self-run school. She explains that the people of her community have difficulty convincing themselves—let alone anyone else—to invest in water, toilets, or any sort of improvement. Why bother if the neighborhood could be bulldozed the next day? Indeed, a central obstacle to any sort of "self-help" in many slums is that the residents do not belong on the land where they live in the eyes of the law.

If governments were to grant people in informal settlements legal recognition or titles to the property where they live, it could open up new opportunities for development, and even credit. Buildings without titles are "dead capital," says Peruvian economist Hernando de Soto. They are useful only for whatever shelter they provide. Buildings with titles, in contrast, can have a second "life" in capital markets, where their owners can leverage them.

"In India if you want to mobilize people, you first go to the public toilets," says Jockin Arputham, president of National Slum Dwellers Federation of India. In May 2002, his organization built a toilet on the front lawn of the UN headquarters in Nairobi to publicize the lack of public toilets and sewers in slums.

De Soto was instrumental in prompting Peru to undertake a massive titling program, which formalized some 1

million urban land parcels between 1996 and 2000, first in the *pueblos jovenes* of Lima, and then in other cities. In his recent book, *The Mystery of Capital,* de Soto suggests that titling programs could have a huge global impact. He estimates that the value of real estate not legally owned in the developing world and former Soviet bloc nations is $9.3 trillion.

Granting titles to residents in much of Lima and some other Latin American cities has been fairly straightforward, as a number of informal settlements arose after groups of settlers planned "invasions" of unused public lands. But in places like Kenya, many slums are on private land or on public land given—often under the table—to large-scale shack builders, who rent out their tenement housing. Sorting out ownership can be further complicated by a confusing mix of English land laws and African customary laws. One new innovation in Kenya is a "community land trust," which allows a neighborhood to collectively own its property, while each household retains some individual property rights.

The issue of secure land tenure is gaining in prominence. Heads of state meeting in New York for the UN's Millenium Summit in 2000 pledged to improve the lives of 100 million slum dwellers by 2020. The two measures of "improvement" are to be access to sanitation and security of tenure. When asked how improved security will be measured, Billy Cobbett of the Cities Alliance acknowledges that "it's tricky." Many governments don't count slum dwellers in their censuses, let alone measure their sense of security.

2) EMPLOYMENT OPPORTUNITIES

Most people come to cities seeking jobs. And the slums that many of these people end up living in—with rickety homes, mounds of refuse, and inadequate water supplies—could become key sources of employment. At little cost, municipal authorities could employ slum dwellers to build sewers, collect trash, compost organic waste, or otherwise improve their communities. If organic waste is composted, it can be used to nourish urban agriculture, which can provide both food and jobs. Cities could also revamp their policies on transportation, land use, and small-scale credit to improve the ability of poor people to make a living.

In 2000, the Kenyan government committed itself to working with the slum dwellers federation, local authorities, and the UN on a seven year slum-upgrading initiative. This program aims to make physical improvements—to extend roads and services into slums to connect them to the rest of the city. "We're looking at all possible sources of job generation," says UN-HABITAT's Chris Williams, including providing housing, water, electricity, and other services.

Schemes to collect and compost organic waste—such as paper, food scraps, and even human excrement—can

EVEN IN A SLUM, LIFE

While no two slums in the world are alike, and conditions vary greatly, getting by in the "informal" world is a daily struggle.

MONEY
Without property to leverage, you are hard pressed to get a reasonable loan. But a number of slums have started "revolving loan funds," where residents pool their savings for local use in small-scale loans.

HOME
The perpetual threat of eviction—because you do not have legal claim to the land where you live—reduces your incentive to improve your surroundings. Peru has begun to address this dilemma by issuing 1 million property titles to slum residents.

COOKING
Unless you have access to propane or electricity, you must cook your meals over a smoky fire pit burning charcoal or scraps of wood.

GARBAGE
Instead of curbside garbage pickup, you live with curbside garbage dumps. Burning the garbage—and creating noxious fumes—is how you keep levels of trash down.

TRANSPORTATION
Your main mode of transportation is your feet, so hopefully you can afford a pair of shoes. Bicycles, buses, and jitneys can greatly expand your employment opportunities by getting you closer to workplaces.

EDUCATION
If you are lucky, your community has cobbled together a school for your children to attend, and collectively supports a teacher.

WATER
Getting clean water can be difficult or expensive for you. Local waterways often double as sewers, and water piped in or brought in tanks by truck can eat up much of your day's meager earnings.

TOILETS
If you think sharing a toilet with your family is frustrating, try sharing with 1,000 families. Infectious diseases spread quickly when there are few sewers, and open waterways are often fouled by waste.

earnings in small-scale enterprises in Nairobi, and the third highest in all of urban Kenya.

High transportation costs limit poor people's access to jobs. Zoning laws that separate homes from businesses discriminate against the poor, as do decisions to invest in infrastructure for private cars, rather than dedicated bus lanes, cheap transit, safe pedestrian walkways, or bicycle paths. "More than 95 percent of money that is meant to tackle transport issues in Kenya goes to motorization, while less than 5 percent of Kenyans actually own cars," says Jeff Maganya of the Nairobi office of the global Intermediate Technology Development Group (ITDG). Today more than 40 percent of Nairobi's residents can't afford to pay bus fares.

Most people would benefit if governments were to shift their priorities towards cheaper forms of transportation, including information jitneys (small buses called *matatus* in Nairobi) and bicycles. For many years, high luxury taxes on bicycles and a large fee for registering bicycles prevented poor people from buying and keeping them in Nairobi. Isaac Mburu, a bicycle mechanic who lives in Mtumba, had his bicycle confiscated by local authorities because he could not pay the fee. When Kenya reduced its tax on bicycles from 80 percent to 20 percent between 1986 and 1989, bicycle sales surged by 1,500 percent.

Governments can also take steps to open up lines of credit in informal communities, not only for home improvement, but for small-business development. Even in the poorest neighborhoods, there are buildings and money-making activities that could be leveraged to increase economic opportunities and strengthen communities. Nairobi's *jua kali*, or "hot sun," workers—street hawkers selling vegetables, motor parts, and all manner of goods and services—act as a crucial source of income for many poor people.

3) GOVERNMENT REPRESENTATION

A number of factors can contribute to silencing the voices of the poor and limiting public scrutiny of key decisions about how resources are allocated: collusion between politicians and real estate developers; government influence over or control of the press; or a weak civil society, for example. The wealthy, even if a small minority, simply have greater political power.

help nurture urban gardens and reduce the problems and costs of waste management while producing food and money. The UN Development Programme estimates that 800 million urban farmers harvest 15 percent of the world's food supply—and the share could grow if governments promoted, rather than discouraged, the practice. Agriculture provides the highest self-employment

MANILA

More than one-half of Asia's urban residents live in slums, according to a recent article in the Harvard International Review. In the Philippines, poor people have settled in slum dwellings along the banks of Manila's Pasig River, where they risk being washed away by floods.

Government corruption also takes a disproportionate toll on slum residents. "When you take a complaint to a local authority employed by the government," says Isaac Mburu, who lives in Nairobi's Mtumba slum, "if you go without cash, you won't be served." While 67 percent of all Kenyans surveyed recently by Transparency International-Kenya said that interactions with public officials required bribes, 75 percent of the poorest and least educated said they were forced to pay bribes. An independent fact-finding team visited Kenya in March 2000 and concluded that "the land and housing situation is characterized by forced evictions, misallocation of public land, and rampant land grabbing through bureaucratic and political corruption." According to Transparency International's Michael Lippe, "corruption is a tax on the poor."

In some parts of the world, however, corruption is being thwarted by community organizers and committed leaders. Porto Alegre, Brazil has become famous for a municipal budgeting experiment started in 1989 that invites citizens to engage in setting public priorities and shows people how funds are allocated. A survey done after the first year of participatory budgeting in Porto Alegre revealed that the process had amplified the voices of the city's poor. Most of that city's slum population had indicated that clean water and toilets were their highest priority, whereas the government previously assumed that public transport was at the top of their list.

Today, more than 200 cities in Latin America have introduced participatory budgeting. In July 2001, Brazil enacted a national "City Statute" that requires municipalities to include citizens in urban planning and management, through participatory budgeting, among other measures. While only a small share of a city budget is usually up for grabs, the process does get important issues on the agenda and helps thwart corruption.

In Bombay, both the municipality and poor neighborhoods have gained as a result of the evolving partnership between local authorities and the national slum dwellers federation. "Fifteen years ago, we were just trying to get poor people to be part of the city," said Sheela Patel, director of the India-based Society for the Promotion of Area Resource Centres. "Now there's a realization that this is a key component of good governance." For example, she says, "when hawking is illegal, the municipality loses 170 million rupees ($3.5 million) per month by not giving the hawkers licenses."

In Nairobi, citizens convened the first ever Nairobi Civic Assembly in January 2002 to demand that the government open itself up to all citizens, including the poor majority. "We have a city without citizens because most of them have no voice," said Davinder Lamba, the head of the local human rights group, Manzigira Institute. Participants discussed how they might tackle a number of specific problems, from the city council's failure to provide water in poor neighborhoods to corrupt "land-grabbing" by public officials.

NEIGHBORHOOD BY NEIGHBORHOOD, THINGS ARE beginning to change. For years, whenever residents of a Nairobi slum called Huruma Ghetto tried to repair their homes, the city council blocked them, forcing them to pay bribes or forbidding their efforts on the grounds that they were squatters on public land. The community's initial efforts to organize themselves to overcome these obstacles met with failure. Once, when the community collectively refused to pay the tribes, their houses were set ablaze.

Banding together, and fortified by allies, Huruma Ghetto's residents are getting local authorities to work with them, rather than against them. In May 2002, I watched as the Huruma Ghetto held a ground-breaking ceremony for a model home paid for by its locally organized savings group and approved for construction by the Nairobi slums, as well as activist friends from all over the world (including Jockin Arputham of Slum Dwellers International), came to Huruma Ghetto to take part.

"With the savings scheme, we are not only collecting money, we are collecting people," says David Mwaniki, a 37-year old father of five who makes a living hawking utensils. He also serves as the assistant to the secretary of Huruma's community council, which organized the savings group. "We want to eradicate poverty, and we want people living in informal settlements all over the world to join us, so we can wipe out slums."

Molly O'Meara Sheehan *is a senior researcher at the Worldwatch Institute.*

FOR FURTHER INFORMATION

Jorge Hardoy, Diana Mitlin, and David Satterthwaite, *Environmental Problems in an Urbanizing World* (London: Earthscan, 2001)

Cities in a Globalizing World: Global Report on Human Settlements 2001 (Nairobi: UN-HABITAT, 2001)

Environment & Urbanization journal: www.iied.org/eandu

Habitat Debate journal: www.unhabitat.org/hd

World Bank, "Upgrading Urban Communities," www.worldbank.org/html/fpd/urban//urb_pov/up_body.htm

For information about Shack/Slum Dwellers International (SDI) go to: www.homeless-international.org

For information on Nairobi's Pamoja Trust, contact landrite@wananchi.com

From *World Watch*, November/December 2002, pp. 20–32. © 2002 by Worldwatch Institute, www.worldwatch.org.

UNIT 3
Energy: Present and Future Problems

Unit Selections

Key Points to Consider

- What are some of the differences between energy conservation produced by government regulation and that arising from market forces? Are there elements in the energy market that make it a particularly useful mechanism for directing energy conservation?

- According to recent and still controversial projections on oil supplies, when will the world begin to run out of oil (that is, reach peak production after which the supply of oil will no longer be sufficient to meet the demand)? How can the world adjust to this potential shortfall within such a short period of time?

- What are some of the major benefits of such alternate energy sources as solar power and wind power? Do these energy alternatives really have a chance at competing with fossil fuels for a share of the global energy market?

- In the context of the California energy crisis of 2000, assess the debate between those who claim energy shortages produced high energy costs and those who claim that the energy shortages were created to produce artificially high costs. Does the nature of the debate suggest something about the future ownership and control of energy resources and delivery systems?

 **Links: www.dushkin.com/online/**
These sites are annotated in the World Wide Web pages.

Alliance for Global Sustainability (AGS)
http://www.global-sustainability (AGS)

Alternative Energy Institute, Inc.
http://www.altenergy.org

Communications for a Sustainable Future
http://csf.colorado.edu

Energy and the Environment: Resources for a Networked World
http://zebu.uoregon.edu/energy.html

Institute for Global Communication/EcoNet
http://www.igc.org/igc/gateway/

Nuclear Power Introduction
http://library.thinkquest.org/17658/pdfs/nucintro.pdf

U.S. Department of Energy
http://www.energy.gov

There has been a tendency, particularly in the developed nations of the world, to view the present high standards of living as exclusively the benefit of a high-technology society. In the "techno-optimism" of the post–World War II years, prominent scientists described the technical-industrial civilization of the future as being limited only by a lack of enough trained engineers and scientists to build and maintain it. This euphoria reached its climax in July 1969 when American astronauts walked upon the surface of the Moon, an accomplishment brought about solely by American technology—or so it was supposed. It cannot be denied that technology has been important in raising standards of living and permitting Moon landings, but how much of the growth in living standards and how many outstanding and dramatic feats of space exploration have been the result of technology alone? The answer is few—for in many of humankind's recent successes, the contributions of technology to growth have been no more important than the availability of incredibly cheap energy resources, particularly petroleum, natural gas, and coal.

As the world's supply of recoverable (inexpensive) fossil fuels dwindles and becomes more important as a factor in international conflict, it becomes increasingly clear that the energy dilemma is the most serious economic and environmental threat facing the Western world and its high standard of living. With the exception of the specter of global climate change, the scarcity and cost of conventional (fossil fuel) energy is probably the most serious threat to economic growth and stability in the rest of the world as well. The economic dimensions of the energy problem are rooted in the instabilities of monetary systems produced by and dependent on inexpensive energy. The environmental dimensions of the problem are even more complex, ranging from the hazards posed by the development of such alternative sources as nuclear power to the inability of developing world farmers to purchase necessary fertilizer produced from petroleum, which has suddenly become very costly, and to the enhanced greenhouse effect created by fossil fuel consumption. The only answers to the problems of dwindling and geographically vulnerable, inexpensive energy supplies are conservation and sustainable energy technology. Both require a massive readjustment of thinking, away from the exuberant notion that technology can solve any problem. The difficulty with conservation, of course, is a philosophical one that grows out of the still-prevailing optimism about high technology. Conservation is not as exciting as putting a man on the Moon. Its tactical applications—caulking windows and insulating attics-are dog—paddle technologies to people accustomed to the crawl stroke. Does a solution to this problem entail the technological fixes of which many are so enamored? Probably not, as it appears that the accelerating energy demands of the world's developing nations will most likely be first met by increased reliance on the traditional (and still relatively cheap) fossil fuels. Although there is a need to reduce this reliance, there are few ready alternatives available to the poorer developing countries. It would appear that conservation is the only option.

But market forces operate in intriguing patterns, and perhaps the picture for the immediate energy future is not as bleak as we might have imagined just a few years ago. In the first article in this section, the editors of *The Economist*, in a major survey of energy note that there may be "Energy: A Brighter Future?" for energy conservation and, hence, for energy supply. While conservation mechanisms have been driven by government intervention (tax breaks, and so on) since the 1970s, it has become increasingly apparent that there are powerful market forces at work in energy conservation as well. Indeed, one of the authorities cited in the article suggests that the best thing that governments can do now in terms of conservation is "get out of the way" of those market forces. But conservation is probably not enough to solve the environmental problems related to fossil fuel use. In "Beyond Oil: The Future of Energy," a team of science writers from *Newsweek* magazine suggests that controversial projections, which show global oil production peaking in the next couple of years, may well be accurate, making alternate energy strategies more than just attractive—they become necessary. Even if we adopted all the available alternative energy sources immediately, there would still probably be significant shortfalls. Such a relatively pessimistic outlook could be tempered if promising new experimental strategies such as the hydrogen fuel cell prove useful in tests now being carried out in, among other places, Iceland. Where the *Newsweek* article stresses the availability of alternative renewable energy sources, the next article in this section, assembled by a research team from the University of California at Berkeley's Renewable and Appropriate Energy Laboratory, focuses more on the public policies that are necessary to implement new energy systems. Their conclusion is that the United States has an unprecedented opportunity to build a sustainable energy future "by engaging and stimulating the tremendous innovative and entrepreneurial capacity of the private sector." The next article in this unit on energy stresses the importance of the need to combine conservation strategies with fuel-saving technologies. In "Fossil Fuels and Energy Independence," engineering professor B. Samuel Tanenbaum notes that ultimately we will probably derive most of our energy from solar sources. But until we learn to harness that energy efficiently and cost-effectively, we will need to rely on a number of other alternatives to fossil fuels. These other alternatives already exist and most of them—such as wind power—are already being used, albeit not widely enough. But Tanenbaum reminds us that enormous savings in energy expenditures could be accomplished by the simplest method of all: conservation.

In the final article in this section on energy, author Harvey Wasserman (The Last Energy War) takes on the difficult task of analyzing the recent energy crisis in California and what it means for the future of electricity as a public utility. While the conventional explanation for the supposed shortages that created "rolling blackouts" during California's summer of 2000 was that demand for electricity exceeded supply, Wasserman contends that the actual cause was a withholding of supply to create shortages that could, in turn, be used as an excuse for price/rate hikes. While industry experts will dispute Wasserman's contention, the debate does raise an intriguing question: should energy be viewed as a public or private commodity? Answers to energy questions and issues are as diverse as the world's geography.

ENERGY:
A brighter future?

The world of energy is being turned upside down. The best thing governments can do is to get out of the way, says **Vijay Vaitheeswaran**

FOR most of the past two decades, the world has enjoyed exceptionally low and stable energy prices, but for the past couple of years or so, world oil markets have been on an unnerving roller-coaster ride: prices collapsed to around $10 a barrel two years ago, and soared to a ten-year high of over $35 last year. It was those peaks that set off a political crisis over petrol prices and shortages in America's mid-western states last summer, and that provoked the fuel riots which paralysed several European countries in September. Now oil prices are lower, but remain volatile.

Controversy over the environmental impact of fossil-fuel use has added an extra layer of complication. Last November a ministerial summit on the Kyoto Protocol, a UN treaty among industrial countries to curb global warming, broke up in rancorous disarray. Just a few weeks earlier, the California Air Resources Board had delighted greens and outraged car manufacturers by unanimously upholding its controversial "zero-emissions" mandate, which requires 10% of the new cars sold in the state by big manufacturers to meet the state's definition of zero emissions by 2004. Greenery surfaced as a big factor in the European fuel protests too, this time at the opposite end: rather than blaming the price-fixers at OPEC, most rioters—especially in Britain—attacked their own governments for levying hefty fuel taxes in the name of protecting the environment.

Most recently, attention has focused on the turmoil in the gas and power markets. Thanks to under-investment in gas production in recent years, low stocks and soaring demand, gas prices have skyrocketed in the past year. A number of energy gurus think the natural-gas market may be facing a decade-long problem.

Such experts usually point to California, where utilities have been brought to the verge of bankruptcy by a botched deregulation of the power industry (of which more later). This has left the firms exposed to spot prices for electricity fired by natural gas which have been as much as ten times as high as a year ago. The resulting debts, and the utilities' attempts to recover them from unwilling ratepayers, have caused a political crisis in the country's biggest and richest state and raised fears of a possible recession there.

But despite the recent volatility in energy markets, comparisons with the oil shocks of recent decades are vastly overblown. For one thing, the causes of recent energy crises are quite different from those that produced the oil shocks of the 1970s and the lesser upsets during the Gulf war a decade ago. Today's woes come at a time of peace in energy-producing regions. OPEC has been working with oil-consuming nations to try to stabilise energy prices. The spikes in natural-gas prices are causing short-term pain, but the price signals they send are already encouraging the development of more gas fields.

No need to panic

Also, in a reversal of the conventional wisdom of two decades ago, most experts now believe that oil and especially natural gas will remain plentiful for decades hence, and that the means of converting those fuels into useful energy, such as internal combustion engines and combined-cycle gas turbines, will grow ever more efficient. What is more, the world has become much less vulnerable to oil shocks: thanks to conservation, fuel switching and improvements in efficiency, oil's share of industrial countries' imports, and their economies' reliance on it, has shrunk significantly.

Three powerful factors are now combining to shape the future of the energy industry: market forces, greenery and technological innovation. None of them is new, but together they are exerting strong pressure for change. Yet the industry's incumbents tend to resist change because they have much to lose from it; and given the sector's enormous and long-lived stock of fixed assets, a turnaround is bound to take time. And, confusingly, some of the forces for change pull in opposite directions: rising environmental standards may favour renewable energy, for example, but market reforms may choke off subsidies for it at the same time.

The [*Economist*'s] survey will argue that energy is indeed on the cusp of dramatic change. The sections [that are in the *Economist's* Survey: Energy, *The Economist*, February 10, 2001], show how the cross-currents at work today are reshaping the en-

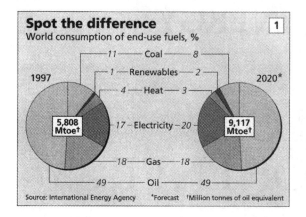

Spot the difference
World consumption of end-use fuels, %

1997 — 5,808 Mtoe†

2020* — 9,117 Mtoe†

	1997	2020*
Coal	11	8
Renewables	1	2
Heat	4	3
Electricity	17	20
Gas	18	18
Oil	49	49

Source: International Energy Agency *Forecast †Million tonnes of oil equivalent

ergy world, from the liberalisation of power markets to the greening of the world's oil giants and the advance of disruptive innovations such as fuel cells and distributed power. These changes will force energy companies to think hard about what they are truly good at, where tomorrow's competitive threats may come from, and what their customers really want. They will also help bring modern energy to poor countries that so far have been left out. Ultimately, they may even propel the world towards a cleaner and more sustainable technology: hydrogen energy.

Hand over to the market

However, whether the world realises the full potential of these prospects depends crucially on one factor: government. The invisible hand may be ascendant, but that does not mean the visible one has become irrelevant. On the contrary, during the transition to liberalised energy markets the role of regulators and officials is vitally important. And as California's sad example shows, governments can make a big difference by getting it wrong.

All three main forces of change closely involve government. In pushing for deregulation, regulators must be willing to trust market forces: they must make the rules of the game clear and refrain from arbitrary interference during the transition, yet remain on the look-out for market abuses. And ultimately they must yield most of their powers to the market.

In advocating greenery to meet their citizens' rising expectations, governments must be careful to avoid the distorting effects of such measures as excessive petrol taxes and flip-flopping environmental standards. There is a good case for some government support for renewable energy and other alternatives to fossil fuels as an insurance policy against the possibility of distant hazards such as global warming and oil depletion. However, the final test for all such technologies must remain the marketplace.

When it comes to clean technology, the most effective boost that bureaucrats can give to a sustainable energy future is to avoid picking winners. Instead, they would do better to provide a level playing field by scrapping the huge and usually hidden subsidies for fossil fuels, and by introducing measures such as carbon taxes so that the price of fossil fuels reflects the costs they impose on the environment and human health. Governments should also ensure that incumbents do not obstruct the entry of nimble newcomers, and keep open a range of options for producing energy, including running existing nuclear plants to the end of their useful life. They should provide strong incentives for firms to invest in today's creaking electricity grids, but also remove barriers to the spread of distributed generation.

Lastly, the governments of the rich world should do much more to help the poorer part meet its energy needs by leapfrogging to clean technologies. Most of the growth in both energy demand and in emissions will soon come from developing countries. If they invest in yesterday's dirty and inefficient technologies, they will be locked into them for decades to come— and the whole world will suffer the consequences. Lack of energy, especially modern fuels, in the developing world is likely to depress the productivity of billions of its workers, and so hold back future global economic growth.

Taken together, these prescriptions suggest that successful reform will be a tricky balancing act. But without it, the future would look much dimmer.

Beyond Oil:
THE FUTURE OF ENERGY
When Wells Go Dry

ENERGY: The rate of global oil production will start to fall in just a few years, says a controversial geologist. And alternative technologies aren't ready yet.

BY FRED GUTERL

As KENNETH DEFFEYES walks the five blocks from the Princeton University campus to his home, he veers sharply through a parking lot and then without warning takes a diagonal path across a side street. He doesn't seem to be paying any particular attention to where his sneaker-clad feet are taking him. His hands are tucked firmly in his parka, his eyes are looking up at a cloudless blue sky and his mind is where it usually is: on the world's supply of oil. In particular, Deffeyes is trying to explain why anybody should believe that the entire human enterprise of oil exploration—the search for reserves, the drilling of wells, the extraction of crude and all the attendant calculations of supply and demand—why this whole messy business should obey a simple but elegant piece of mathematics.

Deffeyes has reached a conclusion with far-reaching consequences for the entire industrialized world. So far-reaching that many of his colleagues in the field of petroleum geology dismiss it. The conclusion is this: in somewhere between two and six years from now, worldwide oil production will peak. After that, chronic shortages will become a way of life. The 100-year reign of King Oil will be over. And there will be nothing that President George W. Bush or Saudi princes or the invisible hand of the marketplace will be able to do about it. "There's nothing we could conceivably do now that would have much of an effect on the oil supply for at least 10 years," says Deffeyes.

This news isn't all bad. For the past decade, climate scientists have lobbied for drastic reductions in the use of fossil fuels, which release carbon dioxide into the air when burned, creating the "greenhouse effect." And

since September 11, some political leaders have called for reducing Western dependence on oil from the Middle East. At the same time, alternative technologies have advanced considerably in recent years. Farms of slender windmills have sprung up in California and Europe, generating electrical power at prices nearly competitive with conventional fuels. Photovoltaic panels that convert sunlight to electricity have fallen in price and are being built into roofing shingles. Hydrogen-fuel cells, another "clean" energy technology, offers an increasingly attractive long-term alternative not only to electrical power but to replace the internal combustion engine in automobiles; President Bush has made fuel cells the centerpiece of his long-term energy plan.

> ## "There's nothing we could do now that would have much effect on the oil supply for at least 10 years."
>
> –KENNETH S. DEFFEYES, Princeton geologist

Despite the progress in alternative fuels, however, the world will have to depend on hydrocarbons such as oil, coal and natural gas for a while yet. Without the political will to pursue alternatives with the wartime vigor of those who created the Manhattan Project to build the atom bomb, it would take decades to replace old power plants and phase out the ubiquitous gas station. Deffeyes, by conjuring up the bad years of the 1970s, when the Organization of Petroleum Exporting Countries restricted oil supplies and sent Western economies reeling, is injecting a large dose of urgency to the debate.

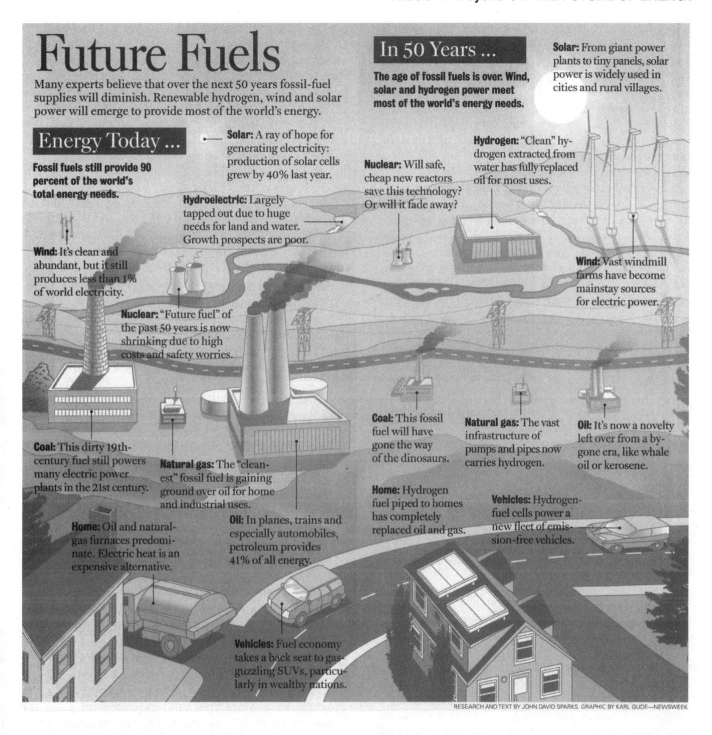

Future Fuels

Many experts believe that over the next 50 years fossil-fuel supplies will diminish. Renewable hydrogen, wind and solar power will emerge to provide most of the world's energy.

Energy Today ...

Fossil fuels still provide 90 percent of the world's total energy needs.

Solar: A ray of hope for generating electricity: production of solar cells grew by 40% last year.

Hydroelectric: Largely tapped out due to huge needs for land and water. Growth prospects are poor.

Wind: It's clean and abundant, but it still produces less than 1% of world electricity.

Nuclear: "Future fuel" of the past 50 years is now shrinking due to high costs and safety worries.

Coal: This dirty 19th-century fuel still powers many electric power plants in the 21st century.

Natural gas: The "cleanest" fossil fuel is gaining ground over oil for home and industrial uses.

Home: Oil and natural-gas furnaces predominate. Electric heat is an expensive alternative.

Oil: In planes, trains and especially automobiles, petroleum provides 41% of all energy.

Vehicles: Fuel economy takes a back seat to gas-guzzling SUVs, particularly in wealthy nations.

Nuclear: Will safe, cheap new reactors save this technology? Or will it fade away?

In 50 Years ...

The age of fossil fuels is over. Wind, solar and hydrogen power meet most of the world's energy needs.

Solar: From giant power plants to tiny panels, solar power is widely used in cities and rural villages.

Hydrogen: "Clean" hydrogen extracted from water has fully replaced oil for most uses.

Wind: Vast windmill farms have become mainstay sources for electric power.

Coal: This fossil fuel will have gone the way of the dinosaurs.

Natural gas: The vast infrastructure of pumps and pipes now carries hydrogen.

Oil: It's now a novelty left over from a bygone era, like whale oil or kerosene.

Home: Hydrogen fuel piped to homes has completely replaced oil and gas.

Vehicles: Hydrogen-fuel cells power a new fleet of emission-free vehicles.

RESEARCH AND TEXT BY JOHN DAVID SPARKS. GRAPHIC BY KARL GUDE—NEWSWEEK

How did Deffeyes arrive at his gloomy forecast? He likes to say that he grew up in the Oklahoma oil patch: his father was a petroleum engineer, and during summers off from high school and college the younger Deffeyes worked odd jobs in the oil fields. But the real genesis of his thinking was a brief stint he did at Shell Oil's research labs in the late 1950s, at the very start of his career. There he worked under the late geophysicist M. King Hubbert, who in 1956 made a startling prediction. At the time, oil reserves were being discovered at an accelerating pace, and the oil business was booming, with no apparent end in sight. But Hubbert had done groundbreaking work by applying a combination of geophysics and statistical analysis to the base of U.S. oil reserves, and from there projecting domestic production rates for years to come. The result was a bell curve: production was then going up, but it would inevitably peak and then start to decline. The math told Hubbert that the peak would come in the early 1970s, which clashed with the conventional wisdom of the time. In fact, that is more or less what happened: U.S. oil production hit its high in 1970, then started down just in time for the OPEC-induced oil crisis. But at the time Deffeyes was one of the few people who took Hubbert's prediction seriously. "I realized that a contracting oil industry was not a good career prospect," he says, "so I decided to get out and go into academia."

After spells at the University of Minnesota and Oregon State, Deffeyes pitched up at Princeton (and served as John McPhee's guide and mentor for his memorable writings on geology, collected as "Annals of the Former World"). There he applied Hubbert's methodology to global oil supplies. The results shocked him: the bell curve would peak sometime between 2004 and 2008, depending on how you crunched the numbers.

But why should oil production follow a bell curve? Deffeyes offers the analogy of a hunter shooting at a target. Each time he's going to miss by a certain amount due to wind or air temperature or faulty aim or hand tremors. If he fires 1,000 shots, you would measure how far each shot falls from the center of the target and plot how many shots fall at each interval of distance. The resulting graph will take the shape of a bell. The fat part of the graph will sit over the target, where most of the shots landed, and the two downward slopes will correspond to those distances on either side of the bull's-eye. Oil exploration is a similarly hit-and-miss affair. Despite all the high-tech methods geologists use nowadays, nine out of every 10 exploration wells turn out to be dry. It stands to reason, Deffeyes argues, that as more of the world's oil is discovered, the smaller the new finds are going to get. Deffeyes's bell curve is a plot of the volume of oil discovered each year for most of the 20th century. The curve rises slowly at first, as geologists picked off the easy-to-find fields—the ones that announced themselves with telltale tar pits and oil slicks. The curve shoots up during the '50s and '60s, when geologists discovered many of the bigger, deeper reserves, such as those in the North Sea, the Bass Strait and Saudi Arabia. As the century draws to a close, the pace of discovery actually accelerates, but the finds are smaller. The curve begins to flatten out.

Already, Saudi Arabia is the only oil-producing country that doesn't sell as much as it can pump. That fact has allowed the Saudis to dominate global markets, but they can't offset the shortages yet to come. Pumping at maximum capacity, they would add less than 4 percent to the world supply. Nor is technology likely to offer a way out. Energy companies, says Deffeyes, invested billions of dollars in the past 30 years to improve their ability to discover and extract oil, and it's unlikely that any breakthroughs in the offing would significantly change the equation. Technology would help eke out the hardest-to-reach reserves, but the curve would still fall. "What would it take to get another hump in the curve, with a peak farther out?" says Deffeyes. "There would have to be another kind of oil field altogether. And there's no evidence that such a thing exists."

Deffeyes, who reported his findings last October in a book, "Hubbert's Peak," has his detractors. Ronald Charpentier, a geologist at the U.S. Geological Survey, wrote recently in the journal Science that Deffeyes's estimates are based on a "questionable methodology" that is attractive in large part because it requires "modest data and human resources." The USGS's 2000 survey of world oil supply—which took 100 man-years to prepare, as opposed to the six months Deffeyes spent running his numbers—shows a worldwide oil supply of more than 3 trillion barrels (a trillion more than Deffeyes estimates). And Charpentier argues that additional discoveries, such as deep-water oil now being extracted in the South Atlantic, as well as untapped reserves in the Caspian Sea, Siberia and Africa, could change the outlook considerably. "There's a lot of oil and gas out there," he says, "but it's not necessarily where you want it to be and at a price you want it to be." Deffeyes says he doesn't dispute the possibility of finding new reserves, but he insists they're likely to be less fruitful than the USGS thinks.

What if Deffeyes is right? How will the world satisfy its growing demand for energy? Carmakers (not to mention President Bush) have pinned their hopes on hydrogen-fuel cells, which would emit no carbon and wouldn't entail drilling in unfavorable parts of the world. General Motors has stepped up its research budget for fuel cells from $1 million a year in 1990 to $100 million this year. "We believe hydrogen is the long-term answer," says Matt Fronk, GM's chief engineer for fuel cells. Photovoltaics (solar panels) have been hobbled by the high cost of semiconductors used to make them, but cheaper alternatives are in the offing. The National Renewable Energy Laboratory in Colorado, for instance, is developing an inexpensive way to deposit ultrathin layers of semiconductors on glass, steel or flexible plastic sheets. Says Christoph Frei of the World Economic Forum: "Everyone expects the energy mix to change."

But developments such as these, as exciting as they are, will not be of much use in the next decade. In that time, natural gas will probably be the most attractive hydrocarbon substitute for oil, at least for electricity generation and heating. It is relatively clean and, by all estimates, plentiful. Using it in cars, though, would entail a whole new fuel infrastructure that couldn't be built in a hurry. The only other real alternative would be conservation. According to conservation advocate Amory Lovins, improving the average fuel efficiency of vehicles in the United States by 2.7 miles per gallon would equal all U.S. oil imports from the Persian Gulf.

There's always the chance, of course, that the next decade will come and go without so much as a wobble in the supply of oil. Deffeyes allows for the possibility that he is wrong. "You've got to give Hubbert credit for getting it right," he says. "Then you ask yourself if he was lucky or if he really knew." Deffeyes sighs. "And you just don't know." With global warming, that brings to two the number of urgent but tentative reasons to pursue alternatives to fossil fuels.

With ADAM PIORE *and* SANDY EDRY

Hot Springs Eternal

HYDROGEN POWER: People mocked Bragi Arnason's vision of producing energy from the H in H_2O. Now the first test is about to be launched in his native Iceland. Next: the world?

BY ADAM PIORE

IT's A LITTLE BEFORE NOON, AND an anemic winter sun rises slowly over an Icelandic wasteland of jagged lava rocks. At a steaming pool of milky blue water, Bragi Arnason is talking about the future. A ruddy-faced chemistry professor with a white pompadour and a heavy maroon coat, Arnason has a staccato voice that barely rises above the hiss of steam surging up from the earth's core. For almost three decades, it has been Arnason's dream to harness this volcanic power to split the H from H_2O, and create a society free of fossil fuels: the first "hydrogen economy." "I will see the first steps," says Arnason, raising a gloved hand to the horizon, where a geothermal power plant gushes white steam. "My children, they will watch the whole transformation. My grandchildren, they will live with this new energy economy."

Oblivious to the frigid wind whipping across the plain just west of Reykjavik, Arnason spells out the vision that earned him the moniker "Professor Hydrogen." First, take all of Iceland's cars and fishing trawlers and gradually replace their internal combustion engines with electric motors that run on hydrogen-fuel cells, just like American space shuttles. Meanwhile, harness the power of Iceland's active volcanoes and raging rivers to begin producing pure hydrogen gas on a mass scale. Arnason has even inspired some compatriots to imagine Iceland as the "Kuwait of the North," a major source of energy in a world where all nations have decided to replace hydrocarbons with hydrogen. "There were not many people who really listened to him" at first, says Valgerdur Sverrisdottir, Iceland's minister of Industry and Commerce, throwing back her head and laughing. "It was so far away—to use water for cars?"

It's not far out anymore. The roster of experts who see hydrogen as the most likely replacement for oil in the long term now includes not only futurists and environmentalists but also the oilmen of the Bush administration and researchers at General Motors and Ford. Iceland's plan is now backed by DaimlerChrysler, Shell and the European Union, which plan to spend tens of millions of euros to create the first societal lab test of a hydrogen economy. In the coming months, Iceland will roll out three hydrogen-powered buses and begin constructing a filling station where hydrogen gas will be produced on-site. If all goes according to plan, this demonstration will expand to cars and fishing vessels in

2005, and all vehicles within 30 to 40 years. Other nations are likely to follow. The only question is when, says Margaret Mann, an engineer at the U.S. National Renewable Energy Laboratory. "In the long term we have to move to hydrogen. It's the only way to really divorce ourselves from fossil fuels."

Even now, the vision of a hydrogen economy sounds too good to be true. Hydrogen occurs naturally in water, so the supply is as close to endless as the oceans. And pure hydrogen is a harmless gas, not a toxic liquid, so spills would dissipate in the air. Hydrogen-fuel cells emit only water vapor, and the electric motors make nary a sound. In the case of Iceland, the process will rely on the island nation's abundant steam and water to generate the power to make hydrogen fuel. The process involves bombarding H_2O with ions to split off the hydrogen, which is recombined with oxygen in a fuel cell that makes charged ions to run a motor, and water molecules as a byproduct. In short, Arnason plans to use natural energy to make a powerful hydrogen fuel that emits only water mist as waste, promising a bottomless well of clean energy that produces no greenhouse gases and no threat of global warming.

ARNASON THOUGHT it was an obvious idea. In the 1970s, he was living on top of a glacier and mapping Iceland's reservoirs of hot groundwater as part of his doctoral thesis in chemistry. The reservoirs were no secret in a nation where cooks have been known to bake bread by burying boxes of dough in the ground, and farmers drilling in the backyard are likely to set off geysers of boiling-hot vapor. But Arnason was the first to map them, revealing huge reserves of natural geothermal energy.

It was a portentous discovery. Iceland grew up energy-poor, a harsh reality that rankled an independent-minded nation founded in the 900s by Norwegian Vikings. With no fossil fuels of its own, and huge energy needs due to the cold weather and the economy's reliance on gas-guzzling fishing vessels, Iceland was heavily dependent on foreign oil. It reeled from recurrent oil shocks, and its inflation rate averaged a stagger-ing 17.6 percent between 1944 and 1995. Driving down the frigid North Atlantic coast, Arnason recently recalled that it was "quite natural" for a nation in this state of energy servitude to ask, "How can we change this?"

Clean Energy: Electricity With No Waste

Like a battery, a fuel cell passes electrical energy between positive and negative poles. Unlike a battery, a fuel cell generates current rather than simply storing it. Chemical reactions between platinum catalysts and the highly charged poles start the juice flowing.

HOW A FUEL CELL WORKS

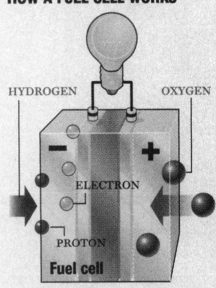

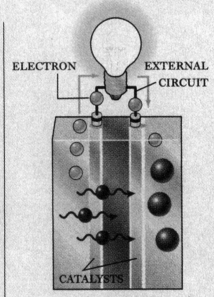

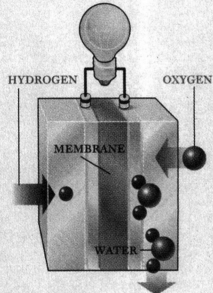

Ⓐ Hydrogen, pumped to the cell's negatively charged side (-), divides into protons and electrons. Oxygen is pumped into the positively charged side (+), which attracts electrons.

Ⓑ A platinum membrane blocks the electrons from moving directly to the positive side of the cell, forcing them through an **external circuit,** where they form a usable current.

Ⓒ Protons pass through the membrane between the two sides. They join with the oxygen to form water. When the water is expelled, oxygen refills the cell, and the process begins again.

TEXT BY JOHN D. SPARKS, GRAPHIC BY TONIA COWAN

From the beginning, Arnason saw hydrogen as the answer, knowing full well it appeared a bit nuts. In the 1970s and 1980s, hydrogen-fuel cells were huge and expensive, and the fact that they were used by the U.S. space program only made Arnason's vision seem farther out there. Arnason delivered his first paper on the subject in 1978 with trepidation. "I was not quite sure whether I should publish it. I thought that maybe people would think I was crazy," Arnason recalls. "[My mentor] read it and said, 'If you really believe this could be realized in 20 to 30 years, you should start talking about it now'."

He did, mostly to Rotary clubs, his classes and other small groups in Iceland, where he soon gained a reputation. "He was always called a guru, not by the general public but by the energy people," says Johann Mar Mariusson, vice president of Landsvirkjun, the National Power Company, and a longtime friend. "He didn't like the name Professor Hydrogen. He thought people were mocking him."

Then the world caught up. Arnason's work began to pay off in the 1990s, when global oil and auto companies started to take seriously the idea of hydrogen as the "next oil." The first breakthrough came in 1992, when Ballard Power Systems of Vancouver, British Columbia, demonstrated the first hydrogen-powered bus—the precursor to the models that will roll onto the streets of Reykjavik. The big surprise: Ballard's bus produced 150 kilowatts, or 15 times more power than most engineers expected from a hydrogen motor. "People kept looking around and asking, 'Where's the rest of it hidden?'" recalls Firoz Rasul, Ballard's CEO. "They never thought it could be done." Daimler-Benz stepped up and invested $250 million, and together they turned to Iceland.

Iceland just made sense. It has extreme weather conditions to test the durability of new automotive models, and a small population of 280,000, for whom high energy costs were reason enough to experiment with alternative energies. Moreover, Iceland's Professor Hydrogen was already well acquainted with the head of fuel-cell research at Daimler, Ferdinand Panik, who had been seeing Arnason at obscure conferences for years. The support of this big-shot scientist from a major multinational silenced the skeptics. In 1998, when officials from Daimler (now merged with Chrysler) arrived in Reykjavik to cut the deal, it was such a big local story that talks were moved to a secret site outside town. "We had to hide," recalls parliamentarian Hjalmar Arnason (no relation to Professor Hydrogen), a leader of Iceland's negotiating team. "Everything went crazy."

The interest of an auto giant drew other global players. The result is Icelandic New Energy, owned 51 percent by Iceland and 49 percent by private interests including DaimlerChrysler, Shell Oil and Norway's Norsk Hydro. The European business executives were stunned at how fast business got done in Reykjavik, a tiny capital where everyone knows everyone. It also didn't take long to interest the European Union, which so far has allocated €2.85 million to the bus phase of the project. The EU plans to follow with similar bus projects in Britain, Germany, Spain and at least four other nations. "If it works in Iceland, it'll work in other countries," says Panik. "It's the perfect place to start."

ICELAND IS ALSO UNUSUALLY AMBITIOUS. Cities from Vancouver to Palm Springs have deployed hydrogen buses, but Iceland represents the first attempt to phase out fossil fuels entirely, and thus become the first society to eliminate greenhouse emissions. Icelandic New Energy will be doing extensive market research to refine the design of hydrogen stations and cars, and to make the whole idea of the hydrogen economy "socially acceptable." One key hurdle is cost: even under the most favorable estimates, Icelandic New Energy expects the hydrogen to eventually cost about twice as much as gasoline, though it would generate twice as many miles per gallon.

Iceland is a model in the making. The International Code Council in Virginia is holding a public hearing in April on draft fire, electrical and industrial-fuel codes to govern a hydrogen economy. This June the International Organization for Standardization will consider safety guidelines for tanks, containers and fueling stations that hold hydrogen, which is about as flammable as gasoline. "Iceland is small enough that you can actually test and validate codes and standards and so on," says Rasul. "Iceland is going to be able to show where the rest of the world can go."

The question is when, and at what cost. Shell Hydrogen figures it would cost at least $19 billion to build hydrogen fuel plants and stations in the United States, $1.5 billion in Britain and $6 billion in Japan. That's compared with a "matter of millions of dollars in tiny Iceland," says CEO Don Huberts. "Also, in Iceland people do not drive their cars off an island, so we won't have to wait for the infrastructure to develop elsewhere."

Iceland may prove unique in other ways, too. Its huge reserves of natural geothermal energy are unrivaled in Europe, and some dream of exporting the hydrogen to the continent and creating a booming new industry (though first they'll have to figure out a way to get it there). Whatever happens, when the new energy age dawns, Arnason will have the satisfaction of knowing that the first seeds of the hydrogen economy grew in his Viking land of snowy moonscapes and steaming water holes. And Professor Hydrogen will no longer have reason to hate his nickname.

With STEFAN THEIL *in Berlin*

Taking the Breeze

ELECTRICITY: New technologies make Europe take another look at wind as a power source

BY WILLIAM UNDERHILL

THE NATIVES OF LEWIS KNOW wind—sometimes too well. Every winter the Atlantic gales come blasting across the northern tip of Scotland's Outer Hebrides. The wind hardly slows down after striking land; in the island's marshy interior, gusts regularly exceed 100 miles an hour. Everyone stays indoors but the sheep. Tourists arrive in summer, lured by mild temperatures and wide expanses of unspoiled countryside; even so, there's rarely a calm day. "The weather here is changeable," says Nigel Scott, a spokesman for the local government. "But the wind is constant."

The brutal climate could finally be Lewis's salvation. The place has been growing poorer and more desolate for generations, as young people seek sunnier prospects elsewhere. But now the energy industry has discovered the storm-swept island. The multinational engineering and construction giant AMEC and the electricity generator British Energy are talking about plans to erect some 300 outsize wind turbines across a few thousand acres of moorland and peat bog. If the $700 million project goes through, the array will be Europe's largest wind farm, capable of churning out roughly 1 percent of Britain's total electrical needs—and generating some badly needed jobs and cash for the people of Lewis. "We have been slowly bleeding to death," says Iain McIver of the Stornoway Trust, the proposed site's owners. "The benefits from this project will continue to flow for as long as the wind blows over the Western Isles."

It sounds like the answer to a lot of prayers—and not only on Lewis. Enthusiasts around the world call wind a perfect alternative to fossil fuels and nuclear power: safe, inexhaustible and free. "This is simply one of the cheapest ways of reducing our output of greenhouse gases," says Christian Kjaer of the Brussels-based European Wind Energy Association. Still, not everyone is such a fan. "I find it incredible that organizations which describe themselves as 'Green' or 'Friends of the Earth' can contemplate the ravage of our hills with these industrial installations," says Margaret Thatcher's former press secretary Sir Bernard Ingham, a champion of nuclear power. Even environmentalists confess to a few reservations. "The wind industry is as capable of environmental insensitivity as any other," says Roger Higman, a senior campaigner for Friends of the Earth.

The energy industry isn't bothered by such quibbles. For the last seven years the world market for wind turbines has grown by an average of 40 percent annually. Last year alone, generating capacity worldwide jumped by almost a third. The more wind plants you build, the cheaper and more powerful you can make them. Turbine makers are now mass-producing giant machines—the rotors' 211-foot diameter is wider than the wingspan of a jumbo jet—that once existed only in theory. Today one standard-issue turbine can produce at least 1 megawatt of power, more than double the typical model's output of 20 years ago and enough to provide electricity for as many as 800 modern households. The next generation, capable of more than twice that output, is already emerging.

The new turbines are not just bigger; they're smarter. The basic design, a triple-vaned rotor atop a vertical shaft, hasn't changed much in 50 years. The big difference is that the towers are taller; the new ones rise almost 300 feet above the ground. The higher they go, the stronger and steadier are the winds they catch. But scientists keep tweaking the specs in subtler ways, too. Tough, light structural materials have been borrowed from the aeronautics industry. Likewise for "vortex generators," tiny fins added to the surfaces of wind rotors and aircraft wings. They induce turbulence that helps prevent stalling at low speeds. Best of all for people who live nearby, im-

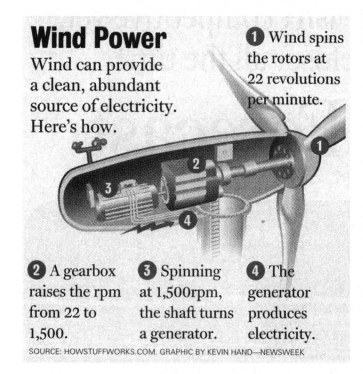

Wind Power

Wind can provide a clean, abundant source of electricity. Here's how.

1 Wind spins the rotors at 22 revolutions per minute.

2 A gearbox raises the rpm from 22 to 1,500.

3 Spinning at 1,500rpm, the shaft turns a generator.

4 The generator produces electricity.

SOURCE: HOWSTUFFWORKS.COM. GRAPHIC BY KEVIN HAND—NEWSWEEK

proved design on the latest models has cut noise to a relative whisper.

Still, some nature lovers hate wind power. Turbines seem to hold the same fatal attraction for birds that bug zappers have for mosquitoes, although no one is quite sure why. And the best sites for wind farms are often previously unspoiled hilltops. "We don't think esthetics are an ecological criterion," says Sven Tiske of Greenpeace. "If we opposed a nuclear power station just because it didn't look good, everyone would laugh." Not everyone. Ask Robert Woodward, a British art historian who campaigns against the spread of wind farming. His holiday residence is in Wales, at the edge of the Cambrian Mountains, and since the early 1990s the view from the hillside above his house has encompassed more than 100 turbines, all flailing out of sync. "A staggeringly beautiful landscape is being devastated," says Woodward, who used to support environmentalist groups—until wind power blew him away. "This is the first time in the history of the conservation movement that concern for the landscape has become a dirty concept."

Such gripes are one reason the industry is experimenting with offshore facilities. Viewed from land, even the tallest offshore farm is a tiny thing. Besides, sea breezes tend to be stiff, and steady enough to raise a turbine's output at least 20 percent. Denmark is already proving the idea can work. Take a walk on the beach in Copenhagen's northern suburbs, and you'll notice that the horizon is broken by a distant line of 20 turbines. They're built on a submerged limestone reef, almost two miles out to sea. Billed as the world's largest offshore wind farm when it opened last year, the installation generates enough power for 30,000 city households. Now vast tracts of open water are being staked out, from western Ireland to the Baltic. Nick Goodall, director of the British Wind Energy Association, says: "The great thing about offshore wind power is that you are limited only by your imagination and the amount of money that the banks will lend you."

Bankers aren't getting too swept away. Building and running a farm at sea costs up to 40 percent more than onshore. Consider just the price of laying undersea cable to deliver electricity from turbine to grid. Construction is tricky, usually requiring a pile driver to sink the turbine's concrete foundation deep into the seabed, and maintenance is inconvenient and expensive. "Offshore is a luxury," says Per Krogsgaard of the Danish consultancy BTM. "And it will be for a very, very long time."

Which leaves one question: can the wind industry undersell its competitors? The answer seems easy. After all, the fuel is literally as free as the wind. The chief expense is setting up a turbine farm. That's still too high a price to drive fossil-fuel plants out of business. "It is still cheaper to put more coal into an existing power station than to build a new wind farm," admits Kjaer. If you built a new coal-fired plant today, wind enthusiasts insist they could put up their own power installation and equal your rates. But no one is staging any such contest: unlike America, Europe has a chronic surplus of generating capacity.

Even so, the question is not academic. Sometimes there's no choice but to build new generating facilities. Conventional plants get old and obsolete, and environmental laws keep getting tougher. A recent British government report on energy policy includes a projection that by 2020, wind is likely to beat nuclear's prices and roughly match those of natural-gas power stations. But the wind industry seems unable to offer a solid cost-comparison against other power sources right now. Why?

"That's really a political question," says Andrew Garrad, of the wind-power consultants Garrad Hassan. "Scratch anywhere beneath the surface of energy economics, and you will find politics." It's officeholders, not engineers, who set subsidy levels for different energy sources, and their decisions can be far more inscrutable than the wind. Not that wind advocates are complaining. So far wind power is thriving only in countries—notably Denmark, Spain and Germany—where governments force utility companies to pay their suppliers premium rates for wind energy.

Wind still has one big drawback: sometimes it refuses to blow. Users need a dependable backup power supply for days when the turbines won't turn. Hard-core enviros say wind is only a stopgap; in the long run, they argue, there's no alternative to the tough conservation measures that are too painful for any politician to espouse. At present, though, wind has no eco-OK rivals. Environmentalists have fallen out of love with big hydroelectric plants, and Europe has few suitable sites left anyway. Solar technology remains prohibitively expensive and space-consuming. And biomass—burning fuel crops such as wood chips or agricultural waste in steam-powered generators—is a messy business that requires the right equipment on the spot. You can't just burn the stuff in a conventional coal-fired plant, for example.

The people of Lewis are putting their money on wind. "This is all about preserving the environment," says Nigel Scott, who moved here seven years ago. "If we don't go down the wind-energy road, in the long run there won't be any habitats to protect." No one seems too worried about turbines' spoiling the peat bog's vistas—or about what might happen if the winter gales should ever turn gentle.

RENEWABLE ENERGY: A VIABLE CHOICE

By Antonia V. Herzog, Timothy E. Lipman, Jennifer L. Edwards, and Daniel M. Kammen

Renewable energy systems—notably solar, wind, and biomass—are poised to play a major role in the energy economy and in improving the environmental quality of the United States. California's energy crisis focused attention on and raised fundamental questions about regional and national energy strategies. Prior to the crisis in California, there had been too little attention given to appropriate power plant siting issues and to bottlenecks in transmission and distribution. A strong national energy policy is now needed. Renewable technologies have become both economically viable and environmentally preferable alternatives to fossil fuels. Last year the United States spent more than $600 billion on energy, with U.S. oil imports climbing to $120 billion, or nearly $440 of imported oil for every American. In the long term, even a natural gas–based strategy will not be adequate to prevent a buildup of unacceptably high levels of carbon dioxide (CO_2) in the atmosphere. Both the Intergovernmental Panel on Climate Change's (IPCC) recent Third Assessment Report and the National Academy of Sciences' recent analysis of climate change science concluded that climate change is real and must be addressed immediately—and that U.S. policy needs to be directed toward implementing clean energy solutions.[1]

Renewable energy technologies have made important and dramatic technical, economic, and operational advances during the past decade. A national energy policy and climate change strategy should be formulated around these advances. Despite dramatic technical and economic advances in clean energy systems, the United States has seen far too little research and development (R&D) and too few incentives and sustained programs to build markets for renewable energy technologies and energy efficiency programs.[2] Not since the late 1970s has there been a more compelling and conducive environment for an integrated, large-scale approach to renewable energy innovation and market expansion.[3] Clean, low-carbon energy choices now make both economic and environmental sense, and they provide the domestic basis for our energy supply that will provide security, not dependence on unpredictable overseas fossil fuels.

Energy issues in the United States have created "quick fix" solutions that, while politically expedient, will ultimately do the country more harm than good. It is critical to examine all energy options, and never before have so many technological solutions been available to address energy needs. In the near term, some expansion of the nation's fossil fuel (particularly natural gas) supply is warranted to keep pace with rising demand, but that expansion should be balanced with measures to develop cleaner energy solutions for the future. The best short-term options for the United States are energy efficiency, conservation, and expanded markets for renewable energy.

Traditional power plants based on fossil fuels emit pollutants that contribute substantially to climate change. Renewable energy sources are becoming economically viable and environmentally preferable alternatives.

For many years, renewables were seen as energy options that—while environmentally and socially attractive—occupied niche markets at best, due to barriers of cost and available infrastructure. In the last decade, however, the case for renewable energy has become economically compelling as well. There has been a true revolution in technological innovation, cost improvements, and our understanding and analysis of appropriate applications of renewable energy resources and technologies—notably solar, wind, small-scale hydro, and biomass-based

Figure 1. Capital cost forecasts for renewable energy technologies

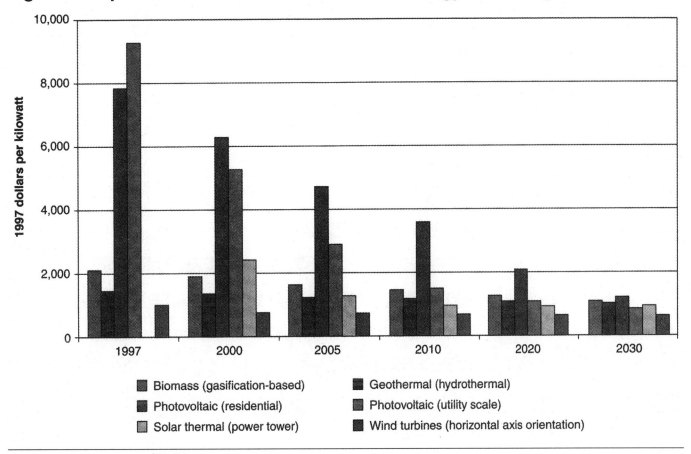

SOURCE: U.S. Department of Energy (DOE), *Renewable Energy Technology Characterizations*, Topical Report prepared by DOE Office of Utility Technologies and EPRI, TR-109496, December 1997.

energy, as well as advanced energy conversion devices such as fuel cells.[4] There are now a number of energy sources, conversion technologies, and applications that make renewable energy options either equal or better in price and services provided than the prevailing fossil-fuel technologies. For example, in a growing number of settings in industrialized nations, wind energy is now the least expensive option among all energy technologies—with the added benefit of being modular and quick to install and bring on-line. In fact, some farmers, notably in the U.S. Midwest, have found that they can generate more income per hectare from the electricity generated by a wind turbine than from their crop or ranching proceeds.[5] Also, photovoltaic (solar) panels and solar hot water heaters placed on buildings across the United States can help reduce energy costs, dramatically shave peak-power demands, produce a healthier living environment, and increase the overall energy supply.

The United States has lagged in its commitment to maintain leadership in key technological and industrial areas, many of which are related to the energy sector.[6] The United States has fallen behind Japan and Germany in the production of photovoltaic systems, behind Denmark in wind and cogeneration system deployment, and behind Japan, Germany, and Canada in the development of fuel-cell systems. Developing these indus-

tries within the United States is vital to the country's international competitiveness, commercial strength, and ability to provide for its own energy needs.

Renewable Energy Technologies

Conventional energy sources based on oil, coal, and natural gas have proven to be highly effective drivers of economic progress, but at the same time, they are highly damaging to the environment and human health. These traditional energy sources are facing increasing pressure on a host of environmental fronts, with perhaps the most serious being the looming threat of climate change and a needed reduction in greenhouse gas (GHG) emissions. It is now clear that efforts to maintain atmospheric CO_2 concentrations below even double the pre-industrial level cannot be accomplished in an oil- and coal-dominated global economy.

Theoretically, renewable energy sources can meet many times the world's energy demand. More important, renewable energy technologies can now be considered major components of local and regional energy systems. Solar, biomass, and wind energy resources, combined with new efficiency measures

Figure 2. Actual electricity costs in 2000

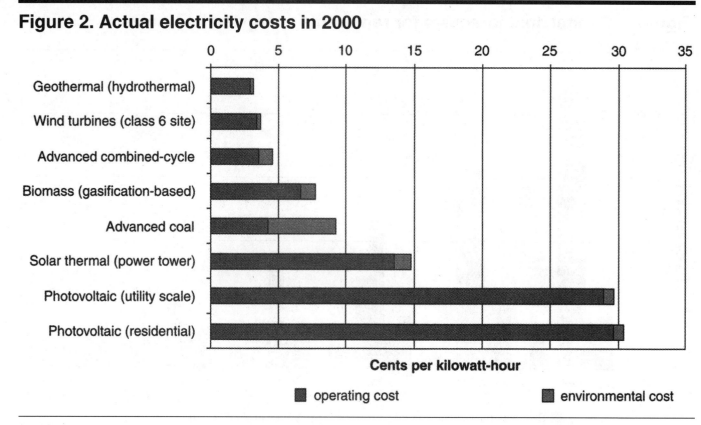

Cents per kilowatt-hour

■ operating cost   ■ environmental cost

SOURCE: R.L. Ottinger et al., *Environmental Costs of Electricity* (New York: Oceana Publications, Inc., 1991); and U.S. Department of Energy, *Annual Energy Outlook 2000*, DOE/EIA-0383, Energy Information Administration, Washington, D.C., December 2000.

available for deployment in California today, could supply half of the state's total energy needs. As an alternative to centralized power plants, renewable energy systems are ideally suited to provide a decentralized power supply that could help to lower capital infrastructure costs. Renewable systems based on photovoltaic arrays, windmills, biomass, or small hydropower can serve as mass-produced "energy appliances" that can be manufactured at low cost and tailored to meet specific energy loads and service conditions.These systems have less of an impact on the environment, and the impact they do have is more widely dispersed than that of centralized power plants, which in some cases contribute significantly to ambient air pollution and acid rain.

There has been significant progress in cost reductions made by renewable technologies (see Figure 1).[7] In general, renewable energy systems are characterized by low or no fuel costs, although operation and maintenance costs can be considerable. Systems such as photovoltaics contain far fewer mechanically active parts than comparable fossil fuel combustion systems, and are therefore likely to be less costly to maintain in the long term.

Costs of solar and wind power systems have dropped substantially in the past 30 years and continue to decline. For decades, the prices of oil and natural gas have been, as one research group noted, "predictably unpredictable."[8] Recent analyses have shown that generating capacity from wind and solar energy can be added at low incremental costs relative to

additions of fossil fuel–based generation. Geothermal and wind can be competitive with modern combined-cycle power plants-and geothermal, wind, and biomass all have lower total costs than advanced coal-fired plants, once approximate environmental costs are included (see Figure 2).[9] Environmental costs are based, conservatively, on the direct damage to the terrestrial and river systems from mining and pollutant emissions, as well as the impacts on crop yields and urban areas. The costs would be considerably higher if the damage caused by global warming were to be estimated and included.

The push to develop renewable and other clean energy technologies is no longer being driven solely by environmental concerns; these technologies are becoming economically competitive. According to Merrill Lynch's Robin Batchelor, the traditional energy sector has lacked appeal to investors in recent years because of heavy regulation, low growth, and a tendency to be cyclical.[10] The United States' lack of support for innovative new companies sends a signal that U.S. energy markets are biased against new entrants. The clean energy industry could, however, become a world-leading industry akin to that of U.S. semi-conductors and computer systems.

Renewable energy sources have historically had a difficult time breaking into markets that have been dominated by traditional, large-scale, fossil fuel–based systems. This is partly because renewable and other new energy technologies are only now being mass produced and have previously had high capital costs relative to more conventional systems, but also because coal-, oil-, and gas-powered systems have benefited from a range of subsidies over the years. These include military expenditures to protect oil exploration and production interests overseas, the costs of railway construction to enable economical delivery of coal to power plants, and a wide range of tax breaks.

One disadvantage of renewable energy systems has been the intermittent nature of some sources, such as wind and solar. A solution to this problem is to develop diversified systems that maximize the contribution of renewable energy sources but that also use clean natural gas and/or biomass-based power generation to provide base-load power (energy to meet the daily needs of society, leaving aside the peak in energy use associated, for example, with afternoon and evening air-conditioner or heating demands).

Solar energy can be harnessed by both industrial-scale and smaller residential solar power systems. Solar and other renewable energy sources could take over for fossil fuel–based power plants while saving energy and money.

Renewable energy systems face a situation confronting any new technology that attempts to dislodge an entrenched technology. For many years, the United States has been locked in to nuclear- and fossil fuel–based technologies, and many of its secondary systems and networks have been designed to accommodate only these sources. The U.S. administration's recent National Energy Policy plan focused on expanding the natural gas supply, without any attention to the benefits of building a diverse energy system.[11] The plan would add one to two new power plants each week for the next several years. The majority of these plants would be fired by natural gas, making the country far more dependent on natural gas than it ever was on oil—even at the height of the OPEC crisis in the 1970s.

Renewable energy technologies are characterized by low environmental costs, but many of these environmental costs are termed "externalities" and are not reflected in market prices. Only in certain areas and for certain pollutants do these environmental costs enter the picture. The international effort to limit the growth of GHG emissions through the Kyoto Protocol may lead to some form of carbon-based tax, which continues to face stiff political opposition in the United States. It is perhaps more likely that concern about emissions of particulate matter and ozone formation from fossil-fuel power plants will lead to expensive mitigation efforts by the plant operators, and this would help to tip the balance toward cleaner renewable systems.

There are two principal rationales for government support of R&D to develop clean energy technologies. First, conventional energy prices generally do not reflect the social and environmental cost of pollution. Second, private firms are generally unable to appropriate all the benefits of their R&D investments. The social rate of return for R&D exceeds the returns captured by individual firms, so they do not invest enough in R&D to maximize social benefits.[12] Public investment, however, would help spread innovation among clean energy companies, which would benefit the public.

Publicly funded market transformation programs (MTPs) for desirable clean energy technologies would provide an initial subsidy and incentive for market growth, thus stimulating long-term demand. A principal reason for considering MTPs is inherent in the production process. When a new technology is first introduced, it is more expensive than established substitutes. The unit cost of manufactured goods then tends to fall as a function of cumulative production experience. Cost reductions are typically very rapid at first and then taper off as the industry matures—resulting in an "experience curve". Gas turbines, photovoltaic cells and wind turbines have all exhibited this expected price-production relationship, with costs falling roughly 20 percent for each doubling of the number of units produced.[13]

If producers of clean energy consider the experience-curve effect when deciding how much to produce, they will "forward-price," producing at a loss initially to bring down their costs and thereby maximize profit over the entire production period. In practice, however, the benefits of production experience often spill over to competitor producers, and this potential problem discourages private firms from investing in bringing new products down the experience curve. Publicly funded MTPs can help correct the output shortfall associated with these experience effects.[14]

This suggests an important role for MTPs in national and international technology policies. MTPs are most effective with emerging technologies that have steep industry experience curves and a high probability of major long-term market penetration once subsidies are removed. The condition that these technologies be clean mitigates the risk of poor MTP performance, because the investments will alleviate environmental problems whose costs were not taken into account for the older, dirtier energy technologies. Renewable energy products are ideal candidates for support through MTPs, via federal policies that reward the early production of clean energy technologies.

Energy Efficiency

Energy efficiency improvements have contributed a great deal to economic growth and increased standard of living in the United States over the past 30 years, and there is much potential for further improvements in the decades to come. According to the U.S. Department of Energy (DOE), increasing energy efficiency could cut national energy use by 10 percent or more by 2010 and about 20 percent by 2020. The recent Interlaboratory Working Group study Scenarios for a Clean Energy Future estimates that cost-effective end-use technologies could reduce electricity consumption by about 1,000 billion kilowatt-hours (kWh) by 2020, almost entirely offsetting the projected growth in electricity use.[15] This level of energy savings would reduce U.S. carbon emissions by approximately 300 million metric tons, and many of these changes can actually be accomplished with an increase in profits. Still more benefits can be had for investments of only a few cents per kilowatt-hour, far less than the energy cost of new power plants.

Energy efficiency is the single greatest way to improve the U.S. energy economy. Based on data published by the Energy Information Administration (EIA), the American Council for an Energy Efficient Economy (ACEEE) estimates that total energy use per capita in the United States in 2000 was almost identical to that in 1973, while over the same period, economic output (measured by Gross Domestic Product or GDP) per capita increased 74 percent. Furthermore, national energy intensity (energy use per unit of GDP) fell 42 percent between 1973 and 2000, with about 60 percent of this decline attributable to real improvements in energy efficiency and about one-quarter due to structural changes and fuel switching. Between 1996 and 2000, GDP increased 19 percent while primary energy use increased just 5 percent. These statistics clearly indicate that energy use and GDP do not have to grow or decline in lock step with each other, but GDP can, in fact, increase while energy use does not.[16]

The federal government's energy efficiency programs have been a resounding success. Last year, DOE documented the results of 20 of its most successful energy efficiency and renewable energy technology initiatives over the past two decades.[17] These programs have already saved the nation 5.5 quadrillion Btu (British Thermal Units) of energy, equivalent to the amount of energy needed to heat every household in the United States for about a year, and worth about $30 billion in avoided energy costs. Over the last decade, the cost to taxpayers for these 20 programs has been $712 million, less than 3 percent of the energy bill savings that the programs created.[18]

In 1997, the President's Committee of Advisors on Science and Technology (PCAST), a panel that consisted mainly of distinguished academics and private-sector executives, conducted a detailed review of DOE's energy efficiency R&D programs. PCAST concluded, "R&D investments in energy efficiency are the most cost-effective way to simultaneously reduce the risks of climate change, oil import interruption, and local air pollution, and to improve the productivity of the economy." PCAST recommended that the DOE energy efficiency budget be doubled between the fiscal years of 1998 and 2003. They estimated that this could produce a 40-to-1 return on investment for the nation, including reductions in fuel costs of $15 billion to $30 billion by 2005 and $30 billion to $45 billion by 2010.[19] Funding for these DOE programs in the last several years has fallen far short of the PCAST recommendations.

Increasing the efficiency of homes, appliances, vehicles, businesses, and industries must be an important part of a sound national energy and climate-change policy. Increasing energy efficiency reduces energy waste (without forcing consumers to cut back on energy services or amenities), lowers GHG emissions, saves consumers and businesses money (because the energy savings more than pay for any increase in initial cost), protects against energy shortages, reduces energy imports, and reduces air pollution. Furthermore, increasing energy efficiency does not create a conflict between enhancing national security and energy reliability, on the one hand, and protecting the environment on the other.

Climate Change

The threat of global climate change is finally producing a growing understanding and acknowledgement by some in U.S. industry and government that a responsible national energy policy must include a sound global climate-change mitigation strategy. President George W. Bush has rejected the Kyoto Protocol, but the U.S. Congress, in particular the Senate, appears poised to take action to reduce domestic GHG emissions. For example, Senators Jim Jeffords (I-Vt.) and Joe Lieberman (D-Conn.) and Representatives Sherwood Boehlert (R-N.Y.) and Henry Waxman (D-Calif.) recently introduced legislation in Congress to reduce the emission of four pollutants from electricity generation. The legislation puts a national cap on power plants' emissions of nitrogen oxides, sulfur dioxide, mercury, and carbon dioxide, and requires every power plant to meet the most recent emission control standards. It allows market-oriented mechanisms such as the trading of emissions credits, which is widely seen as a way to control pollution and stimulate innovation at the lowest cost. In emissions trading, total emissions are capped and then a market is created involving those firms that have excess credits to sell (resulting from decreased emissions due to efficiency and other improvements) and those firms needing to purchase credits due to emissions exceeding an allocated baseline. In the United States, nitrogen oxides and sulfur dioxide markets have been highly successful. The CO_2 reductions required by the legislation would bring emissions back to 1990 levels by 2007, and the costs of implementing such measures would likely be dwarfed by the resulting benefits of industrial innovation.[20]

A tax credit that rewards fuel-efficient vehicles, including hybrid vehicles now on the market, could be used along with a tax penalty for inefficient vehicles to encourage purchases of fuel-efficient vehicles and the development of new technologies.

Legislation that controls the four major pollutants from power plants in an integrated package will help reduce regulatory uncertainties for electric generators and will be less costly than separate programs for each pollutant. Although voluntary action by companies is an attractive idea, in the last 10 years, voluntary actions have failed to reduce carbon dioxide emissions in the United States. Instead, emissions have increased by 15.5 percent since 1990, with an annual average increase of 1.5 percent since 1990, and they continue to increase.[21] EIA recently released data showing an increase of 2.7 percent in U.S. carbon dioxide emissions from 1999 to 2000. Solutions will become more costly and difficult if mandatory emissions reductions are not enacted now.

Policy Options

The ultimate solutions to meeting the nation's energy needs cost-effectively and reducing GHG emissions must be based on private-sector investment bolstered by well-targeted government R&D and incentives for emerging clean energy technologies. The United States now has the opportunity to build a sustainable energy future by engaging and stimulating the tremendous innovative and entrepreneurial capacity of the private sector. Advancing clean energy technologies requires a stable and predictable economic environment.

Research and Development Funding

Federal funding and leadership for renewable energy and energy efficiency projects has resulted in several notable successes, such as the U.S. Environmental Protection Agency's (EPA) Energy Star and Green Lights Programs, which have been emulated in a number of countries. Fifteen percent of the public-sector building space in the country has now signed up for the Energy Star buildings program, saving more than 21 billion kWh of energy in 1999 and reducing carbon emissions by about 4.4 million metric tons, which according to EPA, has resulted in $1.6 billion in energy bill savings. Despite these achievements, funding in this area has been scant and so uneven as to discourage private sector involvement. By increasing funding for these EPA programs, their scope could be considerably expanded.

The Bush administration's proposed cuts in its 2002 fiscal year budget for DOE's renewable energy and energy efficiency programs would harm existing public-private partnerships as well as R&D. This budgetary roller coaster harms all investments and sends mixed signals to industry.[22] Steadily increasing funding would transform the clean energy sector from a good idea to a pillar of the new economy.

Tax Incentives

The R&D tax credit, which goes to companies based on their R&D expenditures, has proven remarkably effective and popular with private industry, so much so that there is a strong consensus in both Congress and the administration to make this credit permanent. To complement this support of private-sector R&D, tax incentives directed toward those who use the technologies would provide the "demand pull" needed to accelerate the technology transfer process and the rate of market development.

Currently, non-R&D federal tax expenditures aimed at the production and use of energy have an unequal distribution across primary energy sources, distorting the market in favor of conventional energy technologies. Renewable fuels make up 4 percent of the United States' energy supply, yet they receive only 1 percent of federal tax expenditures and direct fiscal spending combined (see Table 1).[23] The largest single tax credit in 1999 was the Alternative Fuel Production Credit, which totaled more than $1 billion.[24] This income tax credit, which has gone primarily to the natural gas industry, was designed to reduce dependence on foreign energy imports by encouraging the production of gas, coal, and oil from unconventional sources (such as tight gas formations and coalbed methane) within the United States. Support for the production and further development of renewable fuels, all found domestically, would have a greater long-term effect on the energy system than any expansion of fossil-fuel capacity.

A production tax credit (PTC) of 1.7 cents/kWh now exists for electricity generated from wind power and "closed loop" biomass (biomass from dedicated energy crops and chicken litter). The wind power credit, in particular, has proven successful in encouraging strong growth of U.S. wind energy over the last several years—with a 30-percent increase in 1998 and a 40-percent increase in 1999. Approximately 2,000 megawatts (MW) of wind energy will be under development or proposed for completion before the end of 2001 (a 40-percent increase from 2000), when the federal wind energy PTC is scheduled to expire. Currently, Germany has twice the U.S. installed wind energy capacity, and the major wind-turbine manufacturers are now in Europe.[25]

This production credit should be expanded to include electricity produced by "open loop" biomass (including agricultural and forestry residues but excluding municipal solid waste), solar energy, geothermal energy, and landfill gas. The extension and expansion of PTC has recently been garnering strong and consistent support in the U.S. Congress. Investment tax incentives are also needed for smaller-scale renewable energy systems, such as residential photovoltaic panels and solar hot-water heaters, as well as small wind systems used in commercial and farm applications. In these cases, an investment credit in capital or installation expenditures is preferable to a production credit based on electricity generated, due to the relatively high capital cost of these smaller-scale renewable technologies and the fact that the electricity and heat produced is used directly.

Many new energy-efficient technologies have been commercialized in recent years or are nearing commercialization. Tax incentives can help manufacturers justify mass marketing and help buyers and manufacturers offset the relatively high initial capital and installation costs for new technologies. A key element in designing the credits is for only high-efficiency products to be eligible. If eligibility is set too low, there may not be enough energy savings to justify the credits. These tax credits should have limited duration and be reduced in value over time, because once these new technologies become widely available, costs should decline. In this manner, the credits will help innovative technologies get established in the marketplace but will not become permanent subsidies.

Recent federal tax credit legislation to encourage the use of high-efficiency technologies includes incentives for highly efficient clothes washers, refrigerators, and new homes; innovative building technologies such as furnaces, stationary fuel cells, gas-fired pumps, and electric heat-pump water heaters; and investments in commercial buildings that have reduced heating and cooling costs. The incentives currently being proposed in Congress and by the administration will have a relatively modest direct impact on energy use and CO_2 emissions. Savings may only amount to 0.3 quadrillion Btu of energy and 5 million metric tons of carbon emissions per year by 2012. How-

Table 1. U.S. energy consumption and federal expenditures

Fuel source	Primary energy supply 1998 consumption		Direct expenditures and tax expenditures (1999)	
	Value (quads, quadrillion Btu)	Percent	Value (millions of dollars)	Percent
Oil	36.57	40	263	16
Natural gas	21.84	24	1,048	65
Coal	21.62	24	85	5
Oil, gas, and coal combined			205	13
Nuclear	7.16	8	0	–
Renewables	3.48	4	19	1
Total	90.67	100	1620	100

NOTE: The Alternative Fuels Credit accounted for $1,030 of the $1,048 in expenditures for natural gas. Oil, gas, and coal combined includes expenditures that were not allocated to any one of the three individual fuels. Research and development are not included in direct expenditures and tax expenditures. Btu = British thermal units.

SOURCE: Energy Information Administration, *Federal Financial Interventions and Subsidies in Energy Markets 1999: Primary Energy*, U.S. Department of Energy (Washington, D.C., 1999).

ever, if these proposed tax credits help to establish innovative products in the marketplace and reduce the first-cost premium so that the products are viable after the credits are phased out, then the indirect impacts could be many times greater than the direct impacts. It has been estimated that total energy savings could reach 1 quadrillion Btu by 2010 and 2 quadrillion Btu by 2015 if these credits are successfully implemented.[26]

Vehicle Fuel Economy Standards

New vehicles with hybrid gasoline-electric power systems are now produced commercially, and fuel cell–electric vehicles are being produced in prototype quantities. These vehicles combine high-efficiency electric motors with revolutionary power systems to produce a new generation of motor vehicles that are vastly more efficient than today's simple cycle combustion systems. The potential for future hybrid and fuel-cell vehicles to achieve up to 100 miles per gallon (mpg) is believed to be both technically and economically viable in the near future. In the long term, fuel-cell vehicles running directly on hydrogen promise to allow motor vehicle use with very low fuel-cycle emissions.

The improvements in fuel economy that these new vehicles offer will help to slow growth in petroleum demand, reducing our oil import dependency and trade deficit. While the Partnership for a New Generation of Vehicles helped generate some vehicle technology advances, an increase in the Corporate Average Fuel Economy (CAFE) standard, which has been stagnant for 16 years now, is required to provide an incentive for companies to bring these new vehicles to market rapidly.

Recent analyses of the costs and benefits of motor vehicles with higher fuel economy have been conducted by the Union of Concerned Scientists, Massachusetts Institute of Technology, the Office of Technology Assessment, and Oak Ridge National Lab/ACEEE.[27] These studies have generally concluded that

with longer-term technologies, motor vehicle fuel economy can be raised to 45 mpg for cars with a retail price increase of $500 to $1,700 per vehicle, and to 30 mpg for light trucks with a retail price increase of $800 to $1,400 per vehicle.[28] These improvements could be the basis for a new combined fuel economy standard of 40 mpg for both cars and light trucks. The combined standard could be accomplished between 2008 and 2012. The net cost would be negligible once fuel savings are factored in, if the auto industry is given adequate time to retool for this new generation of vehicles. A lower combined standard could be implemented sooner and then raised incrementally each year to achieve the 40-mpg goal by 2012.

Tax credits for hybrid-electric vehicles, battery-electric vehicles, and fuel-cell vehicles are an important part of the puzzle. These funds could, in principle, be raised through a revision of the archaic "gas guzzler" tax, which does not apply to a significant percentage of the light-duty car and truck fleet. The tax penalty and tax credit in combination could be a revenue-neutral "fee-bate" scheme—similar to one recently proposed in California—that would simultaneously reward economical vehicles and penalize uneconomical ones.

Efficiency Standards

A critical strategy for effectively promoting energy efficiency is implementing new standards for buildings, appliances, and equipment. Tax credits do not necessarily remove all the market barriers that prevent clean energy technologies from spreading throughout the marketplace. These barriers include lack of awareness, rush purchases when an existing appliance breaks down, and purchases by builders and landlords who do not pay utility bills.

Significant advances in the efficiency of heating and cooling systems, motors, and appliances have been made in recent years, but more improvements are technologically and economically feasible. A clear federal statement of desired improvements in system efficiency would remove uncertainty about and reduce costs of implementing these changes. Under such a federal mandate, efficiency standards for equipment and appliances could be gradually increased, helping to expand the market share of existing high-efficiency systems.[29]

Extensive transmission and distribution networks waste a significant percentage of electricity generated by traditional power plants. Smaller-scale systems can be located closer to where the energy is used.

Standards remove inefficient products from the market and still leave consumers with a full range of products and features from which to choose. Building, appliance, and equipment efficiency standards have proven to be one of the federal government's most effective energy-saving programs. Analyses by DOE and others indicate that in 2000, appliance and equipment efficiency standards saved 1.2 quadrillion Btu of energy (1.3 percent of U.S. electric use) and reduced consumer energy bills by approximately $9 billion, with energy bill savings far exceeding any increase in product cost. By 2020, standards already enacted will save 4.3 quadrillion Btu per year (3.5 percent of projected U.S. energy use) and reduce peak electric demand by 120,000 MW (more than a 10-percent reduction). ACEEE estimates that by 2020, energy use could be reduced by 1.0 quadrillion Btu by quickly adopting higher standards for equipment that is currently covered under federal law, such as central air conditioners and heat pumps, and by adopting new standards for equipment not covered, such as torchiere (halogen) light fixtures, commercial refrigerators, and appliances that consume power while on standby.[30] Energy bills would decline by approximately $7 billion per year by 2020.[31]

A Renewable Portfolio Standard

The Renewable Portfolio Standard (RPS) is akin to the efficiency standards for vehicles and appliances that have proven successful in the past. A gradually increasing RPS is designed to integrate renewables into the marketplace in the most cost-effective fashion, and it ensures that a growing proportion of electricity sales is provided by renewable energy. An RPS provides the one true means to use market forces most effectively—the market picks the winning and losing technologies.

A number of studies indicate that a national renewable-energy component of 2 percent in 2002, growing to 10 percent in 2010 and 20 percent by 2020, that would include wind, biomass, geothermal, solar, and landfill gas, is broadly good for business and can readily be achieved.[32] States that decide to pursue more aggressive goals could be rewarded through an additional federal incentive program. In the past, federal RPS legislation has been introduced in Congress and it was proposed by the Clinton administration, but it has yet to be re-introduced by either this Congress or the Bush administration.

Including renewables in the United States' power-supply portfolio would protect consumers from fossil fuel price shocks and supply shortages by diversifying the energy options. A properly designed RPS will also create jobs at home and export opportunities abroad. To achieve compliance, a federal RPS should use market dynamics to stimulate innovation through a trading system. National renewable energy credit trading will encourage development of renewables in the regions of the country where they are the most cost-effective, while avoiding expensive long-distance transmission.

The coal, oil, natural gas, and nuclear power industries continue to receive considerable government subsidies, even though they are already well established. Without RPS or a similar mechanism, many renewables will not be able to survive in an increasingly competitive electricity market focused on producing power at the lowest direct cost. And while RPS is designed to deliver renewable that are most ready for the market, additional policies will still be needed to support emerging renewable technologies, like photovoltaics, that have enormous potential to become commercially competitive.

RPS is the surest market-based approach for securing the public benefits of renewables while supplying the greatest amount of clean power at the lowest price. It creates an ongoing incentive to drive down costs by providing a dependable and predictable market. RPS will promote vigorous competition among renewable energy developers and technologies to meet the standard at the lowest cost.

Analysis of the RPS target for 2020 shows renewable energy development in every region of the country, with most coming from wind, biomass, and geothermal sources. In particular, the Plains, Western, and mid-Atlantic states would generate more than 20 percent of their electricity from renewables.[33] Texas has become a leader in developing and implementing a successful RPS that then-Governor Bush signed into law in 1999. The Texas law requires electricity companies to supply 2,000 MW of new renewable resources by 2009, and the state is actually expected to meet this goal by the end of 2002, seven years ahead of schedule. Nine other states have signed an RPS into law: Arizona, Connecticut, Maine, Massachusetts, Nevada, New Jersey, New Mexico, Pennsylvania, and Wisconsin. Minnesota and Iowa have a minimum renewables requirement similar to RPS, and legislation that includes RPS is pending in several other states.

While the participation of 12 states signals a good start, this patchwork of state policies would not be able to drive down the costs of renewable energy technologies and move these technologies fully into the marketplace. Also, state RPS policies have differed substantially from each other thus far. These differences could cause significant market inefficiencies, negating the cost savings that a more comprehensive, streamlined, market-based federal RPS package would provide.

Small-Scale Distributed Energy Generation and Cogeneration

Small-scale distributed electricity generation has several advantages over traditional central-station utility service. Distributed generation reduces energy losses incurred by sending electricity long distances through an extensive transmission and distribution network (often an 8- to 10-percent loss of energy). In addition, generating equipment located close to the end use allows waste heat to be utilized (a process called cogeneration) to meet heating and hot water demands, significantly boosting overall system efficiency.

Distributed generation has faced several barriers in the marketplace, most notably from complicated and expensive utility interconnection requirements. These barriers have led to a push for national safety and power quality standards, now being finalized by the Institute of Electrical and Electronics Engineers (IEEE). Although the adoption of these standards would significantly decrease the economic burden on manufacturers, installers, and customers, the utilities are allowed discretion in adopting or rejecting them.

In designing credits, highest priority should go to renewable or fossil fuel systems that utilize waste heat through combined heat and power (CHP) designs. While a distributed generation system may achieve 35- to 45-percent electrical efficiency, the addition of heat utilization can raise overall efficiency to 80 percent. Industrial CHP potential is estimated to be 88,000 MW, the largest sectors being in the chemicals and paper industries. Commercial CHP potential is estimated to be 75,000 MW, with education, health care, and office building applications making up the most significant percentages.[36]

A National Public Benefits Fund

Electric utilities have historically funded programs to encourage the development of a host of clean energy technologies. Unfortunately, increasing competition and deregulation have led utilities to cut these discretionary expenditures in the last several years. Total utility spending on demand-side management programs fell more than 50 percent from 1993 to 1999. Utilities should be encouraged to invest in the future through rewards (such as tax incentives) for companies that reinvest profits and invigorate the power sector.[37] A national public benefits fund could be financed through a national, competitively neutral wires charge of $0.002 per kWh.

Cost and Benefit Analyses

A range of recent studies are all coming to the same conclusions: that simple but sustained standards and investments in a clean energy economy are not only possible but would also be highly beneficial to future prosperity in the United States.[37] If energy policies proceed as usual, the nation is expected to increase its reliance on coal and natural gas to meet strong growth in electricity use (42 percent by 2020). To meet this demand, it is estimated that 1,300 300-MW power plants would need to be built, with electricity generated by non-hydro renewables only increasing from 2 percent today to 2.4 percent of total generation in 2020.[38] A set of clean energy polices could meet a much larger share of our future energy needs, with energy efficiency measures projected to almost completely offset the projected growth in electricity use.[39] A clean energy strategy would significantly reduce emissions from utilities. In fact, through a steady shift to clean energy production, power plant carbon dioxide reductions (as proposed in the current legislation before Congress), would not be difficult or expensive to meet.[40]

The United States is becoming increasingly dependent on oil—$120 billion was spent on oil imports last year. Making renewable energy sources a larger part of the energy economy would enable the nation to provide for its own energy needs.

Recent analysis by the Union of Concerned Scientists focused on the costs and environmental impacts of a package of clean energy polices and how the package would affect fossil fuel prices and consumer energy bills. They found that using energy more efficiently and switching from fossil fuels to renewable energy sources will save consumers money by decreasing energy use.[41] A whole-economy analysis carried out by the International Project for Sustainable Energy Paths has also shown that Kyoto-type targets can easily be met, with a net increase of 1 percent in the nation's 2020 GDP, by implementing the right policies.[42]

One of the greatest advantages that energy efficiency and renewable energy sources offer over new power plants, transmission lines, and pipelines is the ability to deploy these technologies very quickly. They can be installed—and benefits can be reaped—immediately.[43] In addition, reductions in CO_2 emissions will have a "clean cascade" effect on the economy because many other pollutants are emitted during fossil fuel combustion.

The renewable and energy-efficient technologies and policies described here have already proven successful and cost-effective at the national and state levels. Supporting them would allow the United States to cost-effectively meet GHG emission targets while providing a sustainable, clean energy future.[44]

We stand at a critical point in the energy, economic, and environmental evolution of the United States. Renewable energy and energy efficiency are now not only affordable, but their expanded use will also open new areas of innovation. Creating opportunities and a fair marketplace for a clean energy economy requires leadership and vision. The tools to implement this evolution are now well known. We must recognize and overcome the current road blocks and create the opportunities needed to put these renewable and energy-efficient measures into effect.

This article is based on testimony provided by D. M. Kammen to the U. S. Senate Commerce, Science and Transportation (July 10, 2001) and U. S. Senate Finance (July 11, 2001) Committees. Antonia V. Herzog and Timothy E. Lipman are postdoctoral researchers and Jennifer L. Edwards is a research assistant at the Renewable and Appropriate Energy Laboratory (RAEL), Energy and Resources Group (ERG), at the University of California at Berkeley. Daniel M. Kammen is a professor of Energy and Society with ERG and a professor of Public Policy with the Goldman School of Public Policy. Address correspondence to D. M. Kammen, 310 Barrows Hall, University of California, Berkeley, CA 94720-3050, or dkammen@socrates.berkeley.edu. Additional material can be found at http://socrates.berkeley.edu/~rael.

NOTES

1. Intergovernmental Panel on Climate Change (IPCC), *Climate Change 2001: The Scientific Basis* (Working Group I of the IPCC, World Meteorological Organization - U.N. Environment Program, Geneva), January 2001; and National Research Council (NRC), *Climate Change Science: An Analysis of Some Key Questions*, Committee on the Science of Climate Change, (National Academy Press, Washington, D.C., 2001). For more information on energy and climate change, see J. P. Holdren, "The Energy-Climate Challenge: Issues for the New U.S. Administration," *Environment*, June 2001, 8–21.

2. D. M. Kammen and R. M. Margolis, "Evidence of Under-Investment in Energy R&D Policy in the United States and the Impact of Federal Policy," *Energy Policy* 27 (1999), 575–84; and R. M. Margolis and D. M. Kammen, "Underinvestment: The Energy Technology and R&D Policy Challenge," *Science* 285 (1999), 690–92.

3. This work appeared in two influential forms that reached dramatically different audiences: A. B. Lovins, "Energy Strategy: The Road Not Taken," *Foreign Affairs* (1976), 65–96; and A. B. Lovins, *Soft Energy Paths: Toward a Durable Peace* (New York: Harper Colophon Books, 1977).

4. A. V. Herzog, T. E. Lipman, and D. M. Kammen, "Renewable Energy Sources," in *Our Fragile World: Challenges and Opportunities for Sustainable Development*, forerunner to the Encyclopedia of Life Support Systems (EOLSS), Volume 1, Section 1 (UNESCO-EOLSS Secretariat, EOLSS Publishers Co. Ltd., 2001).

5. P. Mazza, *Harvesting Clean Energy for Rural Development: Wind,* Climate Solutions Special Report, January 2001.

6. Kammen and Margolis, note 2 above.

7. U.S. Department of Energy (DOE), *Renewable Energy Technology Characterizations*, Topical Report Prepared by DOE Office of Utility Technologies and EPRI, TR-109496, December 1997.

8. B. Haevner and M. Zugel, *Predictably Unpredictable: Volatility in Future Energy Supply and Price From California's Over Dependence on Natural Gas*, CALIPIRG Charitable Trust Research Report, September, 2001).

9. R. L. Ottinger et al., *Environmental Costs of Electricity* (New York: Oceana Publications, Inc., 1991); U.S. Department of Energy (DOE), *Annual Energy Outlook 2000*, DOE/EIA-0383 (00), Energy Information Administration, Washington, D. C., December 2000; and U.S. Department of Energy, 1997.

10. Reuters News Service, "Fuel Cells and New Energies Come of Age Amid Fuel Crisis," 11 September 2000.

11. National Energy Policy, "Reliable, Affordable, and Environmentally Sound Energy for America's Future," Report of the National Energy Policy Development Group, Office of the President, May 2001.

12. Kammen and Margolis, note 2 above.

13. International Institute for Applied Systems Analysis/World Energy Council, *Global Energy Perspectives to 2050 and Beyond* (Laxenburg, Austria, and London, 1995).

14. R. D. Duke and D. M. Kammen, "The Economics of Energy Market Transformation Initiatives," *The Energy Journal* 20 (1999): 15–64.

15. Interlaboratory Working Group, *Scenarios for a Clean Energy Future* (Oak Ridge, Tenn.: Oak Ridge National Laboratory; and Berkeley, Calif.: Lawrence Berkeley National Laboratory), ORNL/CON-476 and LBNL-44029, November 2000.

16. S. Nadel and H. Geller, "Energy Efficiency Polices for a Strong America," American Council for an Energy-Efficient Economy (ACEEE), May 2001 (draft).

17. Clean Energy Partnerships, *A Decade of Success*, Office of Energy Efficiency and Renewable Energy, DOE/EE-0213 (Washington, D.C., 2000).

18. Nadel and Geller, note 16 above.

19. President's Committee of Advisors on Science and Technology (PCAST), *Federal Energy Research and Development for the Challenges of the Twenty-First Century*, Washington, D.C., Energy Research and Development Panel, November 1997.

20. F. Krause, S. DeCanio, and P. Baer, "Cutting Carbon Emissions at a Profit: Opportunities for the U.S." (El Cerrito, Calif.: International Project for Sustainable Energy Paths), May 2001; and A. P. Kinzig and D. M. Kammen, "National Trajectories of Carbon Emissions: Analysis of Proposals to Foster the Transition to Low-Carbon Economies," *Global Environmental Change* 8 (3) (1998): 183–208.

21. Energy Information Administration (EIA), *U.S. Carbon Dioxide Emissions from Energy Sources 2000 Flash Estimate*, based on data from the Monthly Energy Review (May 2001) and the Petroleum Supply Annual 2000, DOE (Washington, D.C.), June 2001.

22. Kammen and Margolis, note 2 above.

23. This does not include revenue outlays for the Alcohol Fuels Excise Tax, which reduces the tax paid on ethanol-blended gasoline. Most ethanol used in the United States is produced from corn, and the GHG emissions impact is uncertain and has been shown to be negligible (M. Delucchi, *A Revised Model of Emissions of Greenhouse Gases from the Use of Transportation Fuels and Electricity*, Institute of Transportation Studies, UCD-ITS-RR-97-22, (Davis, Calif., 1997)); and EIA, *Federal Financial Interventions and Subsidies in*

Energy Markets 1999: Primary Energy, DOE (Washington, D.C., 1999).

24. Established by the Windfall Profit Tax Act of 1980, this tax credit is $3 per barrel of oil equivalent produced, and it phases out when the price of oil rises to $29.50 per barrel (1979 dollars).

25. American Wind Energy Association web site, available at http://www.awea.org, accessed on September 8, 2001.

26. Nadal and Geller, note 16 above.

27. J. Mark, "Greener SUVs: A Blueprint for Cleaner, More Efficient Light Trucks," Union of Concerned Scientists, 1999; M. A. Weiss, J. B. Heywood, E. M. Drake, A. Schafer, and F. F. AuYeung, "On the Road in 2020: A Lifecycle Analysis of New Automobile Technologies," Energy Laboratory, Massachusetts Institute of Technology, MIT EL 00-003 (Cambridge, Mass., October 2000); Office of Technology Assessment, *Advanced Vehicle Technology: Visions of a Super-Efficient Family Car*, U.S. Congress, OTA-ETI-638 (Washington, D.C., September 1995); and D. L. Greene and J. Decicco, "Engineering-Economic Analyses of Automotive Fuel Economy Potential In The United States," *Annual Review of Energy and the Environment*, 25: (2000) 477–536.

28. Greene and Decicco, note 27 above; and Interlaboratory Working Group, note 15 above.

29. S. L. Clemmer, D. Donovan, and A. Nogee, "Clean Energy Blueprint: A Smarter National Energy Policy for Today and the Future, Phase I," Union of Concerned Scientists and Tellus Institute, June 2001.

30. K. B. Rosen and A. K. Meier, *Energy Use of Televisions and Videocassette Recorders in the U.S.*, DOE, LBNL-42393, (Berkeley, Calif.: Lawrence Berkeley National Laboratory), March 1999.

31. Nadal and Geller, note 16 above.

32. Clemmer, Donovan, and Nogee, note 29 above; S. L. Clemmer, A. Nogee, and M. Brower, "A Powerful Opportunity: Making Renewable Electricity the Standard," Union of Concerned Scientists, January 1999; and A.Nogee, S. Clemmer, B. Paulos, and B. Haddad, "Powerful Solutions: 7 Ways to Switch America to Renewable Energy," Union of Concerned Scientists, January 1999.

33. Clemmer, Nogee, and Brower, note 32 above.

34. Distributed generation reflects a new way to manage energy supply and demand. Instead of the old system of large capital-intensive central-station power plants, the improvements in energy efficiency, renewable energy, and small, modular5 "micro-turbines" that burn gas, as well as hydro and other resources, energy supplies could be located closer to users, reducing transmission losses, improving system reliability, and energy security.

35. R. K. Dixon, Office of Power Technologies, U.S. Department of Energy, Second International CHP Symposium, Amsterdam, Netherlands, May 2001.

36. Kammen and Margolis, note 2 above.

37. Interlaboratory Working Group, note 15 above; Krause, DeCanio, and Baer, note 20 above; and Clemmer, Donovan, and Nogee, note 29 above.

38. National Energy Policy, note 11 above.

39. Clemmer, Donovan, and Nogee, note 29 above.

40. Ibid.

41. Ibid.

42. Krause, DeCanio, and Baer, note 20 above.

43. Kinzig and Kammen, note 20 above.

44. P. Baer, et al. "Equity and Greenhouse Gas Responsibility," *Science* 289 (2000), 2287.

Fossil Fuels and Energy Independence

To become self-sufficient in energy resources, the United States needs to combine the conservation strategies learned over the past 30 years with the latest fuel-saving technologies.

B. Samuel Tanenbaum

Following the outbreak of the Arab-Israeli war in 1973, petroleum-producing Arab nations cut back oil production and imposed an embargo that led to an energy crisis in the United States. In response, President Nixon directed Dixy Lee Ray, then-chairman of the Atomic Energy Commission, to "undertake an immediate review of federal and private energy research and development activities" and asked her to submit a national energy plan by December 1 of that year.

Based on the input of hundreds of individuals in workshops, review panels, and government agencies, Ray prepared a comprehensive program, "The Nation's Energy Future." She noted that it was intended to "mobilize the nation's resources toward the attainment of a capacity for energy self-sufficiency by 1980." To achieve that goal, the report recommended five tasks: (1) conserve energy by reducing consumption and adopting processes that use fuel more efficiently; (2) raise domestic production of oil and natural gas; (3) increase the use of coal; (4) greatly expand the use of nuclear power; and (5) promote the use of renewable energy sources.

These recommendations look remarkably similar to suggestions being made by the Bush administration today. Thus it should be useful to review what happened in the 1970s that prevented the country from achieving independence in

energy resources and to examine what steps may be taken at this time.

Historic patterns of energy consumption

In 1972, energy consumption in the United States totaled 72 quadrillion Btu (72 followed by 15 zeros)—or, more simply, 72 Quads. [One Btu (British thermal unit) is the energy required to raise the temperature of one pound of water by one degree Fahrenheit.] Of this amount, 45.5 percent was derived from oil, 32.3 percent from natural gas, 17.2 percent from coal, and the remainder from nuclear reactors and renewable sources. In other words, fossil fuels—oil, natural gas, and coal—supplied 95 percent of this nation's energy needs.

At that time, the United States produced all the coal and natural gas it consumed, and it even exported a substantial amount of coal to other countries. The major problem then (as now) was that the country relied heavily on imported oil to supplement domestic production. Oil imports had been growing rapidly, and during the first half of 1973, the import rate exceeded 6 million barrels per day (MBPD), representing about one-third of the nation's oil consumption. [Note that 1 MBPD of oil yields 2.1 Quads of energy per year.]

The 1973 report to the president projected that if no action were taken, the demand for energy would continue to rise at its historic rate, reaching about 100 Quads in 1980 and 200 Quads in 2000. At the same time, oil imports were predicted to grow from 6.5 MBPD in 1970 to about 12 MBPD in 1980 and 24 MBPD in 2000.

To achieve energy self-sufficiency, this country had to eliminate the demand for imported oil. To do so by 1980, Ray's plan laid out three proposals. First, 4.7 MBPD could be saved through conservation measures—such as extra insulation in buildings, more fuel-efficient cars and trucks, and more efficient industrial processes. Second, 1.5 MBPD of oil would be replaced with energy from coal, nuclear, and renewable sources. Third, domestic oil production would need to be raised by 5.8 MBPD.

The report further noted that "self-sufficiency based on fossil fuels can only be temporary. Though large, these resources are finite." For energy independence over the long term, it recommended incentives for the development of renewable energy sources, but the major replacement of fossil fuels was expected to come from rapid growth in nuclear power. Even before the 1973 energy crisis, the Department of Interior projected that the number of large nuclear power plants would grow from 29 in 1973 to 132 in 1980 and 1,200 in 2000.

So, what happened? The energy crisis of 1973 was short-lived, but it led to a two-year drop in total energy consumption by about 5 percent. Thereafter, energy usage rose again, until the Iranian revolution of 1978–79 disrupted that country's oil production. Those circumstances stimulated additional conservation efforts in the United States, so that energy usage fell from 81 Quads in 1979 to 78 Quads in 1980, reaching a low of 73 Quads in 1982 and 1983.

Fossil fuels provided 95 percent of all the energy consumed in the United States in 1972, and they now supply about 85 percent of the energy used.

Since then, energy consumption increased, but at a slower pace than expected. In 2000, the total usage was 99 Quads—less than half the amount predicted in 1973—reflecting the effectiveness of conservation measures.

The efforts to replace some oil with other energy sources also appear to have been successful. Between 1973 and 2000, the fraction of energy derived from oil dropped from 46 to 39 percent, while the fraction obtained from coal increased from 17 to 23 percent. These changes represent striking reversals of the trends of earlier decades.

On the other hand, the goal to achieve energy independence was not met for several reasons. In particular, despite efforts to encourage exploration in the United States and completion of the 800-mile trans-Alaska pipeline in 1977, domestic oil production fell from its peak value of 10.9 MBPD in 1970 to 9.8 MBPD in 1980 and 8.1 MBPD in 2000. Between 1973 and 1980, overall oil consumption decreased slightly, but oil imports rose from 6.2 to 6.4 MBPD, before dropping to a low of 4.3 MBPD in 1985, thanks to conservation efforts. Thereafter, the situation gradually worsened, as U.S. oil production declined and consumption increased. Consequently, by 2000, imports reached 10.7 MBPD.

Furthermore, the growth of nuclear power had been greatly overestimated. Instead of the 1,200 large nuclear power plants projected for the year 2000, only 103 plants are currently operating in the United States. Although the last U.S. nuclear power plant came on line in 1996, no new plants were ordered after the accident at Three Mile Island (TMI) in Pennsylvania in March, 1979.

No one was killed by the mishap at TMI, but it was followed by the much more serious incident in Chernobyl, Ukraine, in April, 1986. In the latter case, 31 people died immediately, hundreds were hospitalized, and many more were expected to die of cancer from exposure to the giant cloud of radiation released.

While nuclear power plants can be made safer, the main hurdle is their high cost. In the 1980s, the cost of completing a nuclear plant with a capacity of one gigawatt (one billion watts) reached $5–6 billion—more than ten times the price tag for a conventional power plant with equivalent capacity. In addition, the spent, radioactive fuel has been stored at reactor sites, pending agreement on a long-term storage area, and the cost of decommissioning an old plant has grown to be far higher than originally anticipated. Thus, the nuclear power industry blossomed for only about a decade before power companies returned to purchasing fossil fuel plants for electricity.

Fossil fuel reserves

Although the energy policies adopted in the 1970s had a greater effect than most people realize, energy production was (and will continue to be) affected by the availability of fossil fuels. These fuels provided 95 percent of all the energy consumed in the United States in 1972, and they now supply about 85 percent of the energy used. The remainder is obtained from nuclear reactors (8 percent) and renewable sources (7 percent). It is therefore important to examine the reserves of oil, natural gas, and coal, and to look for potential substitutes.

Oil reserves. Following the discovery of huge oil deposits in Alaska in 1970, proven reserves of oil in the United States have steadily declined,

from about 210 Quads in 1973 to about 120 Quads today. During this period, domestic oil companies discovered more than 300 Quads, but they produced over 400 Quads, leading to the decline in reserves.

Domestic reserves of natural gas have fallen from about 300 Quads in 1973 to about 170 Quads today.

By contrast, major new discoveries of oil continue to be made overseas. Over the past 30 years, proven oil reserves worldwide have increased from about 3,800 Quads to about 5,900 Quads. Because the worldwide demand for oil is roughly 150 Quads per year, oil is still plentiful on a global scale, and the cost of imported oil remains low.

Natural gas reserves. As a fuel, natural gas offers many advantages. It is clean-burning, inexpensive, and conveniently delivered by pipeline to consumers. The use of natural gas in the United States soared in the decades following the Second World War, as a nationwide network of pipelines was constructed.

Thus far, this country has been able to meet its needs for natural gas through domestic production and modest imports from Canada. Since 1973, U.S. wells have produced over 600 Quads of natural gas. Even so, as with oil, new discoveries have not kept pace with production. Thus, domestic reserves of natural gas have fallen from about 300 Quads in 1973 to about 170 Quads today.

On the other hand, worldwide reserves of natural gas have doubled from about 2,500 Quads in 1973 to about 5,000 Quads today. Given that the global demand is only about 100 Quads per year, natural gas, like oil, is plentiful worldwide. Consequently, the United States may become increasingly reliant on imports of natural gas in the years ahead.

One problem is that, unlike oil, natural gas cannot be shipped easily in tankers. Before shipment, the gas must be compressed and liquefied; refrigerated ships must be used to keep it cool during transit; and special facilities are needed to receive the liquefied fuel and permit it to expand safely for shipment in a normal pipeline. It is therefore much more expensive to import natural gas than oil from overseas. Thus the huge quantities of natural gas available in Russia and Middle Eastern countries cannot be imported as readily as oil.

The United States has plentiful reserves of coal, but coal mining can adversely affect the surrounding land and water unless special protective measures are taken.

Coal reserves. The United States has roughly 25 percent of the world's coal reserves, corresponding to about 7,000 Quads. Domestic annual consumption of coal is only 22 Quads, so there is no likelihood of a coal shortage anytime soon. Coal is also the cheapest fossil fuel. In recent years, the price of coal used in power plants has been about one dollar per million Btu (MBtu), while that of natural gas has been over two dollars per MBtu. Oil has been even more expensive.

Despite the cost advantage, traditional markets for coal in the United States eroded with time. In the 1950s, coal-fired steam locomotives were replaced by more efficient diesel engines, and coal-based heating systems were replaced by cleaner natural gas heaters wherever pipelines made the fuel available.

Today, about 90 percent of the coal sold in the United States is used to produce electricity. New coal-fired plants continue to be built, although tighter requirements for pollution-control equipment have added to the expense and complications of building them. Moreover, coal mining has to be conducted with improved water-management and land-restoration practices to minimize damage to the environment.

Alternative fuels. If coal is heated in hydrogen, it can produce either an oil substitute known as *syncrude* or a gaseous material known as *coal gas*, which has about half the energy content of conventional natural gas. These products are still too costly to be competitive with available fuels, but they may become important during shortages of oil and natural gas.

Another potential fuel is *shale oil*. Shale is a sedimentary rock rich in the organic compound kerogen. Parts of the western United States have rich deposits of shale oil—30 gallons or more of oil per ton of shale, amounting to several times more than the total domestic reserves of conventional oil. To extract shale oil, the rock is mined, crushed, and heated to about 400 1/4°C (750°F). But it takes about six tons of shale to provide as much energy as one ton of coal, and it's not yet possible to produce large amounts of shale oil in a way that protects the environment or competes in price with oil pumped from a well.

Today's technologies, if widely adopted, could reduce the use of fossil fuels in the United States by up to 50 percent.

An asphaltlike oil, called *bitumen*, can be obtained from tar sands. There are sizable deposits of these sands in the United States, but the world's largest deposits are in Canada. Bitumen can be extracted simply by mixing the mined sand with hot water or steam. Again, the major drawback is that the oil cannot be produced at a price that is competitive with conventional oil.

Reducing our consumption of fossil fuels

The annual worldwide consumption of fossil fuels today represents about one percent of known reserves. This is about the same figure as in 1972, because known reserves continue to increase worldwide. Interpretation of these amounts, however, is complicated by the fact that the newly discovered reserves are mainly in remote locations, including at great depths below the surface of the land and sea. Thus, tapping these supplies would require expensive extraction techniques, which would be justified only when the prices of oil and natural gas become sufficiently high.

In any case, there are several reasons why we need to dramatically reduce our consumption of fossil fuels, replacing them with other energy sources. For a country such as the United States, which is heavily dependent on imported oil and is gradually increasing its imports of natural gas, the near-term benefit would be the achievement of energy self-sufficiency. The long-term benefit is that we would save on fuels that are finite in availability. In addition, we would lessen the production of carbon dioxide, which is produced by burning fossil fuels and contributes to global warming—a phenomenon that has been predicted to cause serious problems in the future.

How, then, can we reduce our dependence on fossil fuels? As long as prices for these fuels remain relatively low, most consumers do not see the need to use energy efficiently or to switch to renewable sources. In the last 25 years, however, the United States has demonstrated that sensible steps to conserve energy can be remarkably effective when mandated, or when adequate incentives are provided. Consider the following examples:

- New standards for insulation and efficiency in heating and air-conditioning systems played a leading role in reducing the use of natural gas by about 20 percent in the early 1980s.

- At the same time, requirements for cars to have better gasoline mileage reduced oil consumption by a similar proportion.

- Tax credits and California's requirement that electric utilities buy power from wind and solar generators at a reasonable rate led to a significant use of these alternative energy sources there.

The Toyota Prius, a gasoline-electric hybrid vehicle, has a mileage rating of about 50 miles per gallon of gasoline. If similar engines were to be installed in most new cars and trucks over the next few years, they would dramatically reduce the fuel consumption.

Today's technologies, if widely adopted, could reduce the use of fossil fuels in the United States by up to 50 percent. For instance, hybrid gasoline-electric cars introduced recently by Toyota and Honda have already become quite popular and show that gasoline mileage can be improved by 50–100 percent without sacrificing performance or comfort. If such engines were required of all new cars and trucks within the next five years, the United States could probably save about one-third of its current consumption of 27 Quads per year for transportation, reducing oil imports by more than 4 MBPD.

New, large wind generators can produce electricity at a price that is competitive with any other means of generating power. If used extensively, they could save enormous amounts of fossil fuel. Alternatively, electrical power can be produced with combined-cycle, gas-turbine generators or cogeneration systems that use natural gas twice as efficiently as conventional steam-turbine generating plants [see "Gaining Power by Thinking Small," THE WORLD & I, March 2002, p. 152].

By replacing electric water heaters, stoves, ovens, and clothes dryers with gas units, up to two-thirds of the energy used for these purposes can be saved. Similarly, replacing old air-conditioning systems, refrigerators, heating systems, and pool pumps with efficient new models can reduce energy use by as much as 50 percent. Fossil fuels can be further saved by using solar hot-water heaters and clothes lines to dry clothes. Moreover, about 75 percent of the energy used in incandescent light bulbs can be saved by replacing them with compact fluorescent bulbs or efficient LED (light emitting diode) lamps.

In warm climates, air-conditioning loads can be significantly reduced by making use of reflective roof surfaces, window coatings, attic exhaust fans, and shade trees. In cold climates, passive solar heating technology can be employed.

Fossil fuel consumption can also be reduced by a number of other recent technologies, but they have not yet gained broad acceptance for reasons of cost or reliability. They include photovoltaic solar panels, which produce electricity directly from sunlight; fuel cells, in which hydrogen or methanol is used to produce electricity at high efficiency, reducing air pollution; and systems that extract energy from waste materials, geothermal sources, or ocean tides. It is also possible that advances in nuclear technology may lead to more extensive use of nuclear energy, by processes of either fission (which splits atomic nuclei) or fusion (which brings atomic nuclei together).

In the long run, it seems likely that solar energy will be used directly to meet most of the energy needs of our world. The amount of sunlight reaching the continental United States averages over 500,000 Btu per square foot per year. A circular area with a 50-mile radius thus receives about 110 Quads per year—enough to meet the energy demands of the entire nation. But until we learn how to harness that energy efficiently and cost-effectively, the above-mentioned technologies are available to reduce this country's energy demand by half, without changing our life-styles. By adopting these approaches, we can almost certainly achieve energy independence.

B. Samuel Tanenbaum is professor of engineering at Harvey Mudd College in Claremont, California.

Power Struggle

California's Engineered Energy Crisis and the Potential of Public Power

By Harvey Wasserman

THE U.S. BARONS OF FOSSIL AND NUCLEAR FUEL have used a contrived energy crisis in California and the nation as a pretext to declare an all-out assault on environmental protection. George Bush, Dick Cheney and their cohorts from the oil industry claim the rolling West Coast blackouts justify rolling over a century of carefully crafted environmental law.

But this rationalization ignores a telling detail: California does not have an energy shortage. And according to some of the state's highest officials, the blackouts have been choreographed for massive price gouging by some of Bush's closest associates and contributors.

The blackouts, rising electricity prices, government subsidies, utility bankruptcies and near-bankruptcies have many causes. But neither a shortfall in supply nor a surge in demand for electricity is among them.

California is in fact among the most energy efficient of states. Though its population and economy have soared, its overall demand for electricity has risen only modestly in recent years. The amounts of electricity it can generate in state or has contracts to purchase out-of-state are more than sufficient to meet overall and peak demand, and have been throughout the state's crisis.

California's electricity crisis was precipitated by a botched deregulatory scheme pushed by the very utilities now screaming about alleged losses, a plan that has immensely profited both the utilities' parent corporations and a few pirate power generators close to George W. Bush.

California's deregulatory disaster is a "failure by design," prompted not by a real shortage, but by a corporate agenda, says Paul Fenn of the Oakland-based American Public Power Project.

In the view of the state's leading consumer and clean power advocates, California's consumers and taxpayers are victims of a massive, complex double-theft, first by the state's biggest electric power utilities, and then by huge oil and gas companies close to George W. Bush.

STRANDED COST CATASTROPHE

There was very little public debate leading up to the Golden State's decision to deregulate its electric utilities in 1996. The early battles were muddied and muzzled. The state legislature deliberated for a scant three weeks. The media barely covered those few hearings that were open to the public. Southern California Edison (SoCalEd) essentially wrote much of the legislation, AB1890, in its corporate offices.

When the legislature unanimously voted for the bill and watched its September signing by a beaming governor Pete Wilson, the utilities and their lobbyists gushed. This is "a great day for us," cheered John Bryson, president of Southern California Edison, widely regarded as the bill's chief architect. "We believe this plan is the best way to facilitate a smooth, timely transition to a competitive electricity market and maximize value for our shareholders and customers." It was "a large achievement and a sound achievement for the state in terms of giving customers choice," he said.

Ostensibly, AB1890 was meant to dismantle the regulated monopolies that had supplied California with electric power since the early twentieth century. Instead, consumers would be able to choose among competing suppliers. The bill presumed the electric market would fill with power companies vying to sell low-cost juice of all varieties, including "green power" from wind and solar.

But the state's three biggest utilities—SoCalEd, Pacific Gas & Electric (PG&E) and San Diego Gas & Electric—made sure that before competition could come they were first reimbursed for their investments in nuclear power, which they argued was inefficient and could not compete in an open market.

In exchange for this payback, they engineered consumer rate caps that were to remain in place until the utilities collected up to $28.5 billion in surcharges for their obsolete generating plants. Once they collected that money, they said, rates would fall as competing generators came into the open market.

MODEST DEMAND, WITHHELD SUPPLY

No one disputes that a mismatch between supply and demand underlies the California energy crisis. But consumer advocates point out that the data show California's energy demand to be growing slowly, not surging as many news reports suggest. And, they say, the electricity shortage reflects not any real limits on supply, but the market manipulations of the independent generators.

A comprehensive San Francisco Chronicle study in March showed that California's energy demand is in fact rising slowly. "The industry has painted the summer of 2000 as the equivalent of a 100-year storm in meteorology—an event so powerful and unexpected that the existing infrastructure was devastated by its force," the Chronicle reported. "The statistics show that 2000, taken in total, was nothing of the sort." Overall electricity usage in California rose approximately 2 percent a year in the 1990s.

Most importantly, peak use—the demand level that actually stresses the system—was lower at the end of 2000 than in the previous year. While peak demand was high in May 2000, in four out of the last six months of 2000—July, August, October and December—peak demand was lower than in 1999, according to an analysis of statistics from California's Independent System Operator (CAISO) by Public Citizen's Critical Mass Energy and Environment Project.

California's available capacity and electricity on contract vastly exceeds its peak demand. In total, California has 55,500 megawatts of power generating capacity and 4,500 megawatts of power on long-term out-of-state contracts—approximately 15,000 megawatts more than peak demand, Public Citizen reports, citing statistics from CAISO.

The problem California is facing isn't supply, say the consumer/green groups, it is that the power supply companies, now a separate industry segment from the utilities, simply refuse to make the supply available. The independent power generators' alleged market manipulation is now the source of numerous lawsuits and investigations.

—*Robert Weissman*

With abundant infusions of utility cash (the utilities spent more than $3.6 million lobbying to win the bill in 1996, and another $4.1 million to promote it in 1997), the state's energy interests promoted AB1890 as a way to save customers money through the magic of the marketplace. In 1995, Bryson trumpeted deregulation as "the best, soundest way to move to a desirable competitive market that will benefit all customers, large and small." SoCalEd, he said, was "committed to a 25 percent rate reduction effective January 1, 2000. As near as we're able to tell, this [legislation] is consistent with our goal."

A broad coalition of consumer and environmental groups knew better. They saw the "crisis" coming right from the start, and bitterly opposed the original AB1890. In 1998, as deregulation was taking effect, Herbert Chao Gunther of the San Francisco-based Public Media Center, Harvey Rosenfield of the statewide Foundation for Taxpayer and Consumer Rights, Ed Maschke and Anna Aurilio of the California and U.S. Public Interest Research Groups, Fenn's American Public Power Project and others fought back. Operating on a shoestring budget, they gathered an astonishing 700,000-plus signatures to put on the fall ballot an initiative—Proposition 9—that would have nipped the crisis in the bud. Among other things, Prop 9 would have restored regulation to the electric power system and prevented the huge "stranded cost" bailout that was AB1890's central feature.

But the supporters of Proposition 9 ran into a hugely funded utility opposition that would not be denied. Still intoxicated by the promises of deregulation, William Hauck, chair of the Concerned Stockholders of California, a SoCalEd front group, spoke for the industry when he warned that returning to public regulation would dismantle "the competitive electricity market and customer choice, and will actually result in higher electric rates."

Big energy steamrolled the campaign with a $40 million counteroffensive. The greens had only $1 million. California voters rejected Proposition 9's proposal to repeal electricity deregulation by a 73-to-27 percent margin. (A parallel Massachusetts campaign was crushed on the same day, by a similar margin.)

It was a grim day for green and consumer advocates who had fought hard to avoid what's now happened. Eugene Coyle, one of the state's most respected energy analysts, and a host of others warned that the electric power industry was a natural monopoly that would never foster true competition. They showed that AB1890 was a cover for a massive bailout for the utilities' bad nuclear investments. They demanded public control. They predicted disaster right from the start.

One early opponent of deregulation was Dan Berman, an energy expert working to win public utility ownership for his hometown of Davis, California. With Boston-based activist John O'Connor, Berman wrote in the 1996 book, *Who Owns the Sun*, "Today deregulation, cheap electricity, and natural gas are all the rage."

"But few people are paying attention to what will happen when the price of natural gas and oil go up, as they most surely will, after falling by 75 percent in the last decade," Berman and O'Connor wrote. "What will happen when the new, unrelated 'independent power producers' of cheap electric power fired by combined-cycle gas turbines pass on whopping rate increases to the public as the price of natural gas soars? Will big industry come weeping to the public, hat in hand, as the savings-and-loan investors did? Are the energy corporations crippling American industry by reinforcing an addiction to cheap fossil fuels and electricity? Will there be a massive ratepayers' revolt

POWER PLAYS: WHERE DID THE CALIFORNIA UTILITIES MONEY GO?

When California lawmakers voted unanimously to deregulate the state's electricity market in 1996, California's three biggest utilities—PG&E, SoCalEdison, and San Diego Gas & Electric (SDGE)—owned most of the state's power generating plants, as well as most of the state's distribution grid and power lines. The companies operated as regional monopoly utilities, with strict regulatory oversight over retail electricity rates.

Once deregulation began, however, the companies rapidly began to restructure. First, they "unbundled" their assets—selling off some of their power generating plants while keeping their distribution grid and power lines and other assets such as nuclear and hydro power generation facilities.

The money from the power plant sales and high utility charges was siphoned off to the utilities' parent companies (e.g., PG&E restructured in 1997, creating a separate holding company called PG&E Corporation), which then passed the money on to other unregulated subsidiaries, which acquired new generating facilities in other states and (in the case of Edison International and SDGE) other countries. The parent companies built a diverse array of new assets and services, including natural gas pipelines, storage and processing plants.

Since the summer of 2000 price spikes, PG&E and SoCalEdison have claimed they are going broke by having to pay the difference between high electricity prices charged by wholesale power suppliers in the newly deregulated market and lower fixed retail rates that the utilities themselves originally negotiated with state regulators.

Claiming they needed relief or else they would be forced to declare bankruptcy, the utilities convinced the state to begin spending $50 million of taxpayer monies per day to purchase wholesale electricity on their behalf. The state began buying super-high-priced electricity on the wholesale market, including on the spot markets where prices were orders of magnitude higher than just a couple years before, and conveying the electricity to the utilities. By May, California taxpayers had spent more than $6 billion, with no end in sight. As the state's budget surplus quickly eroded, Governor Davis proposed floating $13.4 billion in bonds to cover the cost (the bonds are to be paid off over 15 years by ratepayers via higher electricity rates).

Meanwhile, federal regulators (FERC) agreed in January to allow the companies to organize their corporate structure so as to preemptively shield their assets during bankruptcy proceedings.

Less than two months later, on April 6, PG&E—the utility, not the holding company—filed for bankruptcy, leaving consumers and taxpayers in even more of a bind.

But was the company really bankrupt? While the utility was losing $300 million a month in wholesale power purchase costs, its parent company was pulling down a healthy profit. At the end of 2000, the PG&E Corporation (the parent company) reported $30 billion in assets and $20 billion in revenues for the year. It had an ownership interest in 30 independent operating plants in 10 states, including a number under development. In October 2000, it spent nearly $8 billion to acquire 44 new turbines.

Moreover, between 1997 and 1999, PG&E transferred $4 billion to its parent corporation. In the first nine months of 2000, PG&E transferred an additional $632 million. An audit sponsored by the California Public Utility Commission (CPUC) concluded that "historically, cash has flowed in only one direction, from PG&E to PG&E Corp. and then to the unregulated affiliates." Similarly, SoCalEdison transferred $4.8 billion to its parent corporation (Edison International) between 1995 and 2000.

CPUC's audit also found that PG&E Corporation (the parent) is expected to receive an additional federal tax refund of up to $1 billion "largely due to losses sustained by PG&E."

The San Francisco-based Utility Reform Network explains in its report "Cooking the Books" that the utilities' position was "akin to a situation in which one pocket is empty and another is full of cash. A reasonable person would check both pockets before assuming that they are penniless."

PG&E's Jon Tremayne responds that deregulation was set up "to pay the utility investors back money that they invested in power plants that they were now forced under law to sell. So the shareholders essentially recovered their investments and reinvested it. This was done under the direction of the Public Utilities Commission, in synch with the state law."

With support from the FERC, PG&E maintains that its California subsidiary is completely separate from the parent holding company, despite the fact that the two companies have virtually identical boards and file a joint annual report with the Securities and Exchange Commission.

The company may not be entirely in the clear. PG&E admitted in April in a quarterly filing with the U.S. Securities and Exchange Commission that, depending on the terms and conditions of the company's reorganization plan adopted by the company's Bankruptcy Court, creditors of the bankrupt PG&E utility unit may be able to grab some assets from PG&E National Energy Group, a subsidiary the parent had intended to insulate through its restructuring plan.

—*Charlie Cray*

RECENT PURCHASES OR COMMITMENTS OF PG&E CORPORATION (PARENT OF PG&E COMPANY) AND EDISON INTERNATIONAL (PARENT OF SOCAL EDISON)

Subsidiary	What they built/bought/where	Cost ($ billions)	Date
PG&E Corporation			
Nat'l Energy Group	810 MW Southhaven power plant in Mississippi	??	11/00
Nat'l Energy Group	44 turbines, 15 other projects from Societe General	7.8	10/00
PG&E Generating	Okeehobee County, FL, power plant to be completed 2003	0.2	9/00
Nat'l Energy Group	Madison Windpower in New York	0.02	9/00
Nat'l Energy Group	Attala 500 MW power plant, MS	??	9/00
Energy Trading	Tolling rights to peaking plant in suburban Indianapolis	??	9/00
Nat'l Energy Group	Tolling rights to Liberty power plant in suburban Philadelphia	??	6/00
PG&E	Stake in True Quote trading software	??	4/00
PG&E	Aerie broadband pipeline project	??	4/00
PG&E Generating	Pleasant Prairie, Wisconsin power plant	0.5	1999
Nat'l Energy Group	Lake Road power plant in Killingly, CT	0.5	1999
US Gen	New England Electrical Systems hydro, coal, oil, gas generating plants (4,800 MW total capacity; 18 plants)	1.8	9/98
PG&E Corp.	Valero (gas pipelines, processing)	1.5	1997
Edison International			
Citizens Power	P&L Cal Holdings/Boston	0.05	9/00
Edison Capital	Swisscom, telecom network	0.3	9/00
Mission Energy	Italian Vento Power Corp.	0.04	3/00
Mission Energy	Commonwealth Edison's 12 plants in Illinois	5.0	12/99
Mission Energy	Ferrybridge & Fiddler's Ferry power plants in England	2.0	7/99
Mission Energy	40 percent stake in New Zeland's Contact Energy	0.7	5/99
Mission Energy	Homer City power plant, PA	1.8	3/99
EME del Caribe	EcoElectrica co-gen facility in Puerto Rico	0.2	12/98
Mission Energy	1,230 MW coal-fired power plant Paiton, Indonesia (40% stake; under construction since 1994)	.3	1999

Sources: Public Citizen Critical Mass Energy and Environment Program, company annual reports, 10K filings.

when utilities try to stick consumers with doubled and even quadrupled utility bills?"

AB1890 did include measures that appeared to benefit ratepayers. The bill implemented an immediate 10 percent rate cut, and froze it into place for as many as four years. "The same critics who now say the bill was written by the utilities to benefit the utilities were there in Sacramento in 1996 when the legislation was drafted and passed," says Jon Tremayne, a PG&E spokesperson. "Their input, for instance, ensured that residential customers were included in the bill."

But the resulting rates were still 50 percent higher than the national average. And the fixed rate blocked what would have been a natural decline with the onset of new renewables.

Deregulation opponents emphasize that had the natural transition to cheaper and more desirable wind and solar power been allowed to proceed along with efficiency and conservation programs being mandated by the Public Utility Commission (PUC) rates would have drifted downward as green supply increased and demand was held steady.

In the early 1990s, a major program for building renewable-based capacity was eliminated by the utilities through a legal filing in front of the Federal Energy Regulatory Commission. Had that renewable building plan been allowed to proceed, rates would have been on their way down throughout California.

Furthermore, AB1890's so-called rate cut was financed by an elaborate bonding scheme that would force consumers to pay huge sums of long-term interest.

"In effect, the small customers are borrowing to give themselves this rate cut, which is like borrowing money to give yourself a raise," said Coyle at the time. This is a "hidden tax that Californians will have to pay to private utility owners."

GREEN POWER BLOCKED

In the midst of today's crisis, Vice President Dick Cheney, media pundits and others contend that the environmental movement has somehow blocked construction of new power plants that could have helped the state avoid the current crisis.

RECENT SEMPRA ENERGY PURCHASES OR COMMITMENTS

(Sempra was created by the 1998 merger of Pacific Enterprises, the parent of
Southern California Gas and Enova, the parent of San Diego Gas & Electric)

Subsidiary	Purchase/Location	Cost ($ millions)	Date
Sempra Energy International	Sodigas Pampeana S.A./Sodigas Sur S.A. (Argentina natural gas holding co.s)	180	10/00
SEI	Baja natural gas pipeline (joint venture w/PG&E)	230	6/00
SEI	Chilquinta Energia S.A., Chile's third largest electricity distributor; deal includes Energas S.A., gas distributor	830	6/99
SEI	Luz del Sur S.A.A., Peru. (included in above deal)	25	1997
SEI	ECOGAS Mexicali pipeline		
SEI	Termoelectica de Mexicali (Mexico) 600 MW gas-fired power plant (online 2003)	350	2001
SEI	ECOGAS Chihuahua, Mexico (privatized by PEMEX); gas distribution.	50	1997
SEI	DGN De La Laguna-Durango (Mexico) gas pipeline and distribution.	40	2000
SEI	Transportadora de Gas Natural de Baja California, Mexico, pipeline.	??	2000
SEI	Natural gas pipeline, Arizona to Tijuana, Mexico (PG&E will develop US portion).	38	2000
PE	Natural gas/propane distribution (Uruguay) (15 percent interest in Conecta consortium)	160	1998
SEI	Sempra Atlantic Gas (Nova Scotia) exclusive natural gas distribution rights.	800	1999
SEI	Bangor, Maine natural gas distribution	50	2000
SEI	Frontier Energy nat. gas distribution, NC	70	2001
Sempra Energy Trading (SET)	Trading subsidiary purchased from Consolidated Natural Gas, Stamford, CT	36	12/97
SET	Utility.com (Internet utility company)	4	4/00
Sempra Energy Resources (SER)	550 MW Elk Hills power plant Bakersfield, California (online 2003)	360	12/00
SER	1200 MW Mesquite Power plant, Phoenix, AZ	630	12/00
SER	500 MW combined cycle plant, Boulder City, Nevada	280	4/98
Sempra Energy Financial	1200 properties, including housing in Houston, Modesto (CA), Puerto Rico and the Virgin Islands	??	??
Sempra Communications	Aerie Networks nation broadband fiber optic network	??	4/00

Sources: Public Citizen Critical Mass Energy and Environment Program, company annual reports, 10K filings.

Consumer and environmental groups respond that the Federal Electrical Regulatory Commission (FERC) colluded with SoCalEdison to kill a clean generation initiative that would have provided enough additional generating capacity for the state to meet its own needs once deregulation began.

On February 23, 1995, responding to a SoCalEdison petition, FERC blocked a California Public Utilities Commission order that required the utilities to purchase more than 600 megawatts of renewable energy, primarily from wind and geothermal sources. Among other things, the FERC said—with what now seems terrible irony—"we have grave concerns about the need for this capacity," mostly because the state commission was relying on 1990 data, which FERC called "stale."

Throughout the 1990s, California's private utilities resisted the green demands for increased renewables and efficiency.

AB1890 further stifled competition to the benefit of the entrenched utilities that wrote the bill by making it more difficult

HURWITZ'S POWER GRAB

Charles Hurwitz, the man who critics say has made a fortune by pillaging old growth redwoods, busting unions, and engaging in shadowy stock market and savings and loan deals [See "Ravaging the Redwoods: Charles Hurwitz, Michael Milken and the Costs of Greed," *Multinational Monitor*, September 1994], can add another title to his infamous resume: power robber baron.

Hurwitz's Maxxam company controls Kaiser Aluminum, which has limited production at two smelters in Washington state since January 1999. Initially, Kaiser shut down its Washington aluminum smelters as part of the company's nationwide lockout of workers who belong to the United Steelworkers of America. The lockout ended last September, shortly after the federal government threatened to charge Kaiser with violating labor laws.

Then, Kaiser used a novel approach to turn a profit: it started selling the smelters' contracted allotments of electricity back to the Bonneville Power Administration (BPA), the agency that produced the power in the first place.

Under Kaiser's existing contract with the BPA, it purchased power at a rate of $22.50 per megawatt-hour. Kaiser resold the power to BPA for more than 20 times that amount. In late 2000, the spot market price for electricity in the western United States soared to as much as $750 per megawatt-hour.

Kaiser registered $135 million in power sales from October to December 2000.

At year's end, even with its entire U.S. production operations idle, the company's executives funneled some of this easy money into their own wallets. CEO Raymond Milchovich was paid a bonus of $978,000, on top of an annual salary of $630,000. Other executives received bonuses of $250,000 or more.

Aluminum companies have long benefited from generous relationships with the federal BPA, which operates 29 dams in the Columbia and Snake River basins. Smelters aggregate around sources of cheap energy because 45 percent of the cost of aluminum smelting is electricity.

Many Pacific Northwest smelters have been closed since last summer, when their owners found it more profitable to sell power earmarked for their operations to the spot energy market. In June 2000, Alcoa announced that it was halting production at its Troutdale, Oregon smelter. In western Montana, Columbia Falls Aluminum closed its smelter and is reselling the power. Golden Northwest is selling power from its allocation for smelter operations in The Dalles, Oregon, and Kilimat County, Washington.

Some smelter owners recognized the windfall nature of these sales, and pledged to give something back to the workers, local communities and the BPA.

Columbia Falls decided to maintain its 600 workers' salaries and benefits through the year 2001.

Golden Northwest said that it would pass along 25 percent of its estimated $400 million in power sales to the BPA. The company designated another 25 to 50 percent of its electricity revenue for the development of alternative energy, including a wind power plant. It is also converting from the heavily polluting production of primary aluminum to a secondary plant that makes aluminum from scrap, not alumina.

Kaiser has made no such pledges to channel energy resale profits to the BPA, its laid-off workers or alternative power plants. The sales could earn Kaiser a half-billion dollars until the current contract expires in October.

"It's difficult to conceive of a circumstance that would prevent them from coming to terms with the region's other ratepayers and their employees, given the amount of windfall profit," BRA spokesperson Ed Mosey said in January. "There's no way they should be profiteering from reselling federal power and then ask us to draw unemployment," says Steelworkers Local 329 steward Wayne Bentz. Over 900 workers are unemployed due to Kaiser's shutdown.

This winter, Kaiser Vice President Pete Forsyth said the company was keeping its profits to cushion the blow of the new electricity contract, which goes into effect later this year. Initially, BPA and Kaiser reached a deal in which the smelter's power purchase rate would rise by 20 percent. In April, however, BPA officials warned that, due to low water levels and high demand, the agency would not be able to supply enough energy to the region's smelters. BPA advised that the facilities should remain closed for another two years. Kaiser now acknowledges that it is looking to sell its Pacific Northwest smelters.

Like aluminum giants Alcoa and Alcan, Kaiser is slowly but surely moving out of the United States. As energy resources and environmental and labor regulations are tightening, primary aluminum production is shifting to the Third World. Powerful rivers in South America and Africa, coal mines in eastern India, and oil and gas fields of the Middle East are beginning to fuel the global aluminum market.

Ironically, U.S. governmental funding is helping to finance the shift. Kaiser's largest smelter, the 200,000 ton-per-year Valco operation in Ghana, owes its existence to a dam backed by the World Bank and the U.S. government's Overseas Private Investment Corporation (OPIC). The Akosombo dam, according to the International Rivers Network, "flooded more land than any other dam in the world—8,500 square kilometers." Valco consumes most of Akosombo's

(continued)

power. Drought and rising demand have led international financial institutions to back new sources of power for Kaiser's Ghanaian smelter, including a new oil-fired power plant in Takorade.

Hurwitz has focused Kaiser's investments in overseas locations since 1988, when he bought out the corporation with the assistance of fugitive commodities trader Marc Rich. Rich earned the nickname "Aluminum Finger" for his investments in Russia, Iran, and Jamaica, and his stubborn battle against the Steelworkers union at a smelter in Ravenswood, West Virginia.

Under Hurwitz's control, Kaiser has invested in one of the world's largest smelters (Aluminum Bahrain), sold equipment to Russia's notoriously corrupt primary aluminum industry, and has contemplated or made bids to invest in smelters in Guinea, Azerbaijan and Ukraine.

Labor activists say that Kaiser's Pacific Northwest power grab represents a final swindling of the United States before Hurwitz waves goodbye. In April, the president of Steelworkers Local No. 329, Dan Russell, lamented, "I see guys every day that are resigned to the fact that Kaiser is done here. A good number of them are just getting on with their lives… washing their hands of the whole thing."

—Jim Vallette
Jim Vallette is an investigative reporter based in Seawall, Maine. He is working on a report for the Institute of Policy Studies' Sustainable Energy and Economy Network that examines the global structure and social and environmental impacts of the aluminum industry.

for startups to lure new customers away from the established giants. But AB1890 included complex procedural roadblocks that made it virtually impossible for communities to band together to form buying groups that might circumvent the established monopolies.

As Paul Fenn puts it, AB1890 "raises serious questions about the ability of small consumers to exercise any market power. Unless residents and small businesses have the means to purchase power in aggregate through their local governments, 'consumer choice' will mean little more than paying higher rates to a middleman or to your current utility."

PG&E's Tremayne disagrees. "AB1890 provided the opportunity for customers to band together through aggregation and go out on the open market and seek better rates. That was not available prior to deregulation." People are not doing this now, Tremayne says, because "the market is all screwed up. Those lower rates are not out there. Some significant steps need to be taken to stop the out-of-state generators from gouging California customers. Whether it's short-term price caps at the federal level or those generators acting in a responsible fashion, we have to have changes that address the problems of the market."

There were some incentives for conservation and renewables written into AB1890, largely sponsored by the Natural Resources Defense Council, whose chief San Francisco-based energy advocate, Ralph Cavanagh, energetically supported the bill. But the provisions proved marginal at best.

Meanwhile, Cavanagh, with support from the Energy Foundation, supported the utilities at every turn, including acting as a key leader opposing the 1998 grassroots referendum aimed at repealing AB1890.

Cavanagh's role in helping to pass AB1890 and then defending it from repeal has earned him widespread outrage and contempt from the state's green/consumer groups. He still opposes legislation that would take mandated efficiency measures away from utility control and give it to municipalities.

The California experience, warns the state's green/consumer coalition, should stand as a warning to energy activists against accepting the marginal green provisions being tacked onto the Bush/Cheney energy plan.

TOO CHEAP TO METER?

AB 1890's driving force was a utility-sponsored provision to pay the utilities up to $28.5 billion in surcharges for investments in nuclear power, a technology once billed as "too cheap to meter."

In 1996 hearings, SoCalEd and PG&E branded their nuclear reactors at San Onofre and Diablo Canyon as too uneconomical to compete in the competitive free market that deregulation would allegedly bring.

They demanded that ratepayers compensate them for these and other bad investments before deregulation kicked in. That was the rationale for freezing rates at the 1996 established level.

The utility argument, echoed in states around the nation, was that since regulatory agencies had approved the nuclear investments, the public had an obligation to compensate the utilities for their nuclear expenses before opening the electricity market to competition.

Those watching closely argued that not only did this arrangement constitute an outrageous bailout for the utilities, but that the subsidies would be siphoned off for uses that would be of no benefit to Californians. Taken as a whole, warned Gene Coyle at an August 1996 press conference, deregulation and the torrent of cash it would generate for the utilities "will not build infrastructure in California. PG&E and Edison will likely invest it overseas, in places like Indonesia and Australia where both companies are already active. In fact, the entire $27 billion is a liquidation of California assets, with almost all of this ratepayer and taxpayer money likely to flow to foreign investments." That prediction proved prescient.

PG&E's Tremayne responds that the company was under no obligation to reinvest in California. "Under deregulation, the utility's investors—the shareholders of this company—were to get paid back for the investments they made years and years ago. [After the company sold its California generating plants] the shareholders either received the dividends by stock repurchase programs or by investing that money in other investments, including generation facilities outside of California. The way deregulation was set up, [it] intended to do exactly that—

pay the utility investors back money that they had invested in power plants that they were now forced under law to sell. So the shareholders essentially recovered their investments and reinvested them. This was done under the direction of the Public Utilities Commission, in synch with state law."

Both SoCalEd and PG&E say they have suffered huge losses in the last year or two, as the wholesale energy market has spun out of control, though these are losses for the utility subsidiaries, not necessarily for the parent companies which own generating companies as well as the utility subsidiaries' power distribution system. PG&E, the utility subsidiary, not the parent company, has run to bankruptcy protection (handing its executives huge bonuses the day before). In mid-May a federal judge barred consumer groups from participating in those bankruptcy proceedings, meaning the prime suitors in those hearings will be the very power generators who jacked up wholesale prices in the first place.

But by most accounts, the utilities' losses are billions less than the stranded cost bailouts they've laundered to their parent corporations. "It's a mafia operation," says Paul Fenn. "What happened to all that money?"

The Public Media Center's Gunther says the parent companies have spent much of this "rogue cash" as if they were "drunken sailors." Just as Coyle and others warned when AB1890 first became law, Pacific Gas & Electric's owner has made huge investments in power supply networks in New England and New York through its National Energy Group affiliate. SoCalEd's parent, operating through its Mission Energy subsidiary, has been deeply immersed in controversial speculations in Indonesia during the regime of the deposed dictator Suharto.

"The money has not gone to help things in California, that's for sure," says Gunther. "But where is it?"

FOLLOW THE MONEY

While an angry California public increasingly demands to know what its own utilities did with their money, they have also been forced to confront a second band of power magnates—the oil and gas companies close to George W. Bush.

To create competition, the AB1890 deregulation bill of 1996 established a complex scheme by which the utilities divested some, but not all, of their power plants. Those promoting the bill claimed the utilities would become pure distribution companies that would battle one another for the business of small customers.

The transmission wires that delivered the power would remain as regulated monopolies.

And then the generating facilities would, in theory, be bought by dozens of small, entrepreneurial power companies. The magic of the marketplace would drive prices down and service up.

The key to the utilities' deregulation scheme was the assumption that wholesale electric prices would stay low. SoCalEd and PG&E had devised its cap on consumer prices based on the idea that they could dominate supply.

In fact, the utilities did not divest themselves of all their power plants. And, for a series of complex reasons, they failed to enter into long-term contracts with the new generators, thus leaving the utilities dependent on spot markets, where short-term prices could shoot up without notice.

Whereas long-term contracts might have established stable prices over time, the spot market—where energy is sold daily, often in small quantities—is prone to price spikes. Those spikes are supposed to be moderated by the Federal Electrical Regulatory Commission. But over the past year demand on the spot market has regularly been higher than the supply generators make available at any given moment. Prices soar, leaving utilities little choice but to bid prices up dramatically in an effort to procure the electricity to supply their customers.

Thus California was put at the mercy of a handful of out-of-state energy speculators, most notably Duke Power of North Carolina, and Dynergy, Reliant and Enron, all of Texas. These are very big players, who more closely resemble the OPEC cartel than feisty Silicon Valley-type competitors that free market zealots envisioned.

According to Washington, D.C.-based Public Citizen, Enron, Dynergy and Reliant gave in excess of $1.5 million to Bush's campaign and inauguration committee, and to the Republican National Committee. In all, Public Citizen says nine power companies and a trade association with substantial interests in the California energy market gave more than $4 million to Republican candidates and party committees in the 2000 campaign. Bush's new Secretary of Energy, Spencer Abraham, was the energy industry's largest single campaign recipient during his failed U.S. Senate re-election bid in Michigan. Kenneth Lay, president of Enron, the largest U.S. natural gas supplier, is one of George W. Bush's key contributors, and very closest advisers. So is James Baker III, his father's former Secretary of State, and a principal at Reliant.

So while Governor Gray Davis and much of the California legislature received big campaign contributions from the state's utilities, the power companies that manipulated their power supply had their key connections in Washington, which they made good use of.

In a national radio address on May 19, Davis charged that "price gouging" sent the state's annual power bill soaring from $7 billion in 1999 to as much as $60 billion in 2001. Davis blamed it, "pure and simple," on "unconscionable price gouging by the big energy producers—most of them, incidentally, located in Texas."

While Bush has emphatically rejected price caps, widespread charges that his backers' price manipulations are illegal have been given new heft by Loretta Lynch, head of the California Public Utilities Commission. On May 19, the San Francisco Chronicle reported that whistleblower evidence indicated "that generators illegally manipulated prices by deliberately withholding electricity during shortages."

Lynch told a state Senate committee that in at least one instance three power plants simultaneously reduced output, causing prices to spike, and then restored output to cash in. "We certainly see a pattern," she said, while warning of possible criminal prosecutions to come.

By and large the targets of those charges have invested heavily in George Bush, and stand to gain billions if his national energy plan moves ahead. When push came to shove in the California case, they were vastly rewarded by the FERC's crucial refusal to cap prices at which they were selling power to California on the spot market. They also benefited from the FERC's prior opposition to investments in renewables and efficiency, which would have dampened the demand that helped fuel the crisis.

Lynch and the CPUC are not the only state officials warning of legal retribution. On May 2, Lt. Governor Cruz M. Bustamante filed a civil lawsuit against Dynergy, Duke Energy, Mirant, Reliant Energy and Williams Energy Services, alleging that they have systematically engaged in a price-fixing conspiracy to manipulate California's electricity market to "extract" unlawful profits that are draining the state's treasury.

"A cartel of five out-of-state generators has been holding us hostage through a practice of illegal and unfair price-fixing," Bustamante says.

"The energy crisis is not a problem of supply," says Democratic California Assemblywoman Barbara Matthews, who joined Bustamante in filing the suit on behalf of California taxpayers. "It's a problem of manipulation of our supply by out-of-state generators. Generators withhold power, create artificial shortages and play the 'Great American Shell Game' at the public's expense. We will not tolerate this any longer."

While Duke Energy and the others would not comment on the lawsuit, the company says it is "committed to continue playing a major role to help California address its electricity shortfall and the high prices many felt [last] summer." Duke officials say they are working to solve the problem by modernizing and bringing additional generating capacity online at existing power plants as well as signing long-term wholesale electricity contracts with Pacific Gas & Electric (PG&E). But the companies are keeping the terms of those contracts secret.

Public Citizen reports that Enron showed a 42 percent increase in profits last year, Reliant a 55 percent jump and Dynergy a 210 percent rise, all thanks to federal regulators' refusal to cap wholesale prices.

As energy analyst Eugene Coyle puts it: "We've been FERCed."

In April 2001, FERC did finally agree to a price cap scheme that is at best a mixed bag. But in the hot summer of 2000—with Bill Clinton in the White House—FERC stood by while wholesale prices soared.

San Diego Gas & Electric, having collected its final stranded cost money, was allowed by the Public Utilities Commission to unfreeze its consumer prices. The first shockwave of the deregulation disaster hit southern California consumers. SDG&E doubled and tripled its bills.

Consumer rates for SoCalEd and PG&E, however, remained capped. As wholesale prices soared, these large utilities claim to have lost more than $12 billion. When they—and Governor Davis—appealed to President Bush to recap wholesale rates, Bush refused, yielding spectacular profits for his friends at

Enron and Reliant, among others, not to mention the utilities' parent companies, Mission Energy and the National Energy Group.

It is this vise—between skyrocketing wholesale prices combined with frozen retail rates—that has prompted the contention that the problem in California is not a failure of deregulation, but rather that there simply hasn't been enough. The utilities have been desperate to end the rate freeze for consumers that they themselves invented.

In the breach, the utilities convinced Gray Davis to use state funds to buy power, continuing to deliver huge profits for the gas companies. But while escalating his rhetoric against the out-of-state suppliers, Governor Davis has refused widespread consumer demand that he use the state's leverage to take over the assets of the utilities that were still in the process of collecting more than $20 billion in the stranded cost cash bailout, money they promptly laundered to their parent companies.

In all, the double rip-off has yielded at least $20 billion for the utility parent corporations, and another $20 billion for the Bush-related gas companies, all for which California has nothing tangible to show. While PG&E has filed for bankruptcy protection, SoCalEd continues to pressure Davis for concessions, and the gas producers continue to demand that the taxpayers guarantee the purchase of power at rates that appear to fluctuate wildly based not on supply, but on their willingness to sell at uncapped prices.

FOSSIL & NUCLEAR VS. CLEAN AND PUBLIC

In the midst of a convoluted crisis that has so vastly enriched George Bush's supporters, the utility and gas industries are now furthering their agenda on other fronts. Bush and the industry's Congressional allies—most importantly Republican Senator Frank Murkowski of Alaska—want to lift environmental restrictions so that more power plants can be built, as Davis has already done within California. Alaska should be drilled, they say, as well as sensitive protected offshore eco-systems along both coasts and in the Gulf of Mexico, and virtually anywhere else the oil companies think they might make money.

But oil, however, has virtually nothing to do with solving an electricity crisis. Nationwide, less than 4 percent of U.S. electricity is generated from oil. The percentage is even smaller in California.

And then there's the push for more nuclear power plants. Around the April 26, 2001 anniversary of 15 years of fallout from the Chernobyl catastrophe, the mainstream media filled with talk of a revival of atomic energy. Not one of the major stories carried coverage of the huge stranded cost bailouts that had prompted the California crisis in the first place, or the fact that the builders of those nuke plants had labeled them "uncompetitive." Nor was there much mention of the February 3 fire at San Onofre that knocked out a turbine and may keep that nuclear plant off-line for months, at a cost of up to $100 million. In the midst of the state's worst energy crisis, this disaster knocked out, in a flash, fully 25 percent of SoCalEd's generating capacity, and 25 percent of the state's nuclear capacity, enough to supply more than 1.1 million homes.

GIVING IT TO CALIFORNIA CONSUMERS: A REVIEW OF THE DOUBLE RIP-OFF

Thanks to the opposition from utilities and FERC, California has been deprived of the natural-based energy sources the environmental movement tried to win for it.

Thanks to AB1890, the state has been forced to pay surcharges of $20 billion or more to bail the utilities out of their bad nuclear investments.

Thanks to the independent power generators, the state is being gouged for billions to buy electricity it might have otherwise gotten from in-state green sources and increased efficiency.

Thanks to Governor Davis's unwillingness to leverage state money or eminent domain into a takeover of utility assets, the electric grid and production facilities remain owned by utilities and the power producers who caused the crisis in the first place, leaving the state as vulnerable as ever to future manipulations.

And now, thanks to the hysteria whipped up by the rolling blackouts, the state is being inundated by new privately owned, fossil-fuel-burning generators that will escalate pollution levels by rolling over decades of hard-won environmental protections.

—H.W.

Little attention has been devoted to the fact that wind power is now far cheaper than all other sources of new generation except brown coal. At 2.5 cents/kilowatt hour, it is far cheaper than any projected costs for new nuclear capacity, and is even lower than natural gas in many instances. In 2000, Germany, which is moving to shut its 19 nuclear reactors, brought on line 1,300 megawatts of wind, 200 more than was lost at San Onofre. Great Britain has just committed $2.5 billion to offshore turbines. The Public Utilities Commission of Minnesota has deemed wind its "least cost" alternative and ordered at least 300 megawatts of capacity to be added to the 400 already in place. A major wind farm along the Oregon-Washington border began construction in February, and will go on line in December. It is widely known that the Great Plains between the Mississippi and the Rockies are the "Saudi Arabia of wind," with virtually infinite capacity. With current technology, and with additional offshore capacity, wind power could meet much if not all the nation's electric needs long before a new generation of nuclear reactors could be built.

Even atomic energy's staunchest supporters acknowledge it would take at least five years to bring new reactors on line, far more than it would take to bring on vast new supplies of still other alternative technologies, most importantly photovoltaic cells (PV), which transform sunlight directly to electricity. Within five years, PV is expected to plummet far below even

the most optimistic projected costs of atomic power. Increased efficiency and conservation are already far cheaper.

PG&E's Tremayne says conservation is only part of the solution. "PG&E is a leader in conservation efforts and has been for 20 years. Our programs have been looked at from across the country as models of how to institute energy efficiency programs with high customer participation. As a result, Californians are the most energy efficient—in terms of per capita consumption—in the country."

"The problem is that over the past 10 years demand has grown and supply hasn't," Tremayne says. "We've been able to rely upon the Pacific Northwest and the Southwest to supply power to help California meet its needs: As much as 8,000 or 9,000 megawatts on any given summer day. But those two West Coast regions have experienced significant growth as well over the last six years, so they don't have the available power to export any longer. We haven't built any new power plants in California" in a period where 600,000 new residential customers have been added each year.

But as Paul Fenn puts it, the crisis "is not about supply. There's plenty of capacity around. It's a problem of who controls the supply, and the money that pays for it. That's why local control of electricity supply is critical to real solutions and why the idea of gutting environmental laws under the auspices of energy relief is such a horribly impotent gesture."

"We're so afraid to let these companies go under," Fenn complains. "But when all is said and done, the public would be better off letting them go down and using eminent domain to buy their assets. At least then we'll have gotten something tangible out of the deal."

While Governor Davis has talked about possibly buying the utilities' transmission lines, the green/consumer community is increasingly demanding public utility ownership. But Fenn and others are not particularly eager to have the state run a single public-owned utility. Rather, they look to a more local-based solution.

That point of view has gotten powerful backing from James McClatchy, publisher of the Sacramento Bee, and one of the few dissenting voices in the major media's coverage of the California crisis. Beyond the state's taking over the transmission lines, McClatchy wrote in a February 18 editorial, "the next step would be for the state to buy the associated generating facilities."

"Any final solution," says McClatchy, "would have to include public ownership of the generating plants that PG&E and Southern California Edison sold to speculators, as well as the facilities they still own."

The way to deliver the power, McClatchy adds, would be through "existing locally owned and managed utility districts," as well as through ones newly organized to handle the job. "With public ownership of these systems," he says, "would come increased public transparency on all aspects of the operations—where there is little now—and thus less opportunity for sweetheart deals with friendly financiers or broker," leaving less room for "the rapaciousness of speculators and selfish political partisanship."

PG&E's Tremayne denies that public ownership of the companies' generating plants will solve the current crisis.

"Our facilities continue to be regulated under a cost-of-service structure under the Public Utilities Commission. So our generation is being provided to customers here in California at cost and aren't part of the problem. The real problem comes from the out-of-state generators that have been gouging the market for a year. That's where people need to be focusing their efforts."

As for local ownership of the distribution or transmission system, Tremayne says that wouldn't solve the problem either. "Those [distribution] costs are regulated by the CPUC and the FERC. The problem is in the generation market. It doesn't matter who owns the distribution or transmission system—the problem will still exist if we don't get control of the generation market."

THE MUNICIPAL ALTERNATIVE

The greatest of the untold stories of the California crisis is the stunning success of the state's two municipal-owned utilities, the Sacramento Municipal Utility District (SMUD) and the Los Angeles Department of Water and Power (LADWP). Both have not only weathered the storm, but thrived in it.

McClatchy and others point to SMUD and the LADWP for keeping rates stable for their customers while reaping substantial profits selling power into the grid.

In June 1989, Sacramento voted to shut the district's one nuclear reactor, at Rancho Seco. S. David Freeman, who now runs the Los Angeles utility and previously ran the Tennessee Valley Authority, led SMUD into an era powered increasingly—though modestly—by wind and PV facilities, including enough rooftop solar panels for some 6,500 families.

SMUD also inaugurated an unprecedented campaign for increased efficiency. It offered its customers $100 rebates for retiring wasteful old refrigerators (refrigerators account for 20 percent of the average household's electricity consumption). It distributed energy-efficient light bulbs. It promoted solar water heaters and PV panels. It also planted thousands of shade trees to slash summer air conditioning demand.

SMUD's progress occurred amidst a state-wide deregulatory wave that headed in precisely the opposite direction. In the rest of the state, public conservation and efficiency programs were weakened with the deregulation of wholesale supply. The uncertainties of impending consumer price deregulation stalled strong statewide movements to build windmills and install solar panels. And AB1890 freed the utilities from the renewable and efficiency programs mandated by the Public Utilities Commission—which SoCalEd's Bryson had once headed.

This record confirmed public interest advocates' belief that deregulation was incompatible with environmental concerns. Meanwhile, Dan Berman says SMUD was doing "the kinds of things you expect a utility owned by the public to do. It's what we want to see all over California."

Berman has worked for years to win a municipal utility for his home town of Davis. Parallel campaigns have sprung up elsewhere throughout the state. The San Francisco Board of Supervisors has approved a fall referendum for municipal ownership, and five East Bay towns, including Oakland and Berkeley, will hold similar votes. San Diego may also vote on municipal power, and other cities, towns and counties are virtually certain to join in, though they will face massive utility resistance. In mid-May, San Francisco's board of supervisors announced it will take bids on at least 50 megawatts of solar power to be installed in the city and paid for with public bonds—an effort that may be joined by Sacramento.

"SMUD is a role model," says Berman. "If we're to see any progress at all," public power and municipal control is crucial.

Nuclear opponents have long argued that if the hundreds of billions that went into reactors like the ones at the core of the California crisis had gone instead into renewables and efficiency, nothing like California's current crisis would have ever happened. Critics who predicted the disaster that is AB1890 insist that electricity is a service, not a commodity, and that its production and distribution can never be left to an uncertain market that will always be subject to the whims of private utilities and the barons of fossil fuels.

"The last thing the utilities and the independent power producers want is public ownership," says Fenn. "But municipal power, controlled at the local level, is ultimately the key. Until we get it, things are only going to get worse."

"Deregulation of the electricity monopoly is a failure," adds McClatchy. "The monopoly should be returned to the taxpaying consumers who support it and depend on it."

Harvey Wasserman is the author of *The Last Energy War: The Battle Over Utility Deregulation* and a senior advisor to Greenpeace USA.

From *Multinational Monitor*, June 2001, pp. 9-20. © 2001 by Multinational Monitor. Reprinted by permission.

UNIT 4

Biosphere: Endangered Species

Unit Selections

15. **What Is Nature Worth?** Edward O. Wilson
16. **A Fragile Cornucopia: Assessing the Status of U.S. Biodiversity**, Bruce A. Stein
17. **Invasive Species: Pathogens of Globalization**, Christopher Bright

Key Points to Consider

- Are there ways to assess the value or worth of living organisms other than those from whom we derive direct benefits (our domesticated plant and animals species)? What are the relationships between economic assessments of the biosphere and moral or value judgments on the preservation of species?

- Assess the nature of biodiversity in the United States. Why are tropical forests so often mentioned as centers of biodiversity while midlatitude regions like the United States have been ignored as storehouses of biological complexity?

- Why is the spread of invasive species through the global trading network so difficult to control? What suggestions would you make to remedy the problem?

 **Links: www.dushkin.com/online/**
These sites are annotated in the World Wide Web pages.

Endangered Species
http://www.endangeredspecie.com/
Friends of the Earth
http://www.foe.co.uk/index.html
Smithsonian Institution Web Site
http://www.si.edu
World Wildlife Federation (WWF)
http://www.wwf.org

Tragically, the modern conservation movement began too late to save many species of plants and animals from extinction. In fact, even after concern for the biosphere developed among resource managers, their effectiveness in halting the decline of herds and flocks, packs and schools, or groves and grasslands has been limited by the ruthlessness and efficiency of the competition. Wild plants and animals compete directly with human beings and their domesticated livestock and crop plants for living space and for other resources such as sunlight, air, water, and soil. As the historical record of this competition in North America and other areas attests, since the seventeenth century human settlement has been responsible—either directly or indirectly—for the demise of many plant and wildlife species. It should be noted that extinction is a natural process—part of the evolutionary cycle—and not always created by human activity; but human actions have the capacity to accelerate a natural process that might otherwise take millennia.

In the opening article of this unit, one of the world's best-known writers on biological issues asks the central question that will control the future of the biosphere. In "What Is Nature Worth?" Edward O. Wilson notes that there are powerful economic reasons for implementing plans to preserve the world's natural biological diversity. Wilson notes that present losses of Earth's plants and animals from human impact are progressing at rates from 1,000 to 10,000 times those of the average over the last half-billion years. If for no other reason than the fact that many of those lost species have incalculable economic value, humankind needs to fully understand just what it is doing to other inhabitants of Earth. But Wilson notes that there are moral arguments as well as economic ones for slowing the rate of extinction.

The theme of biodiversity is continued in the unit's third article wherein Bruce A. Stein, vice president of the Association for Biodiversity Information, discusses the susceptibility of the United States to disruption of biological systems. In "A Fragile Cornucopia: Assessing the Status of U.S. Biodiversity," Stein notes that the United States has as rich an array of ecological conditions as found anywhere in the world, and a correspondingly high level of biodiversity. In environments stretching from the Alaskan Arctic to the Hawaiian tropics, the lush coastal swamps of Florida to the arid deserts of Nevada, plant and animal species of national and global significance exist. The task of researchers is to identify those species and to fit them into the global pattern of biological complexity in order to under-stand the implications of actions that would seem, at first glance, to have only local relevance. The fourth article in the section also deals with the human impact on environmental systems that begins locally but stretches across the world. Environmental researcher Christopher Bright describes one of the least anticipated results of the growing global economy: the spread of "invasive species." In "Invasive Species: Pathogens of Globalization," Bright notes that thousands of alien species are traveling aboard ships, planes, and railroad cars from one continent to another, often carried in commodities themselves. This consequence of the worldwide global trading network poses enormous hazards for native species in affected areas, and confronting the problem may be as important a challenge to environmental quality as reducing global carbon emissions. As recent events that Bright could not have predicted have shown, when pathogens from one part of the world may be used intentionally as weapons, the challenge becomes even greater.

What Is Nature Worth?

There's a powerful economic argument for preserving our living natural environment:
The biosphere promotes the long-term material prosperity and health of the human race
to a degree that is almost incalculable. But moral reasons, too, should compel us
to take responsibility for the natural world.

by Edward O. Wilson

In the early 19th century, the coastal plain of the southern United States was much the same as in countless millenniums past. From Florida and Virginia west to the Big Thicket of Texas, primeval stands of cypress and flatland hardwoods wound around the corridors of longleaf pine through which the early Spanish explorers had found their way into the continental interior. The signature bird of this wilderness, a dweller of the deep bottomland woods, was the ivory-billed woodpecker, *Campephilus principalis*. Its large size, exceeding a crow's, its flashing white primaries, visible at rest, and its loud nasal call—*kent!… kent!… kent!*—likened by John James Audubon to the false high note of a clarinet, made the ivory-bill both conspicuous and instantly recognizable. Mated pairs worked together up and down the boles and through the canopies of high trees, clinging to vertical surfaces with splayed claws while hammering their powerful, off-white beaks through dead wood into the burrows of beetle larvae and other insect prey. The hesitant beat of their strikes—*tick tick… tick tick tick… tick tick*—heralded their approach from a distance in the dark woods. They came to the observer like spirits out of an unfathomed wilderness core.

Alexander Wilson, early American naturalist and friend of Audubon, assigned the ivorybill noble rank. Its manners, he wrote in *American Ornithology*

(1808–14), "have a dignity in them superior to the common herd of woodpeckers. Trees, shrubbery, orchards, rails, fence posts, and old prostrate logs are all alike interesting to those, in their humble and indefatigable search for prey; but the royal hunter before us scorns the humility of such situations, and seeks the most towering trees of the forest, seeming particularly attached to those prodigious cypress swamps whose crowded giant sons stretch their bare and blasted or moss-hung arms midway to the sky."

A century later, almost all of the virgin bottomland forest had been replaced by farms, towns, and second-growth woodlots. Shorn of its habitat, the ivorybill declined precipitously in numbers. By the 1930s, it was down to scattered pairs in the few remaining primeval swamps of South Carolina, Florida, and Louisiana. In the 1940s, the only verifiable sightings were in the Singer Tract of northern Louisiana. Subsequently, only rumors of sightings persisted, and even these faded with each passing year.

The final descent of the ivorybill was closely watched by Roger Tory Peterson, whose classic *A Field Guide to the Birds* had fired my own interest in birds when I was a teenager. In 1995, the year before he died, I met Peterson, one of my heroes, for the first and only time. I asked him a question common in conversations among American naturalists: What of

the ivory-billed woodpecker? He gave the answer I expected: "Gone."

I thought, surely not gone *everywhere*, not *globally*! Naturalists are among the most hopeful of people. They require the equivalent of an autopsy report, cremation, and three witnesses before they write a species off, and even then they would hunt for it in séances if they thought there were any chance of at least a virtual image. Maybe, they speculate, there are a few ivorybills in some inaccessible cove, or deep inside a forgotten swamp, known only to a few close-mouthed cognoscenti. In fact, several individuals of a small Cuban race of ivorybills were discovered during the 1960s in an isolated pine forest of Oriente Province. Their current status is unknown. In 1996, the Red List of the World Conservation Union reported the species to be everywhere extinct, including Cuba. I have heard of no further sightings, but evidently no one at this writing knows for sure.

Why should we care about *Campephilus principalis*? It is, after all, only one of 10,000 bird species in the world. Let me give a simple and, I hope, decisive answer: because we knew this particular species, and knew it well. For reasons difficult to understand and express, it became part of our culture, part of the rich mental world of Alexander Wilson and all those afterward who cared about it. There is no way to make a full and final valuation of the ivorybill or any other species in the natural world. The measures we use increase in number and magnitude with no predictable limit. They rise from scattered, unconnected facts and elusive emotions that break through the surface of the subconscious mind, occasionally to be captured by words, though never adequately.

We, *Homo sapiens*, have arrived and marked our territory well. Winners of the Darwinian lottery, bulge-headed paragons of organic evolution, industrious bipedal apes with opposable thumbs, we are chipping away the ivorybills and other miracles around us. As habitats shrink, species decline wholesale in range and abundance. They slide down the Red List ratchet, and the vast majority depart without special notice. Over the past half-billion years, the planet lost perhaps one species per million species each year, including everything from mammals to plants. Today, the annual rate of extinction is 1,000 to 10,000 times faster. If nothing more is done, one-fifth of all the plant and animal species now on earth could be gone or on the road to extinction by 2030. Being distracted and self-absorbed, as is our nature, we have not yet fully understood what we are doing. But future generations, with endless time to reflect, will understand it all, and in painful detail. As awareness grows, so will their sense of loss. There will be thousands of ivory-billed woodpeckers to think about in the centuries and millenniums to come.

Is there any way now to measure even approximately what is being lost? Any attempt is almost certain to produce an underestimate, but let me start anyway with macroeconomics. In 1997, an international team of economists and environmental scientists put a dollar amount on all the ecosystems services provided to humanity free of charge by the living natural environment. Drawing from multiple databases, they estimated the contribution to be $33 trillion or more each year. This amount is nearly twice the 1997 combined gross national product (GNP) of all the countries in the world—$18 trillion. *Ecosystems services* are defined as the flow of materials, energy, and information from the biosphere that support human existence. They include the regulation of the atmosphere and climate; the purification and retention of fresh water; the formation and enrichment of the soil; nutrient cycling; the detoxification and recirculation of waste; the pollination of crops; and the production of lumber, fodder, and biomass fuel.

> IF HUMANITY WERE TO TRY TO REPLACE THE FREE SERVICES OF THE NATURAL ECONOMY WITH SUBSTITUTES OF ITS OWN MANUFACTURE, THE GLOBAL GNP WOULD HAVE TO BE INCREASED BY AT LEAST $33 TRILLION.

The 1997 megaestimate can be expressed in another, even more cogent, manner. If humanity were to try to replace the free services of the natural economy with substitutes of its own manufacture, the global GNP would have to be increased by at least $33 trillion. The exercise, however, cannot be performed except as a thought experiment. To supplant natural ecosystems entirely, even mostly, is an economic—and even physical—impossibility, and we would certainly die if we tried. The reason, ecological economists explain, is that the *marginal value*, defined as the rate of change in the value of ecosystems services relative to the rate of decline in the availability of these services, rises sharply with every increment in the decline. If taken too far, the rise will outpace human capacity to sustain the needed services by combined natural and artificial means. Hence, a much greater dependence on artificial means—in other words, environmental prostheses—puts at risk not just the biosphere but humanity itself.

Most environmental scientists believe that the shift has already been taken too far, lending credit to the folk injunction "Don't mess with Mother Nature." The lady is our mother all right, and a mighty dispensational force as well. After evolving on her own for more than three billion years, she gave birth to us a mere million years ago, the blink of an eye in evolutionary time. Ancient and vulnerable, she will not tolerate the undisciplined appetite of her gargantuan infant much longer.

Abundant signs of the biosphere's limited resilience exist all around. The oceanic fish catch now yields $2.5 billion to the U.S. economy and $82 billion worldwide. But it will not grow further, simply because the amount of ocean is fixed and the number of organisms it can generate is static. As a result, all of the world's 17 oceanic fisheries are at or below sustainable yield. During the 1990s, the annual global catch leveled off around 30 million tons. Pressed by ever-growing global demand, it can be expected eventually to drop. Already, fisheries of the western North Atlantic, the Black Sea, and portions of the Caribbean have largely collapsed. Aquaculture, or the farming of fish, crustaceans, and mollusks, takes up part of the slack, but at rising environmental cost. This "fin-and-shell revolution" necessitates the conversion of valuable wetland habitats, which are nurseries for marine life. To feed the captive populations, fodder must be diverted from crop production. Thus, aquaculture competes with other human activities for productive land while reducing natural habitat. What was once free for the taking must now be manufactured. The ultimate result will be an upward inflationary pressure across wide swaths of the world's coastal and inland economies.

Another case in point: Forested watersheds capture rainwater and purify it before returning it by gradual runoffs to the lakes and sea, all for free. They can be replaced only at great cost. For generations, New York City thrived on exceptionally clean water from the Catskill Mountains. The watershed inhabitants were proud that their bottled water was once sold throughout the Northeast. As their population grew, however, they converted more and more of the watershed forest into farms, homes, and resorts. Gradually, the sewage and agricultural runoff adulterated the water, until it fell below Environmental Protection Agency standards. Officials in New York City now faced a choice: They could build a filtration plant to replace the Catskill watershed, at a $6 billion to $8 billion capital cost, followed by $300 million annual running costs, or they could restore the watershed to somewhere near its original purification capacity for $1 billion, with subsequently very low maintenance costs. The decision was easy, even for those born and bred in an urban environment. In 1997, the city raised an environmental bond issue and set out to purchase forested land and to subsidize the upgrading of septic tanks in the Catskills. There is no reason the people of New York City and the Catskills cannot enjoy the double gift from nature in perpetuity of clean water at low cost and a beautiful recreational area at no cost.

There is even a bonus in the deal. In the course of providing natural water management, the Catskill forest region also secures flood control at very little expense. The same benefit is available to the city of Atlanta. When 20 percent of the trees in the metropolitan area were removed during its rapid development, the result was an annual increase in stormwater runoff of 4.4 billion cubic feet. If enough containment facilities were built to capture this volume, the cost would be at least $2 billion. In contrast, trees replanted along streets and in yards, and parking areas are a great deal cheaper than concrete drains and revetments. Their maintenance cost is near zero, and, not least, they are more pleasing to the eye.

In conserving nature, whether for practical or aesthetic reasons, diversity matters. The following rule is now widely accepted by ecologists: The more numerous the species that inhabit an ecosystem, such as a forest or lake, the more productive and stable is the ecosystem. By "production," the scientists mean the amount of plant and animal tissue created in a given unit of time. By "stability," they mean one or the other, or both, of two things: first, how narrowly the summed abundances of all species vary through time; and, second, how quickly the ecosystem recovers from fire, drought, and other stresses that perturb it. Human beings understandably wish to live in the midst of diverse, productive, and stable ecosystems. Who, if given a choice, would build a home in a wheat field instead of a parkland?

Ecosystems are kept stable in part by the insurance principle of biodiversity: If a species disappears from a community, its niche will be more quickly and effectively filled by another species if there are many candidates for the role instead of few. Example: A ground fire sweeps through a pine forest, killing many of the understory plants and animals. If the forest is biodiverse, it recovers its original composition and production of plants and animals more quickly. The larger pines escape with some scorching of their lower bark and continue to grow and cast shade as before. A few kinds of shrubs and herbaceous plants also hang on and resume regeneration immediately. In some pine forests subject to frequent fires, the heat of the fire itself triggers the germination of dormant seeds genetically adapted to respond to heat, speeding the regrowth of forest vegetation still more.

WHO, IF GIVEN A CHOICE, WOULD BUILD A HOME IN A WHEAT FIELD INSTEAD OF A PARKLAND?

A second example of the insurance principle: When we scan a lake, our macroscopic eye sees only relatively big organisms, such as eelgrass, pondweeds, fishes, water birds, dragonflies, whirligig beetles, and other things big enough to splash and go bump in the night. But all around them, in vastly greater numbers and variety, are invisible bacteria, protistans, planktonic single-celled algae, aquatic fungi, and other microorganisms. These seething myriads are the true foundation of the lake's ecosystem and the hidden agents of its stability. They decompose the bodies of the larger organisms. They form large reservoirs of carbon and nitrogen, release carbon dioxide, and thereby damp fluctuations in the organic cycles and energy flows in the rest of the aquatic ecosystem. They hold the lake close to a chemical equilibrium, and, to a point, they pull it back

from extreme perturbations caused by silting and pollution.

In the dynamism of healthy ecosystems, there are minor players and major players. Among the major players are the ecosystems engineers, which add new parts to the habitat and open the door to guilds of organisms specialized to use them. Biodiversity engenders more biodiversity, and the overall abundance of plants, animals, and microorganisms increases to a corresponding degree.

By constructing dams, beavers create ponds, bogs, and flooded meadows. These environments shelter species of plants and animals that are rare or absent in free-running streams. The submerged masses of decaying wood forming the dams draw still more species, which occupy and feed on them.

Elephants trample and tear up shrubs and small trees, opening glades within forests. The result is a mosaic of habitats that, overall, contains larger numbers of resident species.

Florida gopher tortoises dig 30-foot-long tunnels that diversify the texture of the soil, altering the composition of its microorganisms. Their retreats are also shared by snakes, frogs, and ants specialized to live in the burrows.

Euchondrus snails of Israel's Negev Desert grind down soft rocks to feed on the lichens growing inside. By converting rock to soil and releasing the nutrients photosynthesized by the lichens, the snails multiply niches for other species.

TO EVALUATE INDIVIDUAL SPECIES SOLELY BY THEIR KNOWN PRACTICAL VALUE AT THE PRESENT TIME IS BUSINESS ACCOUNTING IN THE SERVICE OF BARBARISM.

Overall, a large number of independent observations from differing kinds of ecosystems point to the same conclusion: The greater the number of species that live together, the more stable and productive the ecosystems these species compose. On the other hand, mathematical models that attempt to describe the interactions of species in ecosystems show that the apparent opposite also occurs: High levels of diversity can reduce the stability of individual species. Under certain conditions, including random colonization of the ecosystem by large numbers of species that interact strongly with one another, the separate but interlocking fluctuations in species populations can become more volatile, thus making extinction more likely. Similarly, given appropriate species traits, it is mathematically possible for increased diversity to lead to decreased production.

When observation and theory collide, scientists turn to carefully designed experiments for resolution. Their motivation is especially strong in the case of biological systems, which are typically far too complex to be grasped by observation and theory alone. The best procedure, as in the rest of science, is first to simplify the system, then to hold it more or less constant while varying the important parameters one or two at a time to see what happens. In the 1990s a team of British ecologists, in an attempt to approach these ideal conditions, devised the *ecotron*, a growth chamber in which artificially simple ecosystems can be assembled as desired, species by species. Using multiple ecotrons, they found that productivity, measured by the increase of plant bulk, rose with an increase in species numbers. Simultaneously, ecologists monitoring patches of Minnesota grassland—outdoor equivalents of ecotrons—during a period of drought found that patches richer in species diversity underwent less decline in productivity and recovered more quickly than patches with less diversity.

These pioneering experiments appeared to uphold the conclusion drawn earlier from natural history, at least with reference to production. Put more precisely, ecosystems tested thus far do not possess the qualities and starting conditions allowed by theory that can reduce production and produce instability as a result of large species numbers.

But—how can we be sure, the critics asked (pressing on in the best tradition of science), that the increase in production in particular is truly the result of just an increase in the number of species?

Maybe the effect is due to some other factor that just happens to be correlated with species numbers. Perhaps the result is a statistical artifact. For example, the larger the number of plant species present in a habitat, the more likely it is that at least one kind among them will be extremely productive. If that occurs, the increase in the yield of plant tissue—and in the number of the animals feeding on it—is only a matter of luck of the draw, and not the result of some pure property of biodiversity itself. At its base, the distinction made by this alternative hypothesis is semantic. The increased likelihood of acquiring an outstandingly productive species can be viewed as just one means by which the enrichment of biodiversity boosts productivity. (If you draw on a pool of 1,000 candidates for a basketball team, you are more likely to get a star than if you draw on a pool of 100 candidates.)

Still, it is important to know whether other consequences of biodiversity enrichment play an important role. In particular, do species interact in a manner that increases the growth of either one or both? This is the process called *overyielding*. In the mid-1990s, a massive study was undertaken to test the effect of biodiversity on productivity that paid special attention to the presence or absence of overyielding. Multiple projects of BIODEPTH, as the project came to be called, were conducted during a two-year period by 34 researchers in eight European countries. This time, the results were more persuasive. They showed once again that productivity does increase with biodiversity. Many of the experimental runs also revealed the existence of overyielding.

Over millions of years, nature's ecosystems engineers have been especially effective in the promotion of overyielding. They have coevolved with other species that exploit the niches they build. The result is a harmony within ecosystems. The constituent species, by spreading out into multiple niches, seize and cycle more materials and energy than is possible in similar ecosystems. *Homo sapiens* is an ecosystems engineer too, but a bad one. Not having coevolved with the majority of life forms we now encounter around the world, we elimi-

nate far more niches than we create. We drive species and ecosystems into extinction at a far higher rate than existed before, and everywhere diminish productivity and stability.

I will grant at once that economic and production values at the ecosystem level do not alone justify saving every species in an ecosystem, especially those so rare as to be endangered. The loss of the ivory-billed woodpecker has had no discernible effect on American prosperity. A rare flower or moss could vanish from the Catskill forest without diminishing the region's filtration capacity. But so what? To evaluate individual species solely by their known practical value at the present time is business accounting in the service of barbarism. In 1973, the economist Colin W. Clark made this point persuasively in the case of the blue whale, *Balaenopterus musculus*. A hundred feet in length and 150 tons in weight at maturity, the species is the largest animal that ever lived on land or sea. It is also among the easiest to hunt and kill. More than 300,000 blue whales were harvested during the 20th century, with a peak haul of 29,649 in the 1930–31 season. By the early 1970s, the population had plummeted to several hundred individuals. The Japanese were especially eager to continue the hunt, even at the risk of total extinction. So Clark asked, What practice would yield the whalers and humanity the most money: Cease hunting and let the blue whales recover in numbers, then harvest them sustainably forever, or kill the rest off as quickly as possible and invest the profits in growth stocks? The disconcerting answer for annual discount rates over 21 percent: Kill them all and invest the money.

Now, let us ask, what is wrong with that argument?

Clark's implicit answer is simple. The dollars-and-cents value of a dead blue whale was based only on the measures relevant to the existing market—that is, on the going price per unit weight of whale oil and meat. There are many other values, destined to grow along with our knowledge of living *Balaenopterus musculus* and as science, medicine, and aesthetics grow and strengthen, in dimensions and magnitudes still unforeseen. What was the value of the blue whale in A.D. 1000? Close to zero. What will be its value in A.D. 3000? Essentially limitless—to say nothing of the measure of gratitude the generation then alive will feel to those who in their wisdom saved the whale from extinction.

No one can guess the full future value of any kind of animal, plant, or microorganism. Its potential is spread across a spectrum of known and as yet unimagined human needs. Even the species themselves are largely unknown. Fewer than two million are in the scientific register, with a formal Latinized name, while an estimated five to 100 million—or more—await discovery. Of the species known, fewer than one percent have been studied beyond the sketchy anatomical descriptions used to identify them.

> OF THE SPECIES KNOWN, FEWER THAN ONE PERCENT HAVE BEEN STUDIED BEYOND THE SKETCHY ANATOMICAL DESCRIPTIONS USED TO IDENTIFY THEM.

Agriculture is one of the vital industries most likely to be upgraded by attention to the remaining wild species. The world's food supply hangs by a slender thread of biodiversity. Ninety percent is provided by slightly more than 100 plant species out of a quarter-million known to exist. Twenty species carry most of the load, of which only three—wheat, maize, and rice—stand between humanity and starvation. For the most part, the premier 20 are those that happened to be present in the regions where agriculture was independently invented some 10,000 years ago, namely the Mediterranean perimeter and southwestern Asia; Central Asia; the Horn of Africa; the rice belt of tropical Asia; and the uplands of Mexico, Central America, and Andean South America. Yet some 30,000 species of wild plants, most occurring outside these regions, have edible parts consumed at one time or other by hunter-gatherers. Of these species, at least 10,000 can be adapted as domestic crops. A few, including the three species of New World amaranths, the carrotlike arracacha of the Andes, and the winged bean of tropical Asia, are immediately available for commercial development.

In a more general sense, all the quarter-million plant species—in fact, all species of organisms—are potential donors of genes that can be transferred by genetic engineering into crop species in order to improve their performance. With the insertion of the right snippets of DNA, new strains can be created that are, variously, cold resistant, pest resistant, perennial, fast growing, highly nutritious, multipurpose, sparing in their consumption of water, and more easily sowed and harvested. And compared with traditional breeding techniques, genetic engineering is all but instantaneous.

The method, a spinoff of the revolution in molecular genetics, was developed in the 1970s. During the 1980s and 1990s, before the world quite realized what was happening, it came of age. A gene from the bacterium *Bacillus thuringiensis*, for example, was inserted into the chromosomes of corn, cotton, and potato plants, allowing them to manufacture a toxin that kills insect pests. No need to spray insecticides; the engineered plants now perform this task on their own. Other transgenes, as they are called, were inserted from bacteria into soybean and canola plants to make them resistant to chemical weed killers. Agricultural fields can now be cheaply cleared of weeds with no harm to the crops growing there. The most important advance of all, achieved in the 1990s, was the creation of golden rice. This new strain is laced with bacterial and daffodil genes that allow it to manufacture beta-carotene, a precursor of vitamin A. Because rice, the principal food of three billion people, is deficient in vitamin A, the addition of beta-carotene is no mean humanitarian feat. About the same time, the almost endless potential of genetic engineering was confirmed by two circus tricks of the trade: A bacterial gene was implanted into a monkey, and a jellyfish bioluminescence gene into a plant.

But not everyone was dazzled by genetic engineering, and inevitably it stirred opposition. For many, human existence was being transformed in a fundamental and insidious way. With little warning, genetically modified organisms (GMOs) had entered our lives and were all around us, changing incomprehensibly the order of nature and society. A protest movement against the new industry began in the mid-1990s and exploded in 1999, just in time to rank as a millennial event with apocalyptic overtones. The European Union banned transgenic crops, the Prince of Wales compared the methodology to playing God, and radical activists called for a global embargo of all GMOs. "Frankenfoods," "superweeds," and "Farmageddon" entered the vocabulary: GMOs were, according to one British newspaper, the "mad forces of genetic darkness." Some prominent environmental scientists found technical and ethical reasons for concern.

As I write, public opinion and official policy toward genetic engineering have come to vary greatly from one country to the next. France and Britain are vehemently opposed. China is strongly favorable, and Brazil, India, Japan, and the United States cautiously so. In the United States particularly, the public awoke to the issue only after the transgenie (so to speak) was out of the bottle. From 1996 to 1999, the amount of U.S. farmland devoted to genetically modified crops had rocketed from 3.8 million to 70.9 million acres. As the century ended, more than half the soybeans and cotton grown, and nearly a third (28 percent) of the corn, were engineered.

> WITH LITTLE WARNING, GENETICALLY MODIFIED ORGANISMS HAD ENTERED OUR LIVES AND WERE ALL AROUND US, CHANGING INCOMPREHENSIBLY THE ORDER OF NATURE AND SOCIETY.

There are, actually, several sound reasons for anxiety over genetic engineering, which I will now summarize and evaluate.

Many people, not just philosophers and theologians, are troubled by the ethics of transgenic evolution. They grant the benefits but are unsettled by the reconstruction of organisms in bits and pieces. Of course, human beings have been creating new strains of plants and animals since agriculture began, but never at the sweep and pace inaugurated by genetic engineering. And during the era of traditional plant breeding, hybridization was used to mix genes almost always among varieties of the same species or closely similar species. Now it is used across entire kingdoms, from bacteria and viruses to plants and animals. How far the process should be allowed to continue remains an open ethical issue.

The effects on human health of each new transgenic food are hard to predict, and certainly never free of risk. However, the products can be tested just like any other new food products on the market, then certified and labeled. There is no reason at this time to assume that their effects will differ in any fundamental way. Yet scientists generally agree that a high level of alertness is essential, and for the following reason: All genes, whether original to the organism or donated to it by an exotic species, have multiple effects. Primary effects, such as the manufacture of a pesticide, are the ones sought. But destructive secondary effects, including allergenic or carcinogenic activity, are also at least a remote possibility.

Transgenes can escape from the modified crops into wild relatives of the crop where the two grow close together. Hybridization has always occurred widely in agriculture, even before the advent of genetic engineering. It has been recorded at one or another time and place in 12 of the 13 most important crops used worldwide. However, the hybrids have not overwhelmed their wild parents. I know of no case in which a hybrid strain outcompetes wild strains of the same or closely related species in the natural environment. Nor has any hybrid turned into a superweed, in the same class as the worst wild nonhybrid weeds that afflict the planet. As a rule, domesticated species and strains are less competitive than their wild counterparts in both natural and human-modified environments, Of course, transgenes could change the picture. It is simply too early to tell.

Genetically modified crops can diminish biological diversity in other ways. In a now famous example, the bacterial toxin used to protect corn is carried in pollen by wind currents for distances of 60 meters or more from the cultivated fields. Then, landing on milkweed plants, the toxin is capable of killing the caterpillars of monarch butterflies feeding there. In another twist, when cultivated fields are cleared of weeds with chemical sprays against which the crops are protected by transgenes, the food supply of birds is reduced and their local populations decline. These environmental secondary effects have not been well studied in the field. How severe they will become as genetic engineering spreads remains to be seen.

Many people, having become aware of the potential threats of genetic engineering in their food supply, understandably believe that yet another bit of their freedom has been taken from them by faceless corporations (who can even name, say, three of the key players?) using technology beyond their control or even understanding. They also fear that an industrialized agriculture dependent on high technology can by one random error go terribly wrong. At the heart of the anxiety is a sense of helplessness. In the realm of public opinion, genetic engineering is to agriculture as nuclear engineering is to energy.

The problem before us is how to feed billions of new mouths over the next several decades and save the rest of life at the same time—without being trapped in a Faustian bargain that threatens freedom and security. No one knows the exact solution to this dilemma. Most scientists and economists who have studied both sides of it agree that the benefits outweigh the risks. The benefits must come from an evergreen revolution that has as its goal to lift food production well above the level attained by the green rev-

olution of the 1960s, using technology and regulatory policy more advanced, and even safer, than that now in existence.

Genetic engineering will almost certainly play an important role in the evergreen revolution. Energized by recognition of both its promise and its risk, most countries have begun to fashion policies to regulate the marketing of transgenic crops. The ultimate driving force in this rapidly evolving process is international trade. More than 130 countries took an important first step in 2000 to address the issue by tentatively agreeing to the Cartagena Protocol on Biosafety, which provides the right to block imports of transgenic products. The protocol also sets up a joint "biosafety clearing house" to publish information on national policy. About the same time, the U.S. National Academy of Sciences, joined by the science academies of five other countries (Brazil, China, India, Mexico, and the United Kingdom) and the Third World Academy of Sciences, endorsed the development of transgenic crops. They made recommendations for risk assessment and licensing agreements and stressed the needs of the developing countries in future research programs and capital investment.

Medicine is another domain that stands to gain enormously from the world's store of biodiversity, with or without the impetus of genetic engineering. Pharmaceuticals in current use are already drawn heavily from wild species. In the United States, about a quarter of all prescriptions dispensed by pharmacies are substances extracted from plants. Another 13 percent originate from microorganisms, and three percent more from animals—making a total of about 40 percent derived from wild species. What's even more impressive is that nine of the 10 leading prescription drugs originally came from organisms. The commercial value of the relatively small number of natural products is substantial. The over-the-counter cost of drugs from plants alone was estimated in 1998 to be $20 billion in the United States and $84 billion worldwide.

But only a tiny fraction of biodiversity has been utilized in medicine, despite its obvious potential. The narrowness of the base is illustrated by the dominance of ascomycete fungi in the control of bacterial diseases. Although only about 30,000 species of ascomycetes—two percent of the total known species of organisms—have been studied, they have yielded 85 percent of the antibiotics in current use. The underutilization of biodiversity is still greater than these figures alone might suggest—because probably fewer than 10 percent of the world's ascomycete species have even been discovered and given scientific names. The flowering plants have been similarly scanted. Although it is likely that more than 80 percent of the species have received scientific names, only some three percent of this fraction have been assayed for alkaloids, the class of natural products that have proved to be among the most potent curative agents for cancer and many other diseases.

There is an evolutionary logic in the pharmacological bounty of wild species. Throughout the history of life, all kinds of organisms have evolved chemicals needed to control cancer in their own bodies, kill parasites, and fight off predators. Mutations and natural selection, which equip this armamentarium, are processes of endless trial and error. Hundreds of millions of species, evolving by the life and death of astronomical numbers of organisms across geological stretches of time, have yielded the present-day winners of the mutation-and-selection lottery. We have learned to consult them while assembling a large part of our own pharmacopoeia. Thus, antibiotics, fungicides, antimalarial drugs, anesthetics, analgesics, blood thinners, blood-clotting agents, agents that prevent clotting, cardiac stimulants and regulators, immunosuppressive agents, hormone mimics, hormone inhibitors, anticancer drugs, fever suppressants, inflammation controls, contraceptives, diuretics and antidiuretics, antidepressants, muscle relaxants, rubefacients, anticongestants, sedatives, and abortifacients are now at our disposal, compliments of wild biodiversity.

Revolutionary new drugs have rarely resulted from the pure insights of molecular and cellular biology, even though these sciences have grown very sophisticated and address the causes of disease at the most fundamental level. Rather, the pathway of discovery has usually been the reverse: The presence of the drug is first detected in whole organisms, and the nature of its activity subsequently tracked down to the molecular and cellular levels. Then the basic research begins.

> THE PROBLEM BEFORE US IS HOW TO FEED BILLIONS OF NEW MOUTHS OVER THE NEXT SEVERAL DECADES AND SAVE THE REST OF LIFE AT THE SAME TIME.

The first hint of a new pharmaceutical may lie among the hundreds of remedies of Chinese traditional medicine. It may be spotted in the drug-laced rituals of an Amazonian shaman. It may come from a chance observation by a laboratory scientist unaware of its potential importance for medicine. More commonly nowadays, the clue is deliberately sought by the random screening of plant and animal tissues. If a positive response is obtained—say, a suppression of bacteria or cancer cells—the molecules responsible can be isolated and tested on a larger scale, using controlled experiments with animals and then (cautiously!) human volunteers. If the tests are successful, and the atomic structure of the molecule is also in hand, the substance can be synthesized in the laboratory, then commercially, usually at lower cost than by extraction from harvested raw materials. In the final step, the natural chemical compounds provide the prototype from which new classes of organic chemicals can be synthesized, adding or taking away atoms and double bonds here and there. A few of the novel substances may prove more efficient than the natural prototype. And of equal importance to the pharmaceutical companies, these analogues can be patented.

Serendipity is the hallmark of pharmacological research. A chance discovery can lead not only to a successful drug but to advances in fundamental science, which in time yield other successful drugs. Routine screening, for example, revealed that an obscure fungus growing in the mountainous interior of Norway produces a powerful suppressor of the human immune system. When the molecule was isolated from the fungal tissue and identified, it proved to be a complex molecule of a kind never before encountered by organic chemists. Nor could its effect be explained by the contemporary principles of molecular and cellular biology. But its relevance to medicine was immediately obvious, because when organs are transplanted from one person to another, the immune system of the host must be prevented from rejecting the alien tissue. The new agent, named cyclosporin, became an essential part of the organ transplant industry. It also served to open new lines of research on the molecular events of the immune response itself.

The surprising events that sometimes lead from natural history to medical breakthrough would make excellent science fiction—if only they were untrue. The protagonists of one such plot are the poison dart frogs of Central and South America, which belong to the genera *Dendrobates* and *Phyllobates* in the family Dendrobatidae. Tiny, able to perch on a human fingernail, they are favored as terrarium animals for their beautiful colors: The 40 known species are covered by various patterns of orange, red, yellow, green, or blue, usually on a black background. In their natural habitat, dendrobatids hop about slowly and are relatively unfazed by the approach of potential predators. For the trained naturalist their lethargy triggers an alarm, in observance of the following rule of animal behavior: If a small and otherwise unknown animal encountered in the wild is strikingly beautiful, it is probably poisonous, and if it is not only beautiful but also easy to catch, it is probably deadly. And so it is with dendrobatid frogs, which, it turns out, secrete a powerful toxin from glands on their backs. The potency varies according to species. A single individual of one (perfectly named) Colombian species, *Phyllobates horribilis*, for example, carries enough of the substance to kill 10 men. Indians of two tribes living in the Andean Pacific slope forests of western Colombia, the Emberá Chocó and the Noanamá Chocó, rub the tips of their blowgun darts over the backs of the frogs, very carefully, then release the little creatures unharmed so they can make more poison.

COLLECTING SAMPLES OF VALUABLE SPECIES FROM RICH ECOSYSTEMS AND CULTIVATING THEM IN BULK ELSEWHERE IS NOT ONLY PROFITABLE BUT THE MOST SUSTAINABLE OF ALL.

In the 1970s a chemist, John W. Daly, and a herpetologist, Charles W. Myers, gathered material from a similar Ecuadorian frog, *Epipedobates tricolor*, for a closer look at the dendrobatid toxin. In the laboratory, Daly found that very small amounts administered to mice worked as an opiumlike painkiller, yet otherwise lacked the properties of typical opiates. Would the substance also prove nonaddictive? If so, it might be turned into the ideal anesthetic. From a cocktail of compounds taken from the backs of the frogs, Daly and his fellow chemists isolated and characterized the toxin itself, a molecule resembling nicotine, which they named epibatidine. This natural product proved 200 times more effective in the suppression of pain than opium, but was also too toxic, unfortunately, for practical use. The next step was to redesign the molecule. Chemists at Abbott Laboratories synthesized not only epibatidine but hundreds of novel molecules resembling it. When tested clinically, one of the products, code-named ABT-594, was found to combine the desired properties: It depressed pain like epibatidine, including pain from nerve damage of a kind usually impervious to opiates, and it was nonaddictive.

ABT-594 had two additional advantages: It promoted alertness instead of sleepiness, and it had no side effects on respiration or digestion.

The full story of the poison dart frogs also carries a warning about the conservation of tropical forests. The destruction of much of the habitat in which populations of *Epipedobates* live almost prevented the discovery of epibatidine and its synthetic analogues. By the time Daly and Myers set out to collect enough toxin for chemical analysis, after their initial visit to Ecuador, one of the two prime rainforest sites occupied by the frogs had been cleared and replaced with banana plantations. At the second site, which fortunately was still intact, they found enough frogs to harvest just one milligram of the poison. From that tiny sample, chemists were able, with skill and luck, to identify epibatidine and launch a major new initiative in pharmaceutical research.

It is no exaggeration to say that the search for natural medicinals is a race between science and extinction, and will become critically so as more forests fall and coral reefs bleach out and disintegrate. Another adventure dramatizing this point began in 1987, when the botanist John Burley collected samples of plants from a swamp forest near Lundu in the Malaysian state of Sarawak, on the northwestern corner of the island of Borneo. His expedition was one of many launched by the National Cancer Institute (NCI) to search for new natural substances to add to the fight against cancer and AIDS. Following routine procedure, the team collected a kilogram of fruit, leaves, and twigs from each kind of plant they encountered. Part was sent to the NCI laboratory for assay, and part was deposited in the Harvard University Herbarium for future identification and botanical research.

One such sample came from a small tree at Lundu about 25 feet high. It was given the voucher code label Burley-and-Lee 351. Back at the NCI laboratories, an extract made from it was tested routinely against human cancer cells grown in culture. Like the majority of such preparations, it had no effect. Then it was run through screens designed to test its potency against the AIDS virus.

The NCI scientists were startled to observe that Burley-and-Lee 351 gave, in their words, "100 percent protection against the cytopathic effects of HIV-I infection," having "essentially halted HIV-I replication." In other words, while the substance the sample contained could not cure AIDS, it could stop cold the development of disease symptoms in HIV-positive patients.

The Burley-and-Lee 351 tree was determined to belong to a species of *Calophyllum*, a group of species belonging to the mangosteen family, or Guttiferae. Collectors were dispatched to Lundu a second time to obtain more material from the same tree, with the aim of isolating and chemically identifying the HIV inhibitor. The tree was gone, probably cut down by local people for fuel or building materials. The collectors returned home with samples from other *Calophyllum* trees taken in the same swamp forest, but their extracts were ineffective against the virus.

Peter Stevens, then at Harvard University, and the world authority on *Calophyllum*, stepped in to solve the problem. The original tree, he found, belonged to a rare strain named *Calopsyllum lanigerum*, variety *austrocoriaceum*. The trees sampled on the second trip were another species, which explained their inactivity. No more specimens of *austrocoriaceum* could be found at Lundu. The search for the magic strain widened, and finally a few more specimens were located in the Singapore Botanic Garden. Thus supplied with enough raw material, chemists and microbiologists were able to identify the anti-HIV substance as (+)-calanolide A. Soon afterward the molecule was synthesized, and the synthetic proved as effective as the raw extract. Additional research revealed calanolide to be a powerful inhibitor of reverse transcriptase, an enzyme needed by the HIV virus to replicate itself within the human host cell. Studies are now underway to determine the suitability of calanolide for market distribution.

The exploration of wild biodiversity in the search for useful resources is called *bioprospecting*. Propelled by venture capital, it has in the past 10 years grown into a respectable industry within a global market hungry for new pharmaceuticals. It is also a means for discovering new food sources, fibers, petroleum substitutes, and other products. Sometimes bioprospectors screen many species of organisms in search of chemicals with particular qualities, such as antisepsis or cancer suppression. On other occasions bioprospecting is opportunistic, focusing on one of a few species that show signs of yielding a valuable resource. Ultimately, entire ecosystems will be prospected as a whole, and all of the species assayed for most or all of the products they can yield.

The extraction of wealth from an ecosystem can be destructive or benign. Dynamiting coral reefs and clearcutting forests yield fast profits but are unsustainable. Fishing coral reefs lightly and gathering wild fruit and resins in otherwise undisturbed forest are sustainable. Collecting samples of valuable species from rich ecosystems and cultivating them in bulk elsewhere, in biologically less favored areas, is not only profitable but the most sustainable of all.

Bioprospecting with minimal disturbance is the way of the future. Its promise can be envisioned with the following matrix for a hypothetical forest: To the left, make a list of the thousands of plant, animal, and microbial species, as many as you can, recognizing that the vast majority have not yet been examined, and many still lack even a scientific name. Along the top, prepare a horizontal row of the hundreds of functions imaginable for all the products of these species combined. The matrix itself is the combination of the two dimensions. The spaces filled within the matrix are the potential applications, whose nature remains almost wholly unknown.

The richness of biodiversity's bounty is reflected in the products already extracted by native peoples of the tropical forests, using local knowledge and low technology of a kind transmitted solely by demonstration and oral teaching. Here, for example, is a small selection of the most common medicinal plants used by tribes of the upper Amazon, whose knowledge has evolved from their combined experience with the more than 50,000 species of flowering plants native to the region: motelo sanango, *Abuta grandifolia* (snakebite, fever); dye plant, *Arrabidaea chica* (anemia, conjunctivitis); monkey ladder, *Bauhinia guianensis* (amoebic dysentery); Spanish needles, *Bidens alba* (mouth sores, toothache); firewood tree, or capirona, species of *Calycophyllum* and *Capirona* (diabetes, fungal infection); wormseed, *Chenopodium ambrosioides* (worm infection); caimito, *Chrysophyllum cainito* (mouth sores, fungal infection); toad vine, *Cissus sicyoides* (tumors); renaquilla, *Clusia rosea* (rheumatism, bone fractures); calabash, *Crescentia cujete* (toothache); milk tree, *Couma macrocarpa* (amoebic dysentery, skin inflammation); dragon's blood, *Croton lechleri* (hemorrhaging); fer-de-lance plant, *Dracontium loretense* (snakebite); swamp immortelle, *Erythrina fusca* (infections, malaria); wild mango, *Grias neuberthii* (tumors, dysentery); wild senna, *Senna reticulata* (bacterial infection).

Only a few of the thousands of such traditional medicinals used in tropical forests around the world have been tested by Western clinical methods. Even so, the most widely used already have commercial value rivaling that of farming and ranching. In 1992 a pair of economic botanists, Michael Balick and Robert Mendelsohn, demonstrated that single harvests of wild-grown medicinals from two tropical forest plots in Belize were worth $726 and $3,327 per hectare (2.5 acres) respectively, with labor costs thrown in. By comparison, other researchers estimated per hectare yield from tropical forest converted to farmland at $228 in nearby Guatemala and $339 in Brazil. The most productive Brazilian plantations of tropical pine could yield $3,184 per hectare from a single harvest.

In short, medicinal products from otherwise undisturbed tropical forests can be locally profitable, on condition that markets are developed and the extraction rate is kept low enough to be sustainable. And when plant and animal food products, fibers, carbon credit trades, and ecotourism are added to the mix, the commercial value of sustainable use can be boosted far higher.

Examples of the new economy in practice are growing in number. In the Petén region of Guatemala, about 6,000 families live comfortably by sustainable extraction of rainforest products. Their combined annual income is $4 million to $6 million, more than could be made by converting the forest into farms and cattle ranches. Ecotourism remains a promising but largely untapped additional resource.

Nature's pharmacopoeia has not gone unnoticed by industry strategists. They are well aware that even a single new molecule has the potential to recoup a large capital investment in bioprospecting and product development. The single greatest success to date was achieved with extremophile bacteria living in the boiling-hot thermal springs of Yellowstone National Park. In 1983 Cetus Corporation used one of the organisms, *Thermus aquaticus*, to produce a heat-resistant enzyme needed for DNA synthesis. The manufacturing process, called *polymerase chain reaction* (PCR), is today the foundation of rapid genetic mapping, a stanchion of the new molecular biology and medical genetics. By enabling microscopic amounts of DNA to be multiplied and typed, PCR also plays a key role in crime detection and forensic medicine. Cetus's patents on PCR technology, which have been upheld by the courts, are immensely profitable, with annual earnings now in excess of $200 million—and growing.

Bioprospecting can serve both mainstream economics and conservation when put on a firm contractual basis. In 1991, Merck signed an agreement with Costa Rica's National Institute of Biodiversity (INBio) to assist the search for new pharmaceuticals in Costa Rica's rainforests and other natural habitats. The first deposit was $1 million dispensed over two years, with two similar consecutive grants to follow. During the first period, the field collectors concentrated on plants, in the second on insects, and in the third on microorganisms. Merck is now working through the immense library of materials it gathered during the field program and testing and refining chemical extracts made from them.

Also in 1991, Syntex signed a contract with Chinese science academies to receive up to 10,000 plant extracts a year for pharmaceutical assays. In 1998, Diversa Corporation signed on with Yellowstone National Park to continue bioprospecting the hot springs for biochemicals from thermophilic microbes. Diversa pays the park $20,000 yearly to collect the organisms for study, as well as a fraction of the profits generated by commercial development. Funds returning to Yellowstone will be used to promote conservation of the unique microbes and their habitat, as well as basic scientific research and public education.

Still other agreements have been signed between NPS Pharmaceuticals and the government of Madagascar, between Pfizer and the New York Botanical Garden, and between the international company GlaxoSmithKline and a Brazilian pharmaceutical company, with part of the profits pledged to the support of Brazilian science.

Perhaps it is enough to argue that the preservation of the living world is necessary to our long-term material prosperity and health. But there is another, and in some ways deeper, reason not to let the natural world slip away. It has to do with the defining qualities and self-image of the human species. Suppose, for the sake of argument, that new species can one day be engineered and stable ecosystems built from them. With that distant prospect in mind, should we go ahead and, for short-term gain, allow the original species and ecosystems to be lost? Yes? Erase Earth's living history? Then also burn the art galleries, make cordwood of the musical instruments, pulp the musical scores, erase Shakespeare, Beethoven, and Goethe, and the Beatles too, because all these—or at least fairly good substitutes—can be re-created.

The issue, like all great decisions, is moral. Science and technology are what we can do; morality is what we agree we should or should not do. The ethic from which moral decisions spring is a norm or standard of behavior in support of a value, and value in turn depends on purpose. Purpose, whether personal or global, whether urged by conscience or graven in sacred script, expresses the image we hold of ourselves and our society. A conservation ethic is that which aims to pass on to future generations the best part of the nonhuman world. To know this world is to gain a proprietary attachment to it. To know it well is to love and take responsibility for it.

EDWARD O. WILSON *is Pellegrino University Research Professor and Honorary Curator in Entomology at Harvard University's Museum of Comparative Zoology. His books include* Sociobiology: The New Synthesis *(1975),* Consilience: The Unity of Knowledge *(1998), and two Pulitzer Prize winners,* On Human Nature *(1978) and* The Ants *(1990, with Burt Holldobler). This essay is taken from his latest book,* The Future of Life *(2002). Published by arrangement with Alfred A. Knopf, a division of Random House, Inc. Copyright: © 2002 By Edward O. Wilson.*

A Fragile Cornucopia: Assessing the Status of U.S. Biodiversity

Bruce A. Stein

A vast land of contrasts, the United States stretches from above the Arctic Circle to below the tropic of Cancer and spans nearly a third of the globe from eastern Maine to the tip of the Aleutian chain. This enormous expanse harbors a wide array of ecological conditions—from the lush forests of the Appalachians to California's thorny chaparral and from Alaska's frigid tundra to the parched desert of Death Valley. Although tropical rainforests come to mind when most people think of biodiversity—the variety of life—the United States itself contains a surprising diversity.

The Death Valley region, for instance, known for its scorching temperatures and austere landscapes, also holds some watery secrets. Devils Hole, an abrupt fissure in the desert floor, shelters the remnant of a huge lake that formed during the Ice Ages. In this 70-by-10-foot pool dwells the entire population of the Devils Hole pupfish (*Cyprinodon diabolis*), a tiny fish that descended from the lake's original inhabitants and has been isolated for more than 20,000 years. Subsisting on the algae that grow on the pool's single sunlit ledge, this fish has the distinction of

being the world's most narrowly distributed vertebrate species.

Such extreme rarity also makes the fish susceptible to even slight changes in its tiny oasis. Lowering the pool's water below the level of the algal-covered ledge would eliminate the fish's sole food source, leading inevitably to its extinction. Indeed, such a threat occurred in the late 1960s when owners of a nearby ranch began pumping irrigation water from the same aquifer that maintains Devils Hole. In a landmark decision concerning the 1973 Endangered Species Act (ESA), the U.S. Supreme Court ruled in favor of protecting the Devils Hole pupfish and limited the ground water extractions.[1]

Since that time, ESA has become a far-reaching legal tool for protecting the nation's threatened living resources. Conversely, ESA has been and continues to be a lightning rod for those concerned about private property rights and the scope and role of government. One of the few things that all sides in this contentious debate can agree on, however, is the need for sound scientific information on which to base decisions.

A National Status Assessment

What is the full scope of the nation's biological inheritance, and how is it faring? Which species and ecosystems are at greatest risk, and what is threatening them? Where are the most biologically significant places, and where should society direct its conservation efforts? To address these questions, the Association for Biodiversity Information (ABI) and The Nature Conservancy carried out a major assessment of the nation's species and ecosystems, recently published as the book *Precious Heritage: The Status of Biodiversity in the United States*.[2] By drawing together a quarter century of information gathered by the state-based natural heritage programs, this study provides the first-ever comprehensive view of the state of the nation's biota. (For information on ABI and its web site, see the boxes "Knowledge to Protect the Diversity of Life" and "NatureServe: An Online Resource for More Information".)

Biodiversity is a word that gained great currency during the 1990s, yet the concept is still foreign to most people. In its most elemental form, biodiversity refers to the full array of life on Earth. Although plant and animal species are the most tangible manifestations of biodiversity, the concept covers the full hierarchy of biological organization recognized by scientists, including genetic, species, and ecosystem levels. The Devils Hole pupfish is just one example of the extraordinary diversity of species in the United States. A host of other biological superlatives can also be found within the 50 states. The redwood of coastal northern California, for example, is the world's tallest tree, with individual specimens rising as high as 35-story office buildings. What is currently thought to be the world's most massive organism is a 107-acre aspen grove in Utah. Because each trunk is connected by a common root system, the entire grove represents a single, genetic individual.[3] Other oddities and peculiarities abound among the national flora and fauna, including yard-long salamanders, frogs with "antifreeze" in their blood that enables them to survive freezing and thawing, and birds that travel 25,000 miles roundtrip on their annual migration.

A Catalog of U.S. Biodiversity

In an age when exploration of distant planets seems routine, many take it for granted that we have done an adequate job of discovering what exists here on Earth. Nothing could be farther from the truth. The mysteries of life on our own living planet are still profound and none more so than determining the most fundamental question of how many species of plants, animals, and microorganisms exist. A more complete knowledge of the diversity of life is essential to understanding how ecosystems function and evaluating the health of our environment. Only about 1.75 million species have been studied well enough to receive a scientific name. Yet estimates of the total number of living species span more than an order of magnitude, from around 3 million species to more than 100 million, although a working estimate hovers around 14 million.[4] (This enormous range in estimates is due to emerging research that points to previously under-appreciated reservoirs of life such as tropical tree canopies, deep sea floors, and soil generally.)

The United States is one of the best studied nations on Earth, but, even here, producing a tally of life is extremely difficult. Surveying the scientific literature and querying taxonomic specialists reveals that in excess of 200,000 native species are currently known to inhabit the United States.[5] This figure is conservative, including only formally named species and leaving out such poorly known groups as bacteria, protists, or viruses. Still, this represents more than one-tenth of all scientifically documented species on Earth. Although these figures accentuate the richness of the U.S. biota, they also call attention to the disparity between cataloging efforts in the temperate regions, where most biologists live, and the tropics, where most of the world's species are thought to reside.

Even taking into account the disparity in inventory efforts, the United States emerges as an exceptionally rich country biologically and a world leader in the diversity of certain groups of organisms (see Table 1). Four out of every ten salamander species, for example, are found in the United States, the most of any country on Earth. Almost one-

Knowledge to Protect the Diversity of Life

The Association for Biodiversity Information (ABI) is a new nonprofit organization working in partnership with the network of state natural heritage programs. Dedicated to providing the knowledge necessary to protect the diversity of life, ABI reflects a continuation of The Nature Conservancy's 25-year commitment to science-based conservation. Representing a unique institutional and scientific collaboration, ABI and its natural heritage program members are the leading source for detailed information about the condition and location of rare and endangered species and threatened ecosystems. Network programs operate in all 50 U.S. states, across Canada, and in a dozen countries of Latin America and the Caribbean, with each center maintaining detailed maps and computer records about the species and ecosystems that are of greatest conservation concern within their state. This information and expertise is provided to thousands of users annually to assist in project planning, environmental review, and conservation efforts. ABI serves as the network's coordinating body, providing scientific and technical support to ensure consistency of information across state boundaries. ABI also provides a single access point for heritage data and expertise, particularly at regional and national scales. To improve society's understanding of and ability to protect biodiversity, ABI provides policy makers and the public with a variety of objective and credible information products, such as *Precious Heritage* and the recently launched NatureServe web site (see "NatureServe: An On-Line Resource for More Information"). (For more information about ABI or natural heritage programs, visit http://www.abi.org.)

NatureServe: An On-line Resource for More Information

A new web site launched by the Association for Biodiversity Information (ABI) provides an easily accessible source for information about the thousands of U.S. plants and animals that are summarized in this article. NatureServe: An On-line Encyclopedia of Life (http://www.natureserve.org) offers access to ABI's comprehensive databases on more than 50,000 U.S. and Canadian species and ecological communities. The site provides in-depth information about rare and endangered species as well as information about common plants and animals. NatureServe can be used to learn about a particular species or can be queried to learn, for instance, which endangered mammals or butterflies are found in a given state. The site includes scientific and common names, conservation and legal status, color-coded distribution maps, and summaries of life histories and conservation needs. This new web site puts details about our rich natural heritage within reach of everyone.

---

third of the world's freshwater mussel species (about 300) reside in the United States. By comparison, all the rivers of Europe have just 10 mussel species. Of all the nations in the world, the United States has the most species of freshwater crayfishes (61 percent), freshwater turtles (22 percent), and freshwater snails (17 percent). Several other groups of organisms are well represented in the United States. U.S. mammals rank sixth in global diversity, due largely to the diverse fauna of the arid lands in the Southwest. Although the tropics have the greatest diversity of freshwater fishes, the United States has by far the highest diversity of these important organisms among temperate countries. Among plants, the number of U.S. gymnosperms—including conifers such as pines, firs, and cypress—is second only to China.

The United States is also remarkably diverse at the scale of ecosystems. Several systems exist for dividing the globe into major ecological zones. To assist the World Conservation Union in targeting the placement of protected areas, in the 1970s, the now late University of California professor Miklos Udvardy mapped out biogeographic provinces worldwide, depicting distinctive floral and faunal assemblages. Each of these provinces was classed according to one of 14 biomes, representing major ecosystem groups such as temperate grasslands or tropical dry forests.[6] In

another effort to classify large-scale global ecosystems, U.S. Forest Service ecologist Robert Bailey identified large ecologically defined areas that share similar vegetation and climate—termed ecoregions.[7] Interestingly, the United States has a larger number of both biomes and ecoregions than any other country on Earth.[8] Certain globally significant large-scale ecosystems are especially well represented in the United States, including temperate broadleaf forest and prairies. Coastal California's shrubby vegetation is particularly distinctive, representing one of only five examples worldwide of Mediterranean-climate vegetation.

The Condition of U.S. Species

How is this extraordinary array of species faring? The Association for Biodiversity Information and its affiliated natural heritage programs assess the conservation status of species based on several factors linked to increases in risk of extinction.[9] These criteria include the total number of individuals of a species, the number of distinct populations across which these individuals are distributed, the viability of these populations, and the short- and long-term trends for the species. Conservation status ranks are assigned on a scale from one to five, ranging from critically imperiled (G1) to demonstrably secure (G5)

(see Table 2 for more details on the ranking system). In general, species classified as vulnerable (G3) or rarer may be considered to be "at risk."

Given the paucity of information about most of the nation's more than 200,000 species, any attempt to characterize the status of the biota overall will necessarily be incomplete. Nonetheless, ABI and natural heritage program scientists have assessed the status of more than 30,000 U.S. species and subspecies. These assessments are comprehensive for 14 of the best-known groups of plants and animals—that is, the conservation status has been evaluated for each and every species in these groups. These 14 groups, which include all vertebrates and vascular plants, represent about 20,900 species (see Table 3). These assessments begin to paint an overall picture of the condition of wildlife in America. And that picture is not particularly pretty.

Overall, a surprisingly high one-third of the native U.S. flora and fauna is at risk, with 16 percent considered vulnerable, 8 percent imperiled, 7 percent critically imperiled, and I percent missing or extinct. The proportion of species at risk varies greatly from one group of plants and animals to another (see Figure 1). Freshwater mussels in the United States show the highest risk levels, with nearly 70 percent in trouble. Indeed, one in ten mussels are already missing or extinct (listed as GX (presumed extinct) or GH (possibly ex-

Table 1. Global significance of selected U.S. plant and animal groups				
Taxonomic group	Number of U.S. species	Number of species worldwide	% of global species in U.S.	U.S. ranking worldwide
Mammals	416	4,600	9	6
Birds	768	9,700	8	27
Reptiles	283	6,600	4	14
Amphibians	231	4,400	5	12
Freshwater fishes	799	8,400	10	7
Freshwater mussels	292	1,000	29	1
Freshwater snails	661	4,000	17	1
Crayfishes	322	525	61	1
Tiger beetles	114	2,000	6	7
Dragonflies/ Damselflies	456	5,800	8	?
Butterflies/ Skippers	620	17,500	4	?
Flowering plants	15,320	235,000	7	>10
Gymnosperms	114	760	15	2
Ferns	556	12,000	5	>15

NOTE: Several species have their highest levels of diversity in the United States, Including Freshwater Mussels, Freshwater Snails, And Crayfishes: Several Other Taxonomic Groups, Such As Freshwater Fishes And Gymnosperms, Are Also Well Represented In The United States.

SOURCE: B. A. Stein, L. S. Kutner, and J. S. Adams, eds., Precious Heritage: The Status of Biodiversity in the United States (New York: Oxford University Press, 2000), Table 3.2, 67.

tinct)). Several other freshwater groups have exceptionally high risk levels, including crayfish (51 percent), stoneflies (43 percent), and freshwater fish (37 percent). In contrast, several vertebrate groups appear to be on relatively secure footing, at least at the full species level measured here. This includes the two groups of animals—birds and mammals—that receive the majority of public conservation attention. Of all groups considered, birds are best off, with only 14 percent of their species at risk.

Focusing on the number of species at risk in each group tells a different story. On a proportional basis, species depending on freshwater habitats appear to be at greatest risk. Based on the actual number of species at risk, however, the largely terrestrial flowering plants dominate by a huge margin, with 33 percent of the more than 15,300 native U.S. species—a sobering 5,090 species—at risk.

The Legacy of Extinctions

The Devils Hole pupfish's brush with extinction was staved off thanks to actions of the nation's highest court in 1976. Another small fish living not far from Devils Hole was not so lucky. The Ash Meadows poolfish (Empetrichthys merriami) was first discovered in 1891. Never abundant, the fish lived in just five isolated desert springs. Probably due to the introduction of two voracious predators—bullfrogs and crayfish—the species slipped into oblivion sometime between 1948, when it was last seen, and 1953, when researchers were unable to locate it. Like the Ash Meadows poolfish, a number of declining U.S. species never had friends in high places to guard over them and consequently have suffered the ultimate demise. Overall, at least 100 U.s. species have gone extinct since Euro-

pean colonization of North America and Hawaii and nearly 440 more are missing and may be extinct.[10]

These extinctions span the gamut of organisms, including vertebrates such as the great auk (Pinguinus impennis), plants like the Santa Catalina monkeyflower (Mimulus traskiae), and invertebrates such as the Wabash riffleshell mussel (Epioblasmna sampsonii). Snails have been particularly hard hit by extinctions; with 26 species presumed extinct and another 106 species missing and possibly extinct, gastropods lead all other groups in this unenviable category. Among vertebrates, birds have been most severely affected by extinctions, with 22 species of birds presumed extinct and another 3 missing. Although several bird species have disappeared from the mainland United States—the passenger pigeon, Carolina parakeet, Labrador duck, and great auk—most extinct birds are Hawaiian. A considerable number of plants—11 extinct and 130 missing—are also gone from the U.S. landscape.

Table 2. Definition of conservation status ranks

GX	Presumed Extinct: not located despite intensive searches
GH	Possibly Extinct: of historical occurrence; missing but still some hope of rediscovery
G1	Critically Imperiled: typically 5 or fewer occurrences or 1,000 or fewer individuals
G2	Imperiled: typically 6 to 20 occurrences or 1,000 to 3,000 individuals
G3	Vulnerable: rare; typically 21 to 100 occurrences and 10,000 individuals
G4	Apparently Secure: uncommon but not rare, some cause for long-term concern; usually more than 100 occurrences and 10,000 individuals
G5	Secure: common; widespread, and abundant

NOTE: The conservation status of a given species is assessed based on several factors using a one to five scale. This table summarizes key assessment criteria. "G" refers to the global or rangewide status of a species. Both national (N) and state (S) status ranks are also assessed.

SOURCE: B. A. Stein, L. S. Kutner, and J. S. Adams, eds., Precious Heritage: The Status of Biodiversity in the United States (New York: Oxford University Press, 2000), Table 4.2, 97.

Although extinctions have touched every state in the nation, certain regions have lost disproportionate numbers of species. States with large numbers of extinctions tend to have either high overall species numbers, an inherently fragile flora and fauna, or intense human alteration of the landscape. Hawaii, not surprisingly, has suffered the gravest losses, with 249 extinct species (29 presumed and 220 possibly extinct). The oceanic isolation of the Hawaiian Islands has produced one of the world's most distinctive biotas and one that evolved largely in the absence of aggressive continental predators. The native Hawaiian species have proven to be extremely susceptible to the kinds of outside influences introduced first by the Polynesian immigration (between 200 and 500 A.D.) and later by European colonists.

On the mainland, Alabama tops the list of extinction-prone states, with 96 species gone (22 presumed and 74 possibly extinct). Home to an exceptionally rich variety of freshwater fauna, many of the waterways in Alabama have been dammed, dredged, or diverted, leading to the loss of numerous snails, mussels, and fish. California ranks third in the nation with 35 extinctions (11 presumed and 24 possibly extinct). The intensive conversion of the state's lands and waters for agriculture, urbanization, and other uses has had a severe impact on the many restricted-range species that have evolved in this ecologically unique state.

State of the States

The ecological complexion of the 50 states varies dramatically, reflected by the composition and character of the plants and animals inhabiting each state. The East is dominated by wide coastal plains and the ancient remains of the north-south trending Appalachian chain. Stretching across the Midwest and the Great Plains, the continent's vast interior exhibits relatively little topographic relief. In contrast, the western third of the country is a welter of mountains and valleys resulting from relatively recent (from a geological perspective) flurries of mountain-building. Topography interacts with and influences climatic factors, which in turn combine with regional evolutionary histories to produce distinctive assemblages of plants and animals.

The areas with the greatest species diversity are found in the topographically and climatically diverse Southwest, with California and Texas leading the nation in the number of species.[11] This is due in part to the enormous size of these two states, in part to their southern location (a general ecological principle is that species diversity increases as one moves toward the equator), and partly because of their ecological complexity. Because of California's benign climate and extraordinary diversity of unique habitats, the state is often referred to as an ecological island, attached to but discrete from the rest of the continent. Centrally situated, Texas straddles several major ecological zones—the Great Plains, the southwestern deserts, the humid southeastern coastal plain, and even a touch of the Mexican subtropics—each contributing their own distinctive species to the state's mix. (For a map of state patterns of diversity, see Figure 2.)

Another way to look at the biological significance of different states is to consider those species that are restricted—or endemic—to a single state (see Figure 2). California, with almost 1,500 endemic species, leads the nation in this category, highlighting the ecological distinctiveness of the Golden State. Hawaii is the other state that stands out for its number of endemic species. Because of the island chain's extreme isolation, most plants and animals native to the archipelago are descended from a relatively few colonists. A mere 15 original colonists, for example, may account for all 53 species of endemic birds, including the famous honeycreepers.[12] As a result, Hawaii has some of the highest levels of endemism of any place on Earth. Forty-three percent of Hawaii's native vertebrates are endemic to the state, as are 87 percent of its vascular plants and an astonishing 97 percent of its insects.

Risk patterns among states—as reflected by the proportion of a state's species that are considered imperiled or vulnerable—also highlight Hawaii and California (see Figure 2). Secondary centers of rarity are found in several other western and southeastern states. The upper Midwest, in turn, shows relatively low levels of rarity, a condition due in part to the lingering effects of Ice

Table 3. Species report card: Status of U.S. plants and animals

		Presumed Extinct (GX)	Possibly Extinct (GH)	Critically Imperiled (G1)	Imperiled (G2)	Vulnerable (G3)	Apparently Secure (G4)	Secure (G5)	Other	Total
Vertebrate animals										
	Mammals	1	0	8	21	36	96	253	1	416
	Birds	22	3	27	20	36	87	572	1	768
	Reptiles	0	0	7	14	30	45	186	1	283
	Amphibians	1	1	21	28	33	42	104	1	231
	Freshwater fishes	16	1	91	87	105	147	351	1	799
Vertebrate totals		40	5	154	170	240	417	1,466	5	2,497
Invertebrate animals										
	Butterflies/Skippers	0	0	8	31	78	101	388	14	620
	Crayfishes	1	2	54	53	55	87	69	1	322
	Freshwater mussels	17	20	73	42	50	44	42	4	292
	Dragonflies/Damselflies	0	2	10	24	45	91	268	16	456
	Stoneflies	1	11	13	71	164	168	178	0	606
	Tiger beetles	0	0	3	5	14	19	62	11	114
Invertebrate totals		19	35	161	226	406	510	1,007	46	2,410
Vascular plants										
	Ferns/Fern allies	0	4	32	24	65	168	242	21	556
	Gymnosperms	0	0	7	8	12	27	60	0	114
	Flowering plants	11	126	1,031	1,309	2,615	4,469	5,618	141	15,320
Vascular plant totals		11	130	1,070	1,341	2,692	4,664	5,920	162	15,990
Totals		70	170	1,385	1,737	3,338	5,591	8,393	213	20,897

SOURCE: B. A. Stein, L. S. Kutner and J. S. Adams, eds., Precious Heritage: The Status of Biodiversity in the United States (New York: Oxford University Press, 2000), Table 4.4, 104.

Age glaciations and the resulting wide ranges of most midwestern species. Rare plants and animals in this region tend to be widespread but spotty in their distribution, often as a result of large-scale agricultural conversion of their habitat. Mead's milkweed (*Asciepias meadii*), for instance, formerly ranged throughout the tallgrass prairie but is now restricted to about 100 sites.

State patterns for specific groups of organisms can be very revealing and may differ significantly from the general patterns. The Southeast has been a center of evolution for freshwater fish, giving rise to a plethora of species like darters and mad toms. Maps of fish diversity reflect the large number of species in states such as Tennessee, Alabama, and Georgia (see Figure 3). Patterns of rarity, however, accentuate the arid states of the Southwest (see Figure 3). Although this region has far fewer rivers, streams, and lakes, their very scarcity has contributed to the evolution of numerous narrowly restricted species in groups such as pupfish and suckers. Most of the major rivers in the Southwest have also been extensively dammed or otherwise altered, leading to serious declines in even once abundant and wide-ranging fish species. The Colorado pikeminnow (*Ptycliocheilus lucius*), a "minnow" capable of reaching six feet in length, was once abundant throughout the major riv-

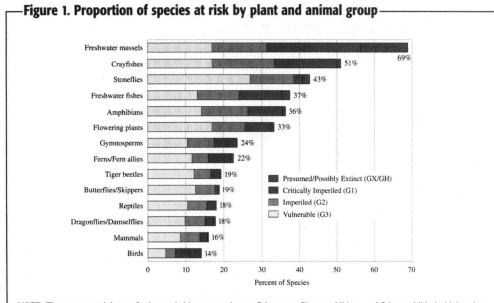

Figure 1. Proportion of species at risk by plant and animal group

Legend:
- Presumed/Possibly Extinct (GX/GH)
- Critically Imperiled (G1)
- Imperiled (G2)
- Vulnerable (G3)

(X-axis: Percent of Species, 0 to 70)

- Freshwater mussels — 69%
- Crayfishes — 51%
- Stoneflies — 43%
- Freshwater fishes — 37%
- Amphibians — 36%
- Flowering plants — 33%
- Gymnosperms — 24%
- Ferns/Fern allies — 22%
- Tiger beetles — 19%
- Butterflies/Skippers — 19%
- Reptiles — 18%
- Dragonflies/Damselflies — 18%
- Mammals — 16%
- Birds — 14%

NOTE: Those groups relying on freshwater habitats—mussels, crayfishes, stoneflies, amphibians, and fishes—exhibit the highest levels of risk.

SOURCE: B. A. Stein, L. S. Kutner, and J. S. Adams, eds., *Precious Heritage: The Status of Biodiversity in the United States* (New York: Oxford University Press, 2000), Figure 4.2, 102 (figure © The Nature Conservancy and Association for Biodiversity Information, reprinted with permission).

ers of the Colorado River basin. Due to changes in water flow and temperature brought about by dam operations, this magnificent fish now occupies barely a quarter of its former habitat.

Hot Spots of Imperilment

State-level assessments are useful for broadly identifying the magnitude of the conservation challenge in different regions, but to accomplish on-the-ground conservation and to minimize inadvertent damage to sensitive resources, it is necessary to have a much finer level of knowledge about the distribution of these species. Conducting field inventories and mapping the precise localities for rare and endangered species is a hallmark of the Association for Biodiversity Information (ABI) and its natural heritage program members. Collectively, the natural heritage programs maintain databases with nearly half a million detailed locality records for rare and endangered species and threatened ecological communities. This information is routinely used by government agencies, industry, landowners, and conservationists to improve the environmental sensitivity of development projects and to target conservation activities. *Precious Heritage* is the first project to bring together this detailed information to create a truly national view of imperiled species in the United States.

A striking picture of biodiversity hot spots in United States emerges from mapping imperiled species against a uniform grid (see Figure 4).[13] Charting the rarest of the rare—approximately 2,800 imperiled and critically imperiled species—this map depicts the number of different imperiled species in each 640,000-acre hexagonal cell, an area about half the size of the typical eastern U.S. county. Most previous maps showing the distribution of U.S. endangered species have been based on county distributions, and therefore suffer distortions from the huge disparity in size among counties, especially in the West.[14] By using this equal-area hexagon grid, size differences arid edge effects are eliminated, leading to a more accurate picture of rarity.

Concentrations of imperiled species are particularly prominent in four major regions: Hawaii, Califor-nia, the southern Appalachians, and Florida. Hawaii, in particular, has extraordinarily high concentrations of imperiled species, including a single hexagon centered on the Alakai Swamp that contains 128 different imperiled species. The highest concentration of imperiled species on the mainland is a hexagon centering on the Clinch and Powell Rivers in southwestern Virginia that contains 27 imperiled species, a large number of which are freshwater mussels.

Mapping these imperiled species by ecoregion provides a complementary view of biodiversity hot spots (see Figure 5).[15] Ecoregions consist of large, ecologically defined zones that share similar climate, vegetation, ecological processes, and suites of species. This view of imperiled species confirms the importance of coastal California and the Appalachian region but also highlights another region—the Great Basin, which stretches across Nevada and into Utah. Many of this region's isolated mountain ranges harbor at least a few highly localized species. Thus, even though few places have large concentrations of imperiled species, in combination these mountain range-restricted rarities elevate

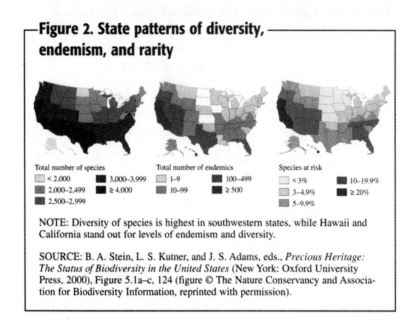

Figure 2. State patterns of diversity, endemism, and rarity

Total number of species
- < 2,000
- 2,000–2,499
- 2,500–2,999
- 3,000–3,999
- ≥ 4,000

Total number of endemics
- 1–9
- 10–99
- 100–499
- ≥ 500

Species at risk
- < 3%
- 3–4.9%
- 5–9.9%
- 10–19.9%
- ≥ 20%

NOTE: Diversity of species is highest in southwestern states, while Hawaii and California stand out for levels of endemism and diversity.

SOURCE: B. A. Stein, L. S. Kutner, and J. S. Adams, eds., *Precious Heritage: The Status of Biodiversity in the United States* (New York: Oxford University Press, 2000), Figure 5.1a–c, 124 (figure © The Nature Conservancy and Association for Biodiversity Information, reprinted with permission).

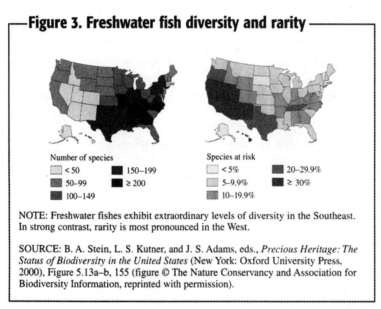

Figure 3. Freshwater fish diversity and rarity

Number of species
- < 50
- 50–99
- 100–149
- 150–199
- ≥ 200

Species at risk
- < 5%
- 5–9.9%
- 10–19.9%
- 20–29.9%
- ≥ 30%

NOTE: Freshwater fishes exhibit extraordinary levels of diversity in the Southeast. In strong contrast, rarity is most pronounced in the West.

SOURCE: B. A. Stein, L. S. Kutner, and J. S. Adams, eds., *Precious Heritage: The Status of Biodiversity in the United States* (New York: Oxford University Press, 2000), Figure 5.13a–b, 155 (figure © The Nature Conservancy and Association for Biodiversity Information, reprinted with permission).

the Great Basin as an ecoregional hot spot.

Implications for the Endangered Species Act

Many of the listed threatened and endangered species that command the highest public attention—and consume the major share of Endangered Species Act funding—are relatively wide-ranging species, such as the grizzly bear, the northern spotted owl, and the red-cockaded woodpecker. Some listed animals, like the desert tortoise, are also still quite abundant and listed primarily because of steep declines in their populations. The process for listing species as threatened or endangered has been criticized by some as not being based on sufficiently clear guidelines, with the potential for poor listing decisions.[16] A legitimate question, then, is to what extent such wide-ranging and relatively abundant species are representative of the endangered species list as a whole?

This relates to a current controversy regarding ESA: how to determine listing priorities. Specifically, how can the listing priorities of the implementing agencies be balanced with court-ordered priorities generated by lawsuits? The environmental community has been very effective at using the courts to focus listing activities on certain species, often with extensive habitat requirements, while both the Clinton and current Bush administrations have argued in favor of allowing the U.S. Fish and Wildlife Service greater flexibility in allocating listing funds towards internally derived priorities.

One approach for determining how rare a species needs to be before making it onto the federal endan-

Figure 4. Hot spots of rarity in the United States

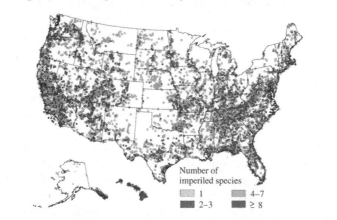

Number of
imperiled species

☐ 1   ☐ 4–7
☐ 2–3   ☐ ≥ 8

NOTE: A striking picture emerges from mapping the rarest of rare species, based on field data gathered by state natural heritage programs. Particular concentrations of imperiled species occur in California, Hawaii, Florida, and the Southern Appalachians.

SOURCE: B. A. Stein, L. S. Kutner, and J. S. Adams, eds., *Precious Heritage: The Status of Biodiversity in the United States* (New York: Oxford University Press, 2000), Figure 6.6, 169 (figure © The Nature Conservancy and Association for Biodiversity Information, reprinted with permission).

Figure 5. Rarity and ecological regions

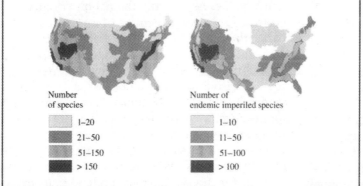

Number
of species

☐ 1–20
☐ 21–50
☐ 51–150
■ > 150

Number of
endemic imperiled species

☐ 1–10
☐ 11–50
☐ 51–100
■ > 100

NOTE: Ecologically defined regions (ecoregions) offer another view of imperiled species patterns, with the Great Basin joining coastal California and the Southern Appalachians as important areas.

SOURCE: B. A. Stein, L. S. Kutner, and J. S. Adams, eds., *Precious Heritage: The Status of Biodiversity in the United States* (New York: Oxford University Press, 2000), Figure 6.5a–b, 168 (figure © The Nature Conservancy and Association for Biodiversity Information, reprinted with permission).

gered species list is to consider how listed species correspond to ABI conservation status assessments (see Table 2). The vast majority (92 percent) of all federally listed species (including subspecies and distinct populations) are categorized by ABI as imperiled, critically imperiled, or historical (G1, G2, or GH, respectively).[17] Only 6 percent of listed species fall into the vulnerable category (G3), with fewer than 2 percent

in the more secure categories of G4 and G5. Those few species that are regarded by ABI as "secure or apparently secure" (G4 or G5) are for the most part wide-ranging vertebrates, for which the listing affects only specific populations. This analysis confirms that most species are already exceptionally rare and restricted in their distributions before making it onto the federal endangered species list.

The entry bar for inclusion on the list, however, seems to be more stringent for plants and invertebrates than for vertebrate animals. Whereas 5 percent of listed vertebrates are ranked as secure or apparently secure (G4 or G5) and 13 percent as vulnerable (G3), no listed plants or invertebrates fall into the secure categories and very few (5 percent and 0.7 percent, respectively) are regarded as vulnerable (G3).[18] These figures are consistent with earlier studies that have found that plants and invertebrates tend to be considerably more rare than vertebrates before they are considered for listing.[19]

This analysis also confirms that a biological basis underlies whether species are listed as endangered (formally defined as "in danger of extinction within the foreseeable future") or in the lesser category of threatened (defined as "likely to become endangered within the foreseeable future").[20] From these definitions one would predict that species listed as endangered should receive higher ABI ranks than those listed as threatened. In fact, this is the case: Almost three-quarters (74 percent) of listed endangered species are regarded as critically imperiled by ABI, while only 37 percent of listed threatened species are in that category.[21]

The distribution of federally listed species relative to different land ownership patterns also has important implications for how ESA is administered. For example, vertebrate animals receive protection from being killed or harmed even on private lands, while plants and invertebrates are fully protected only on federal lands. In documenting the locations for imperiled and endangered species, natural heritage programs keep track of the type of landowner on whose property the species occurs. Analyzing this data from a national perspective indicates that federal lands support examples of about three-fifths (59 percent) of federally listed species. Interestingly, Department of Defense (DOD) lands contain the most federally listed species of any federal agency, supporting examples of about one-fifth (21 percent) of all listed species. This finding is particularly striking, given that DOD lands represent just 3 percent of the overall federal estate. Many military bases turn out to be strategically placed, not just from a military standpoint, but also from a biological perspective. Often found in coastal areas with fast growing human populations, many of DOD land holdings, such as southern Cal-

ifornia's Camp Pendelton Marine Base, are becoming islands of natural habitat in rapidly urbanizing regions.

Federal land management agencies have special responsibilities for protecting federally listed species, and there are numerous legal mechanisms in place to promote such public stewardship. However, given that about 40 percent of federally listed species are nowhere found on the federal lands, protection for these species will necessarily need to focus on private lands and lands managed by state and local governments.

Beyond Fences

Historically, conservation efforts in America have targeted a fairly narrow set of the nation's biological wealth. These efforts were often opportunistic and tended to focus on scenic lands, often of limited economic value. As a result, the nation is well endowed with "rock and ice" parks that offer spectacular vistas and inspiring recreational experiences. These parks are not, however, representative of the biota as a whole. Yellowstone National Park, for instance, was originally set aside for its spectacular geysers and other geological features rather than its equally impressive wildlife and ecosystem functions. Current understanding of biodiversity and particularly the large-scale ecological processes necessary to sustain it, suggests the need for a more biologically rational strategy for targeting conservation efforts and for strategies that address larger geographic scales.

A chain-link fence encircles the tiny oasis sheltering the Devils Hole pupfish. While providing some protection against vandals, this fence is as much psychological as practical, because the most serious threats to the species' existence come from offsite. Fences, no matter how high, are ineffective against regional-scale threats, like the drawdown of aquifers. Conservationists now under-

stand that even for the most rare and localized species—and none are more so than this little fish—protection efforts must go well beyond the fences, whether symbolic or physical, that bound our parks and nature preserves. Indeed, to sustain and restore the nation's biodiversity over the long term, protection efforts will need to be planned for and undertaken at the scale of whole landscapes and regions.

Twenty-five years ago The Nature Conservancy began creating a mechanism to improve its own ability to rationally identify and protect ecologically significant lands. The organization's efforts to establish natural heritage programs in each of the 50 states set in motion a process for informing not just the Nature Conservancy's conservation priorities, but those of society at large. Building on this foundation, the conservation community is now focusing on the need to get ahead of the extinction curve by considering the scale of habitat conservation that will be needed to protect not just individual species but whole ecological systems. Although the Endangered Species Act is a powerful legal tool and a formal testament to U.S. society's concern for its living inheritance, it is designed to function as a safety net rather than the much broader support scaffolding that will be required to protect the full array of the nation's biodiversity.

Building such a robust conservation infrastructure will require identifying those lands and waters that will be most important for maintaining and restoring the nation's species and ecosystems. Such activities are already under way in numerous organizations working at local, state, and ecoregional scales. One thing is already clear though: Approaching biodiversity conservation through land acquisition alone would be extraordinarily costly and probably not possible. Fortunately, many of the lands important to biodiversity are currently in uses that could be compatible with conservation values if they were managed responsibly.

Across the nation, innovative collaborations involving farmers, ranchers, and loggers are demonstrating that land can be managed in a way that produces economic benefits as well as maintains ecosystem values. For example, in the Malpais borderlands region of New Mexico, local ranchers have banded together and are working with conservationists and government agencies to improve the quality of the regional ecosystem so that they can continue to draw their livelihood from ranching. Indeed, the future of biodiversity in the United States may well rely as much on how we manage the working landscape (those lands in some form of economic production) as on how much land is strictly protected. Developing effective incentives for private landowners to manage their lands in biodiversity-friendly ways must be a major policy focus in the next few years. In the debate over endangered species protection, incentive-based approaches are at least one thing on which fairly broad consensus exists.

As the scientific community learns more about how natural ecosystems work, conservationists inevitably must think on broader scales of both space and time. To be successful, though, these grand visions must be translated into action at particular places inhabited by real people. One of the great challenges is to link local biodiversity concerns with national and global priorities. The assessment of U.S. species and ecosystems contained in *Precious Heritage: The Status of Biodiversity in the United States* confirms that there is globally significant biodiversity in our backyard and that protecting this inheritance is an obligation that the nation cannot afford to take lightly. Yet as the Devils Hole pupfish and its tiny range reminds us, the implications of even our most local actions can and do have global resonance.

NOTES:

1. J. E. Deacon and C. D. Williams, "Ash Meadows and the Legacy of the Devils Hole Pupfish," in W. L. Minckley and J. E. Deacon, eds., *Battle against Extinction: Native Fish Management in the American West* (Tucson: University of Arizona Press, 1991), 69–87.

2. B. A. Stein, L. S. Kutner, and J. S. Adams, eds., *Precious Heritage: The Status of Biodiversity in the United States* (New York: Oxford University Press, 2000).

3. M. C. Grant, J. B. Mitton, and Y. B. Linhart, "Even Larger Organisms," *Nature* 360 (1992): 216.

4. P. M. Hammond, "The Current Magnitude of Biodiversity," in V. Heywood. ed., *Global Biodiversity Assessment* (Cambridge: Cambridge University Press, 1995). 113–38.

5. Stein, Kutner, Adams, eds., note 2 above, pages 62–3.

6. M. D. F. Udvardy, "A Classification of the Biogeographic Provinces of the World," *IUCN Occasional Paper*, no. 18 (Gland, Switzerland: IUCN, 1975).

7. R. G. Bailey, *Ecoregions of tire Continents*, Map 1:30,000,000 (Washington D.C.: U.S. Department of Agriculture, Forest Service, 1989).

8. Stein, Kutner, Adams, eds., note 2 above, page 207.

9. For a discussion of species status assessments, see B. A. Stein and S. R. Flack, "Conservation Priorities: The State of U.S. Plants and Animals." *Environment* May 1997, 6–11, 34–9.

10. Stein, Kutner, Adams, eds., note 2 above, pages 112–6.

11. Ibid., pages 123–5. Figures for overall species diversity, endemism, and rarity are based on an analysis of a total of 19,279 species, which includes all native vascular plants, vertebrates, mussels, and crayfish.

12. A. J. Berger. Hawaiian Birdlife (Honolulu: University Press of Hawaii, 1981), 15.

13. Stein, Kutner, Adams, eds., note 2 above, pages 169–70.

14. For two recent county-level analyses of endangered species distributions, see A. P. Dobson, J. P. Rodriguez, W. M. Roberts, and D. S. Wilcove, "Geographic Distribution of Endangered Species in the United States," *Science* 275, 24 January 1997, 550–53; and C. H. Flather, M. S. Knowles, and I. A. Kendall, "Threatened and Endangered Species: Geography Characteristics of Hot Spots in the Conterminous United States," *BioScience* 48, no. 5 (1998): 365–76.

15. Stein, Kutner, Adams, eds., note 2 above, pages 166–8. The ecoregions displayed here are those recognized by The Nature Conservancy, which are based initially on those defined by R. Bailey of the U.S. Forest Service (see note 7 above).

16. A. Easter-Pilcher, "Implementing the Endangered Species Act: Assessing the Listing of Species as Endangered or Threatened," *BioScience* 46, no. 5 (1996): 355–63.

17. Stein, Kutner, Adams, eds., note 2 above, page 109, This analysis is based on the 1,048 U.S. species listed an of April 1998.

18. Ibid., pages 109–10.

19. D. S. Wilcove, M. McMillan, and K. C. Winston, "What Exactly Is an Endangered Species? An Analysis of the U.S. Endangered Species List, 1985–1991," *Conservation Biology* 7, no. 1 (1993): 87–93.

20. U.S. Fish and Wildlife Service, *Endangered Species Act of 1973 as Amended through the 100th Congress* (Washington, D.C., 1988).

21. Stein, Kutner, Adams, eds., note 2 above, pages 109–10.

Bruce A. Stein is vice president for programs with the Association for Biodiversity Information and was formerly a senior scientist with The Nature Conservancy. He was a coauthor and lead editor of *Precious Heritage: The Status of Biodiversity in the United States* (Oxford University Press. 2000), from which this article draws. A tropical botanist by training, Stein's current work focuses on evaluating the condition of the nation's ecosystems and working with the network of natural heritage programs to make biological and ecological information more accessible to the public and to environmental decision makers. He may be reached at the Association for Biodiversity Information, 1101 Wilson Boulevard, 15th floor, Arlington, VA 22209 (telephone: 703-908-1800; email: bruce_stein@abi.org).

Invasive Species: Pathogens of Globalization

by Christopher Bright

World trade has become the primary driver of one of the most dangerous and least visible forms of environmental decline: Thousands of foreign, invasive species are hitch-hiking through the global trading network aboard ships, planes, and railroad cars, while hundreds of others are traveling as commodities. The impact of these bioinvasions can now be seen on every landmass, in nearly all coastal waters (which comprise the most biologically productive parts of the oceans), and probably in most major rivers and lakes. This "biological pollution" is degrading ecosystems, threatening public health, and costing billions of dollars annually. Confronting the problem may now be as critical an environmental challenge as reducing global carbon emissions.

Despite such dangers, policies aimed at stopping the spread of invasive "exotic" species have so far been largely ineffective. Not only do they run up against far more powerful policies and interests that in one way or another encourage invasion, but the national and international mechanisms needed to control the spread of non-native species are still relatively undeveloped. Unlike chemical pollution, for instance, bioinvasion is not yet a working category of environmental decline within the legal culture of most countries and international institutions.

In part, this conceptual blindness can be explained by the fact that even badly invaded landscapes can still look healthy. It is also a consequence of the ancient and widespread practice of introducing exotic species for some tangible benefit: A bigger fish makes for better fishing, a faster-growing tree means more wood. It can be difficult to think of these activities as a form of ecological corrosion—even if the fish or the tree ends up demolishing the original natural community.

The increasing integration of the world's economies is rapidly making a bad situation even worse. The continual expansion of world trade—in ways that are not shaped by any real understanding of their environmental effects—is causing a degree of ecological mixing that appears to have no evolutionary precedent. Under more or less natural conditions, the arrival of an entirely new organism was a rare event in most times and places. Today it can happen any time a ship comes into port or an airplane lands. The real problem, in other words, does not lie with the exotic species themselves, but with the economic system that is continually showering them over the Earth's surface. Bioinvasion has become a kind of globalization disease.

They Came, They Bred, They Conquered

Bioinvasion occurs when a species finds its way into an ecosystem where it did not evolve. Most of the time when this happens, conditions are not suitable for the new arrival, and it enjoys only a brief career. But in a small percentage of cases, the exotic finds everything it needs—and nothing capable of controlling it. At the very least, the invading organism is liable to suppress some native species by consuming resources that they would have used instead. At worst, the invader may rewrite some basic ecosystem "rules"—checks and balances that have developed between native species, usually over many millennia.

Although it is not always easy to discern the full extent of havoc that invasive species can wreak upon an ecosystem, the resulting financial damage is becoming increasingly difficult to ignore. Worldwide, the losses to agriculture might be anywhere from $55 billion to nearly $248 billion annually. Researchers at Cornell University recently concluded that bioinvasion might be costing the United States alone as much as $123 billion per year. In South and Central America, the growth of specialty export crops—upscale vegetables and fruits—has spurred the spread of whiteflies, which are capable of transmitting at least 60 plant viruses. The spread of these viruses has forced the abandonment of more than 1 million hectares of cropland in South America. In the wetlands of northern Nigeria, an exotic cattail is strangling rice paddies, ruining fish habitats, and slowly choking off the Hadejia-Nguru river system. In southern India, a tropical American shrub, the bush morning glory, is causing similar chaos throughout the basin of the Cauvery, one of the region's biggest rivers. In the late 1980s, the accidental release into the Black Sea of *Mnemiopsis leidyi*—a comb jelly native to the east coast of the Americas—provoked the collapse of the already highly stressed Black Sea fisheries, with estimated financial losses as high as $350 million.

Controlling invasive species is difficult enough, but the bigger problem is preventing the machinery of the world trading system from releasing them in the first place.

Controlling such exotics in the field is difficult enough, but the bigger problem is preventing the machinery of the world trading system from releasing them in the first place. That task is becoming steadily more formidable as the trading system continues to grow. Since 1950, world trade has expanded sixfold in terms of value. More important in terms of potential invasions is the vast increase in the volume of goods traded. Look, for instance, at the ship, the primary mechanism of trade—80 percent of the world's goods travel by ship for at least part of their journey from manufacturer to consumer. From 1970 to 1996, the volume of seaborne trade nearly doubled.

Ships, of course, have always carried species from place to place. In the days of sail, shipworms bored into the wooden hulls, while barnacles and seaweeds attached themselves to the sides. A small menagerie of other creatures usually took up residence within these "fouling communities." Today, special paints and rapid crossing times have greatly reduced hull fouling, but each of the 28,700 ships in the world's major merchant fleets represents a honeycomb of potential habitats for all sorts of life, both terrestrial and aquatic.

The most important of these habitats lies deep within a modern ship's plumbing, in the ballast tanks. The ballast tanks of a really big ship—say, a supertanker—may contain more than 200,000 cubic meters of water—equivalent to 2,000 Olympic-sized swimming pools. When those tanks are filled, any little creatures in the nearby water or sediment may suddenly become inadvertent passengers. A few days or weeks later, when the tanks are discharged at journey's end, they may become residents of a coastal community on the other side of the world. Every year, these artificial ballast currents move some 10 billion cubic meters of water from port to port. Every day, some 3,000 to 10,000 different species are thought to be riding the ballast currents. The result is a creeping homogenization of estuary and bay life. The same creatures come to dominate one coastline after another, eroding the biological diversity of the planet's coastal zones—and jeopardizing their ecological stability.

Some pathways of invasion extend far beyond ships. Another prime mechanism of trade is the container: the metal box that has revolutionized the transportation of just about every good not shipped in bulk. The container's effect on invasion ecology has been just as profound. For centuries, shipborne exotics were largely confined to port areas—but no longer. Containers move from ship to harbor crane to the flatbed of a truck or railroad car and then on to wherever a road or railroad leads. As a result, all sorts of stowaways that creep aboard containers often wind up far inland. Take the Asian tiger mosquito, for example, which can carry dengue fever, yellow fever, and encephalitis. The huge global trade in containers of used tires—which are, under the right conditions, an ideal mosquito habitat—has dispersed this species from Asia and the Indo-Pacific into Australia, Brazil, the eastern United States, Mozambique, New Zealand, Nigeria, and southern Europe. Even packing material within containers can be a conduit for exotic species. Untreated wood pallets, for example, are to forest pests what tires are to mosquitoes. One creature currently moving along this pathway is the Asian long-horn beetle, a wood-boring insect from China with a lethal appetite for deciduous trees. It has turned up at more than 30 locations around the United States and has also been detected in Great Britain. The only known way to eradicate it is to cut every tree suspected of harboring it, chip all the wood, and burn all the chips.

As other conduits for global trade expand, so does the potential for new invasions. Air cargo service, for example, is building a global network of virtual canals that have great potential for transporting tiny, short-lived creatures such as microbes and insects. In 1989, only three airports received more than 1 million tons of cargo; by 1996, there were 13 such airports. Virtually everywhere you look, the newly constructed infrastructure of the global economy is forming the groundwork for an ever-greater volume of biological pollution.

The Global Supermarket

Bioinvasion cannot simply be attributed to trade in general, since not all trade is "biologically dirty." The natural resource industries—especially agriculture, aquaculture, and forestry—are causing a disproportionate share of the problem. Certain trends within each of these industries are liable to exacerbate the invasion pressure. The migration of crop pests can be attributed, in part, to a global agricultural system that has become increasingly uniform and integrated. (In China, for example, there were about 10,000 varieties of wheat being grown at mid-century; by

1970 there were only about 1,000.) Any new pest—or any new form of an old pest—that emerges in one field may eventually wind up in another.

The key reason that South America has suffered so badly from white-flies, for instance, is because a pesticide-resistant biotype of that fly emerged in California in the 1980s and rapidly became one of the world's most virulent crop pests. The fly's career illustrates a common dynamic: A pest can enter the system, disperse throughout it, and then develop new strains that reinvade other parts of the system. The displacement of traditional developing-world crop varieties by commercial, homogenous varieties that require more pesticide, and the increasing development of pesticide resistance among all the major pest categories—insects, weeds, and fungi—are likely to boost this trend.

Similar problems pertain to aquaculture—the farming and exporting of fish, shellfish, and shrimp. Partly because of the progressive depletion of the world's most productive fishing grounds, aquaculture is a booming business. Farmed fish production exceeded 23 million tons by 1996, more than triple the volume just 12 years before. Developing countries in particular see aquaculture as a way of increasing protein supply.

But many aquaculture "crops" have proved very invasive. In much of the developing world, it is still common to release exotic fish directly into natural waterways. It is hardly surprising, then, that some of the most popular aquaculture fish have become true cosmopolitans. The Mozambique tilapia, for example, is now established in virtually every tropical and subtropical country. Many of these introductions—not just tilapia, but bass, carp, trout, and other types of fish—are implicated in the decline of native species. The constant flow of new introductions catalogued with such enthusiasm in the industry's publications are a virtual guarantee that tropical freshwater ecosystems are unraveling beneath the surface.

Aquaculture is also a spectacularly efficient conduit of disease. Perhaps the most virulent set of wildlife epidemics circling the Earth today involves shrimp production in the developing world. Unlike fish, shrimp are not a subsistence crop: They are an extremely lucrative export business that has led to the bulldozing of many tropical coasts to make way for shrimp ponds. One of the biggest current developments is an Indonesian operation that may eventually cover 200,000 hectares. A horde of shrimp pathogens—everything from viruses to protozoa—is chasing these operations, knocking out ponds, and occasionally ruining entire national shrimp industries: in Taiwan in 1987, in China in 1993, and in India in 1994. Shrimp farming has become, in effect, a form of "managed invasion." Since shrimp are important components of both marine and freshwater ecosystems worldwide, it is anybody's guess at this point what impact shrimpborne pathogens will ultimately have.

Managed invasion is an increasingly common procedure in another big biopolluting industry: forestry. Industrial round-wood production (basically, the cutting of logs for uses other than fuel) currently hovers at around 1.5 billion cubic meters annually, which is more than twice the level of the 1950s. An increasing amount of wood and wood pulp is coming out of tree plantations (not inherently a bad idea, given the rate at which

the world is losing natural forests). In North America and Europe, plantation forestry generally uses native species, so the gradation from natural forest to plantation is not usually as stark as it is in developing countries, where exotics are the rule in industrial-plantation development.

For the most part, these developing-country plantations bear about as much resemblance to natural forests as corn fields do to undisturbed prairies. And like corn fields, they are maintained with heavy doses of pesticides and subjected to a level of disturbance—in particular, the use of heavy equipment to harvest the trees—that tends to degrade soil. Some plantation trees have launched careers as king-sized weeds. At least 19 species of exotic pine, for example, have invaded various regions in the Southern Hemisphere, where they have displaced native vegetation and, in some areas, apparently lowered the water tables by "drinking" more water than the native vegetation would consume. Even where the trees have not proved invasive, the exotic plantations themselves are displacing natural forest and traditional forest peoples. This type of tree plantation is almost entirely designed to feed wood to the industrialized world, where 77 percent of industrial roundwood is consumed. As with shrimp production, local ecological health is being sacrificed for foreign currency.

There is another, more poignant motive for the introduction of large numbers of exotic trees into the developing world. In many countries severely affected by forest loss, reforestation is recognized as an important social imperative. But the goal is often nothing more than increasing tree cover. Little distinction is made between plantation and forest or between foreign and native species. Surayya Khatoon, a botanist at the University of Karachi, observes that "awareness of the dangers associated with invasive species is almost nonexistent in Pakistan, where alien species are being planted on a large scale in so-called afforestation drives."

Even international agreements that focus specifically on ecological problems have generally given bioinvasion short shrift.

The industrial sources of biological pollution are very diverse, but they reflect a common mindset. Whether it is a tree plantation, a shrimp farm, or even a bit of landscaping in the back yard, the Earth has become a sort of "species supermarket"; if a species looks good for whatever it is that you have in mind, pull it off the shelf and take it home. The problem is that many of the traits you want the most—adaptability, rapid growth, and easy reproduction—also tend to make the organism a good candidate for invasion.

Launching a Counter-Attack

Since the processes of invasion are deeply embedded in the globalizing economy, any serious effort to root them out will run the risk of exhausting itself. Most industries and policymakers are striving to open borders, not erect new barriers to trade. Moreover, because bioinvasion is not yet an established policy category, jurisdiction over it is generally badly fragmented—or even absent—on both the national and international levels. Most countries have some relevant legislation—laws intended to discourage the movement of crop pests, for example—but very few have any overall legislative authority for dealing with the problem. (New Zealand is the noteworthy exception: Its Biosecurity Act of 1993 and its Hazardous Substances and New Organisms Act of 1996 do establish such an authority.) Although it is true that there are many treaties that bear on the problem in one way or another—23 at least count—there is no such thing as a bioinvasion treaty.

Even agreements that focus specifically on ecological problems have generally given bioinvasion short shrift. Agenda 21, for example—the blueprint for sustainable development that emerged from the 1992 Earth Summit in Rio de Janeiro—reflects little awareness of the dangers of exotic forestry and aquaculture. Among international agencies, only certain types of invasion seem to get much attention. There are treaties—such as the 1951 International Plant Protection Convention—that limit the movement of agricultural pests, but there is currently no clear international mechanism for dealing with ballast water releases. Obviously, in such a context, you need to pick your fights carefully. They have to be important, winnable, and capable of yielding major opportunities elsewhere. The following three-point agenda offers some hope of slowing invasion over the near term.

The first item: Plug the ballast water pathway. As a technical problem, this objective is probably just on the horizon of feasibility, making it an excellent policy target. Strong national and international action could push technologies ahead rapidly. At present, the most effective technique is ballast water exchange, in which the tanks of a ship are pumped out and refilled in the open sea. (Coastal organisms, pumped into the tanks at the ship's last port of call, usually will not survive in the open ocean; organisms that enter the tanks in mid-ocean probably will not survive in the next port of call.) But it can take several days to exchange the water in all of a ship's ballast tanks, so the procedure may not be feasible for every leg of a journey, and the tanks never empty completely. In bad weather, the process can be too dangerous to perform at all. Consequently, other options will be necessary—filters or even toxins (that may not sound very appealing, but some common water treatment compounds may be environmentally sound). It might even be possible to build port-side ballast water treatment plants. Such a mixture of technologies already exists as the standard means of controlling chemical pollution.

This objective is drifting into the realm of legal possibility as well. As of July 1 this year, all ships entering U.S. waters must keep a record of their ballast water management. The United States has also issued voluntary guidelines on where those ships can release ballast water. These measures are a loose extension of the regulations that the United States and Canada have imposed on ship traffic in the Great Lakes, where foreign ballast water release is now explicitly forbidden. In California, the State Water Resources Control Board has declared San Francisco Bay "impaired" because it is so badly invaded—a move that may allow authorities to use regulations written for chemical pollution as a way of controlling ballast water. Australia now levies a small tax on all incoming ships to support ballast water research.

Internationally, the problem has acquired a high profile at the UN International Maritime Organization (IMO), which is studying the possibility of developing a ballast management protocol that would have the force of international law. No decision has been made on the legal mechanism for such an agreement, although the most likely possibility is an annex to MARPOL, the International Convention for the Prevention of Pollution from Ships.

Within the shipping industry, the responses to such proposals have been mixed. Although industry officials concede the problem in the abstract, the prospect of specific regulations has tended to provoke unfavorable comment. After an IMO meeting last year on ballast water management, a spokesperson for the International Chamber of Shipping argued that rigorous ballast exchange would cost the industry millions of dollars a year—and that internationally binding regulations should be avoided in favor of local regulation, wherever particular jurisdictions decide to address the problem. Earlier this year in California, a proposed bill that would have essentially prohibited foreign ballast water release in the state's ports provoked outcries from local port representatives, who argued that such regulations might encourage ship traffic to bypass California ports in favor of the Pacific Northwest or Mexico. Of course, any management strategy is bound to cost something, but the important question is: What impact will this additional cost have? It may not have much impact at all. In Canada, for example, the Vancouver Port Authority reported that its ballast water program has had no detectable effect on port revenues.

The second item on the agenda: Fix the World Trade Organization (WTO) Agreement on the Application of Sanitary and Phytosanitary Measures. This agreement, known as the SPS, was part of the diplomatic package that created the WTO in 1994. The SPS is supposed to promote a common set of procedures for evaluating risks of contamination in internationally traded commodities. The contaminants can be chemical (pesticide residues in food) or they can be living things (Asian longhorn beetles in raw wood).

One of the procedures required by the SPS is a risk assessment, which is supposed to be done before any trade-constricting barriers are imposed to prevent a contaminated good from entering a country. If you want to understand the fundamental flaw in this approach as it applies to bioinvasion, all you have to do is recall the famous observation by the eminent biologist E. O. Wilson: "We dwell on a largely unexplored planet." When it comes to the largest categories of living things—insects, fungi, bacteria, and so on—we have managed to name only a tiny fraction of them, let alone figure out what damage they can cause. Consider, for example, the rough, aggregate risk assessments done by the United States Department of Agriculture (USDA) for wood imported into the United States from Chile, Mexico, and New Zealand. The USDA found dozens of "moderate" and "high" risk pests and pathogens that have the potential for doing economic damage on the order of hundreds of millions of dollars at least—and ecological damage that is incalculable. But even with wide-open thoroughfares of invasion such as these, the SPS requirement in its current form is likely to make preemptive action vulnerable to trade complaints before the WTO.

Another SPS requirement intended to insure a consistent application of standards is that a country must not set up barriers against an organism that is already living within its borders unless it has an "official control program" for that species. This approach is unrealistic for both biological and financial reasons. Thousands of exotic species are likely to have invaded most of the world's countries and not even the wealthiest country could possibly afford to fight them all. Yet it certainly is possible to exacerbate a problem by introducing additional infestations of a pest, or by boosting the size of existing infestations, or even by increasing the genetic vigor of a pest population by adding more "breeding stock." The SPS does not like "inconsistencies"—if you are not controlling a pest, you have no right to object to more of it; if you try to block one pathway of invasion, you had better be trying to block all the equivalent pathways. Such an approach may be theoretically neat, but in the practical matter of dealing with exotics, it is a prescription for paralysis.

In the near term, however, any effort to repair the SPS is likely to be difficult. The support of the United States, a key member of the WTO, will be critical for such reforms. And although the United States has demonstrated a heightened awareness of the problem—as evidenced by President Bill Clinton's executive order to create an Invasive Species Council—it is not clear whether that commitment will be reflected in the administration's trade policy. During recent testimony before Congress, the U.S. Trade Representative's special trade negotiator for agricultural issues warned that the United States was becoming impatient with the "increasing use of SPS barriers as the 'trade barrier of choice.'" In the developing world, it is reasonable to assume that any country with a strong export sector in a natural resource industry would not welcome tougher regulations. Some developed countries, however, may be sympathetic to change. The European Union (EU) has sought very strict standards in its disputes with the United States over bans on beef from cattle fed with growth hormones and on genetically altered foods. It is possible that the EU might be willing to entertain a stricter SPS. The same might be true of Japan, which has attempted to secure stricter testing of U.S. fruit imports.

The third item: Build a global invasion database. Currently, the study of bioinvasion is an obscure and rather fractured enterprise. It can be difficult to locate critical information or relevant expertise. The full magnitude of the issue is still not registering on the public radar screen. A global database would consolidate existing information, presumably into some sort of central research institution with a major presence on the World Wide Web. One could "go" to such a place—either physically or through cyberspace—to learn about everything from the National Ballast Water Information Clearinghouse that the U.S. Coast Guard is setting up, to the database on invasive woody plants in the tropics that is being assembled at the University of Wales. The database would also stimulate the production of new media to encourage additional research and synthesis. It is a telling indication of how fragmented this field is that, after more than 40 years of formal study, it is just now getting its first comprehensive journal: *Biological Invasions*.

Better information should have a number of practical effects. The best way to control an invasion—when it cannot be prevented outright—is to go after the exotic as soon as it is detected. An emergency response capability will only work if officials know what to look for and what to do when they find it. But beyond such obvious applications, the database could help bring the big picture into focus. In the struggle with exotics, you can see the free-trade ideal colliding with some hard ecological realities. Put simply: It may never be safe to ship certain goods to certain places—raw wood from Siberia, for instance, to North America. The notion of real, permanent limits to economic activity will for many politicians (and probably some economists) come as a strange and unpalatable idea. But the global economy is badly in need of a large dose of ecological realism. Ecosystems are very diverse and very different from each other. They need to stay that way if they are going to continue to function.

WANT TO KNOW MORE?

Although the scientific literature on bioinvasion is enormous and growing rapidly, most of it is too technical to attract a readership outside the field. For a nontechnical, broad overview of the problem, readers should consult Robert Devine's *Alien Invasion: America's Battle with Non-Native Animals and Plants* (Washington: National Geographic Society, 1998) or Christopher Bright's *Life Out of Bounds: Bioinvasion in a Borderless World* (New York: W.W. Norton & Company, 1998).

If you have a long-term interest in bioinvasion, you will want to get acquainted with the book that founded the field: Charles Elton's *The Ecology of Invasions by Animals and Plants* (London: Methuen, 1958). A historical overview of bioinvasions can be found in Alfred Crosby's book *Ecological Imperialism: The Biological Expansion of Europe, 900–1900* (Cambridge: Cambridge University Press, 1986).

Many studies focus on invasion of particular regions. The focus can be very broad, as in P. S. Ramakrishnan, ed., *Ecology of Biological Invasions in the Tropics*, proceedings of an international workshop held at Nainital, India, (New Delhi: International Scientific Publications, 1989). Generally, however, the coverage is much narrower, as in Daniel Simberloff, Don Schmitz, and Tom Brown, eds., *Strangers in Paradise: Impact and Management of Nonindigenous Species in Florida* (Washington: Island Press, 1997). The other standard research tack has been to look at a particular type of invader. The most accessible results of this exercise are encyclopedic surveys such as Christopher Lever's *Naturalized Mammals of the World* (London: Longman, 1985) and his companion volumes on naturalized birds and fish. In the plant kingdom, the genre is represented by Leroy Holm, et al., *World Weeds: Natural Histories and Distribution* (New York: John Wiley and Sons, 1997).

There are many worthwhile documents available for anyone who is interested not just in the ecology of invasion, but also in its economic, social, and epidemiological implications. Just about every aspect of the problem is discussed in Odd Terje Sandlund, Peter Johan Schei, and Aslaug Viken, eds., *Proceedings of the Norway/UN Conference on Alien Species* (Trondheim: Directorate for Nature Management and Norwegian Institute for Nature Research, 1996). A groundbreaking study of invasion in the United States, with particular emphasis on economic effects, is *Harmful Nonindigenous Species in the United States* (Washington: Office of Technology Assessment, September 1993). An assessment of the ballast water problem is available from the National Research Council's Commission on

Ships' Ballast Operations' *Stemming the Tide: Controlling Introductions of Nonindigenous Species by Ships' Ballast Water* (Washington: National Academy Press, 1996). Readers who are interested in exotic tree plantations as a form of "managed invasion" might look through Ricardo Carrere and Larry Lohmann's *Pulping the South: Industrial Tree Plantations and the World Paper Economy* (London: Zed Books, 1996) and the World Rainforest Movement's *Tree Plantations: Impacts and Struggles* (Montevideo: WRM, 1999). Unfortunately, there are no analogous studies of shrimp farms.

For links to relevant Web sites, as well as a comprehensive index of related FOREIGN POLICY articles, access **www. foreignpolicy.com.**

Christopher Bright is a research associate at the Worldwatch Institute in Washington, DC, and author of Life Out of Bounds: Bioinvasion in a Borderless World (New York: W.W. Norton & Company, 1998).

UNIT 5

Resources: Land, Water, and Air

Unit Selections

18. **Where Have All the Farmers Gone?** Brian Halweil
19. **All the Wild Rivers**, Curtis Runyan
20. **Growing More Food With Less Water**, Sandra Postel
21. **Oceans Are on the Critical List**, Anne Platt McGinn
22. **Feeling the Heat: Life in the Greenhouse**, Michael D. Lemonick

Key Points to Consider

• Why is the number of farmers in the world decreasing? What kinds of social, economic, and cultural impacts will be produced by decreasing the number of family farmers and increasing the amount of land farmed by corporate farmers?

• What are some of the incentives that induce people in such different places in the world as India and the United States to remove dams that have been in place for long periods of time? What benefits are produced and what are the costs of returning rivers to free-flowing conditions?

• Why is the world's freshwater resource in such short supply? Describe the nature of the accelerating demand for water as a resource and discuss the relationship between that demand and water management.

• Why and how has overfishing contributed to a decline in the food supply from the oceans? Has oceanic pollution contributed to this decline?

• What are some of the uncertainties about the future impact of global temperature increase on human social and economic systems? What reasons are there to develop extensive monitoring systems to identify the causes and effects of global warming?

 **Links: www.dushkin.com/online/**
These sites are annotated in the World Wide Web pages.

Agriculture Production Statistics
http://www.wri.org/statistics/fao-prd.html

Global Climate Change
http://www.puc.state.oh.us/consumer/gcc/index.html

National Oceanic and Atmospheric Administration (NOAA)
http://www.noaa.gov

National Operational Hydrologic Remote Sensing Center (NOHRSC)
http://www.nohrsc.nws.gov

Virtual Seminar in Global Political Economy/Global Cities & Social Movements
http://csf.colorado.edu/gpe/gpe95b/resources.html

Websurfers Biweekly Earth Science Review
http://www.mindspring.com/~michaelg2/weeksreviews.html

The worldwide situations regarding reduction of biodiversity, scarcity energy resources, and pollution of the environment have received the greatest amount of attention among members of the environmentalist community. But there are a number of other resource issues that demonstrate the interrelated nature of all human activities and the environments in which they occur. One such issue is the declining quality of agricultural land. In the developing world, excessive rural populations have forced the overuse of lands and sparked a shift into marginal areas, and the total availability of new farmland is decreasing at an alarming rate of 2 percent per year. In the developed world, intensive mechanized agriculture has resulted in such a loss of topsoil that some agricultural experts are predicting a decline in food production. Other natural resources, such as minerals and timber, are declining in quantity and quality as well; in some cases they are no longer usable at present levels of technology. The overuse of groundwater reserves has resulted in potential shortages beside which the energy crisis pales in significance. And the very productivity of Earth's environmental systems—their ability to support human and other lif—is being threatened by processes that derive at least in part from energy overuse and inefficiency and from pollution. Many environmentalists believe that both the public and private sectors, including individuals, are continuing to act in a totally irresponsible manner with regard to the natural resources upon which we all depend.

Uppermost in the minds of many who think of the environment in terms of an integrated whole, as evidenced by many of the selections in this unit, is the concept of the threshold or critical limit of human interference with natural systems of land, water, and air. This concept suggests that the environmental systems we occupy have been pushed to the brink of tolerance in terms of stability and that destabilization of environmental systems has consequences that can only be hinted at, rather than predicted. Although the broader issue of system change and instability, along with the lesser issues such as the quantity of agricultural land, the quality of iron ore deposits, the sustained yield of forests, or the availability of fresh water seems to be quite diverse, all are closely tied to a pair of concepts—that of resource marginality and of the globalization of the economy that has made marginality a global rather than a regional problem. Many of these ideas are brought together in the lead article of this unit. In "Where Have All the Farmers Gone?" Brian Halweil of the Worldwatch Institute discusses the globalization of industry and trade that is creating a uniform approach to all forms of economic management, including management of agricultural resources. Halweil notes that increasing agribusiness and decreasing numbers of family farmers represent a loss of both biological and cultural diversity, as standardized farming and business practices that may work for some areas but not for others are applied uniformly.

In the second article in this unit, Curtis Runyan, a senior editor of World Watch magazine, expresses concerns over global issues related to water supplies. In "All the Wild Rivers," Runyan discusses the growing trend to removing dams from river valleys rather than building them. For centuries, if not millennia, farmers facing problems of water scarcity have viewed impoundment of free-flowing rivers as a viable strategy. We now know that the loss of topsoil to accelerated erosion, the loss of living and farming space to reservoirs, and increasing the salinity in irrigation waters and soil are problems greater than the dryness that prompts irrigation to begin with. Sandra Postel of the Global Water Policy Project who, in "Growing More Food With Less Water," the third article in the unit, discusses one of the most critical problems facing food producers in developing and developed countries alike: the stresses placed on the world's freshwater supply by the demands of irrigation agriculture, a necessary component of increasing global food production. Postel describes the inefficiencies of existing irrigation systems and calls for the development of a sustainable base of irrigation agriculture if both food and environmental problems are to be solved.

The final two articles in this unit deal with other global resource problems: those associated with limits to wise use of the ocean of water that makes up 70 percent of Earth's surface and of the ocean of air—the atmosphere—that surrounds us. In each article, the concept of marginality is relevant. In "Oceans Are on the Critical List," researcher Anne Platt McGinn contends that the primary threats to the health of the world's oceanic ecosystems are human-induced and include overharvesting of fish, pollution of both coastal and deep-water zones, introduction of alien species and the consequent threat to oceanic biodiversity, and climate change, which also poses threats to biodiversity. Efforts to protect the oceans, McGinn claims, lag far behind what is needed. In the final article in this unit, the most critical and controversial of the problems that characterize the global atmosphere is dealt with: the continuing accumulation of greenhouse gases and the consequent trend toward a warmer Earth—or what has been termed "global warming." In a special report from Time, the global warming issue is explored by a team of investigative reporters. This report, "Feeling the Heat: Life in the Greenhouse" discusses the social, political, economic, and scientific aspects of the global warming debate and concludes that, in spite of scientific and nonscientific rhetoric and argument about global warming, the fact that the globe is becoming warmer is indisputable. It is equally indisputable that human activities are playing some role in the process. While a small segment of the scientific community still does not see human influence in short-term climate trends, the majority believes in a greenhouse effect enhanced by human activity. Interestingly, some of the strongest proponents of action to curb human-induced global warming are some of the biggest energy companies.

There are two possible solutions to all these problems posed by the use of increasingly marginal and scarce resources and by the continuing pollution of the global atmosphere. One is to halt the basic cause of the problems—increasing population and consumption. The other is to provide incentives and techniques for the conservation and management of existing resources and for the discovery of alternative resources to eliminate the demand for more marginal resources and the use of heavily polluting ones.

Where Have All the Farmers Gone?

The globalization of industry and trade is bringing more and more uniformity to the management of the world's land, and a spreading threat to the diversity of crops, ecosystems, and cultures. As Big Ag takes over, farmers who have a stake in their land—and who often are the most knowledgeable stewards of the land—are being forced into servitude or driven out.

by Brian Halweil

Since 1992, the U.S. Army Corps of Engineers has been developing plans to expand the network of locks and dams along the Mississippi River. The Mississippi is the primary conduit for shipping American soybeans into global commerce—about 35,000 tons a day. The Corps' plan would mean hauling in up to 1.2 million metric tons of concrete to lengthen ten of the locks from 180 meters to 360 meters each, as well as to bolster several major wing dams which narrow the river to keep the soybean barges moving and the sediment from settling. This construction would supplement the existing dredges which are already sucking 85 million cubic meters of sand and mud from the river's bank and bottom each year. Several different levels of "upgrade" for the river have been considered, but the most ambitious of them would purportedly reduce the cost of shipping soybeans by 4 to 8 cents per bushel. Some independent analysts think this is a pipe dream.

Around the same time the Mississippi plan was announced, the five governments of South America's La Plata Basin—Bolivia, Brazil, Paraguay, Argentina, and Uruguay—announced plans to dredge 13 million cubic meters of sand, mud, and rock from 233 sites along the Paraguay-Paraná River. That would be enough to fill a convoy of dump trucks 10,000 miles long. Here, the plan is to straighten natural river meanders in at least seven places, build dozens of locks, and construct a major port in the heart of the Pantanal—the world's largest wetland. The Paraguay-Paraná flows through the center of Brazil's burgeoning soybean heartland—second only to the

United States in production and exports. According to statements from the Brazilian State of Mato Grasso, this "Hidrovía" (water highway) will give a further boost to the region's soybean export capacity.

Lobbyists for both these projects argue that expanding the barge capacity of these rivers is necessary in order to improve competitiveness, grab world market share, and rescue farmers (either U.S. or Brazilian, depending on whom the lobbyists are addressing) from their worst financial crisis since the Great Depression. Chris Brescia, president of the Midwest River Coalition 2000, an alliance of commodity shippers that forms the primary lobbying force for the Mississippi plan, says, "The sooner we provide the waterway infrastructure, the sooner our family farmers will benefit." Some of his fellow lobbyists have even argued that these projects are essential to feeding the world (since the barges can then more easily speed the soybeans to the world's hungry masses) and to saving the environment (since the hungry masses will not have to clear rainforest to scratch out their own subsistence).

Probably very few people have had an opportunity to hear both pitches and compare them. But anyone who has may find something amiss with the argument that U.S. farmers will become more competitive versus their Brazilian counterparts, at the same time that Brazilian farmers will, for the same reasons, become more competitive with their U.S. counterparts. A more likely outcome is that farmers of these two nations will be pitted against each other in a costly race to maximize production, resulting in short-cut practices that essentially strip-mine their soil and throw long-term investments in the land to the wind. Farmers in Iowa will have stronger incentives to plow up land along stream banks, triggering faster erosion of topsoil. Their brethren in Brazil will find themselves needing to cut deeper into the savanna, also accelerating erosion. That will increase the flow of soybeans, all right—both north and south. But it will also further depress prices, so that even as the farmers are shipping more, they're getting less income per ton shipped. And in any case, increasing volume can't help the farmers survive in the long run, because sooner or later they will be swallowed by larger, corporate, farms that can make up for the smaller per-ton margins by producing even larger volumes.

So, how can the supporters of these river projects, who profess to be acting in the farmer's best interests, not notice the illogic of this form of competition? One explanation is that from the advocates' (as opposed to the farmers') standpoint, this competition isn't illogical at all—because the lobbyists aren't really representing farmers. They're working for the commodity processing, shipping, and trading firms who want the price of soybeans to fall, because these are the firms that buy the crops from the farmers. In fact, it is the same three agribusiness conglomerates—Archer Daniels Midland (ADM), Cargill, and Bunge—that are the top soybean processors and traders along both rivers.

Welcome to the global economy. The more brutally the U.S. and Brazilian farmers can batter each-other's prices (and standards of living) down, the greater the margin of profit these three giants gain. Meanwhile, another handful of companies controls the markets for genetically modified seeds, fertilizers, and herbicides used by the farmers—charging oligopolistically high prices both north and south of the equator.

In assessing what this proposed digging-up and reconfiguring of two of the world's great river basins really means, keep in mind that these projects will not be the activities of private businesses operating inside their own private property. These are proposed public works, to be undertaken at huge public expense. The motive is neither the plight of the family farmer nor any moral obligation to feed the world, but the opportunity to exploit poorly informed public sentiments about farmers' plights or hungry masses as a means of usurping public policies to benefit private interests. What gets thoroughly Big Muddied, in this usurping process, is that in addition to subjecting farmers to a gladiator-like attrition, these projects will likely bring a cascade of damaging economic, social, and ecological impacts to the very river basins being so expensively remodeled.

What's likely to happen if the lock and dam system along the Mississippi is expanded as proposed? The most obvious effect will be increased barge traffic, which will accelerate a less obvious cascade of events that has been underway for some time, according to Mike Davis of the Minnesota Department of Natural Resources. Much of the Mississippi River ecosystem involves aquatic rooted plants, like bullrush, arrowhead, and wild celery. Increased barge traffic will kick up more sediment, obscuring sunlight and reducing the depth to which plants can survive. Already, since the 1970s, the number of aquatic plant species found in some of the river has been cut from 23 to about half that, with just a handful thriving under the cloudier conditions. "Areas of the river have reached an ecological turning point," warns Davis. "This decline in plant diversity has triggered a drop in the invertebrate communities that live on these plants, as well as a drop in the fish, mollusk, and bird communities that depend on the diversity of insects and plants." On May 18, 2000, the U.S. Fish and Wildlife Service released a study saying that the Corps of Engineers project would threaten the 300 species of migratory birds and 12 species of fish in the Mississippi watershed, and could ultimately push some into extinction. "The least tern, the pallid sturgeon, and other species that evolved with the ebbs and flows, sandbars and depths, of the river are progressively eliminated or forced away as the diversity of the river's natural habitats is removed to maximize the barge habitat," says Davis.

The outlook for the Hidrovía project is similar. Mark Robbins, an ornithologist at the Natural History Museum at the University of Kansas, calls it "a key step in creating a Florida Everglades-like scenario of destruction in the Pantanal, and an American Great Plains-like scenario in the Cerrado in southern Brazil." The Paraguay-Paraná feeds the Pantanal wetlands, one of the most diverse habitats on the planet, with its populations of woodstorks, snailkites, limpkins, jabirus, and more than 650 other species of birds, as well as more than 400 species of fish and hundreds of other less-studied plants, mussels, and marshland organisms. As the river is dredged and the banks are built up to funnel the surrounding wetlands water into the navigation path, bird nesting habitat and fish spawning grounds will be eliminated, damaging the indigenous and other traditional societies that depend on these resources. Increased barge traffic will suppress river species here just as it will on the Mississippi. Meanwhile, herbicide-intensive soybean monocultures—on farms so enormous that they dwarf even the biggest operations in the U.S. Midwest—are rapidly replacing diverse grasslands in the fragile Cerrado. The heavy plowing and periodic absence of ground cover associated with such farming erodes 100 million tons of soil per year. Robbins notes that "compared to the Mississippi, this southern river system and surrounding grassland is several orders of magnitude more diverse and has suffered considerably less, so there is much more at stake."

Supporters of such massive disruption argue that it is justified because it is the most "efficient" way to do business. The perceived efficiency of such farming might be compared to the perceived efficiency of an energy system based on coal. Burning coal looks very efficient if you ignore its long-term impact on air quality and climate stability. Similarly, large farms look more efficient than small farms if you don't count some of their largest costs—the loss of the genetic diversity that underpins agriculture, the pollution caused by agro-chemicals, and the dislocation of rural cultures. The simultaneous demise of small, independent farmers and rise of multinational food giants is troubling not just for those who empathize with dislocated farmers, but for anyone who eats.

An Endangered Species

Nowadays most of us in the industrialized countries don't farm, so we may no longer really understand that way of life. I was born in the apple orchard and dairy country of Dutchess County, New York, but since age five have spent most of my life in New York City—while most of the farms back in Dutchess County have given way to spreading subdivisions. It's also hard for those of us who get our food from supermarket shelves or drive-thru windows to know how dependent we are on the viability of rural communities.

Whether in the industrial world, where farm communities are growing older and emptier, or in developing nations where population growth is pushing the number of farmers continually higher and each generation is inheriting smaller family plots, it is becoming harder and harder to make a living as a farmer. A combination of falling incomes, rising debt, and worsening rural poverty is forcing more people to either abandon farming as their primary activity or to leave the countryside altogether— a bewildering juncture, considering that farmers produce perhaps the only good that the human race cannot do without.

Since 1950, the number of people employed in agriculture has plummeted in all industrial nations, in some regions by more than 80 percent. Look at the numbers, and you might think farmers are being singled out by some kind of virus:

- In Japan, more than half of all farmers are over 65 years old; in the United States, farmers over 65 outnumber those under 35 by three to one. (Upon retirement or death, many will pass the farm on to children who live in the city and have no interest in farming themselves.)
- In New Zealand, officials estimate that up to 6,000 dairy farms will disappear during the next 10 to 15 years—dropping the total number by nearly 40 percent.
- In Poland, 1.8 million farms could disappear as the country is absorbed into the European Union—dropping the total number by 90 percent.
- In Sweden, the number of farms going out of business in the next decade is expected to reach about 50 percent.

- In the Philippines, Oxfam estimates that over the next few years the number of farm households in the corn–producing region of Mindanao could fall by some 500,000—a 50 percent loss.
- In the United States, where the vast majority of people were farmers at the time of the American Revolution, fewer people are now full-time farmers (less than 1 percent of the population) than are full-time prisoners.
- In the U.S. states of Nebraska and Iowa, between a fifth and a third of farmers are expected to be out of business within two years.

Of course, the declining numbers of farmers in industrial nations does not imply a decline in the importance of the farming sector. The world still has to eat (and 80 million more mouths to feed each year than the year before), so smaller numbers of farmers mean larger farms and greater concentration of ownership. Despite a precipitous plunge in the number of people employed in farming in North America, Europe, and East Asia, half the world's people still make their living from the land. In sub-Saharan Africa and South Asia, more than 70 percent do. In these regions, agriculture accounts, on average, for half of total economic activity.

Some might argue that the decline of farmers is harmless, even a blessing, particularly for less developed nations that have not yet experienced the modernization that moves peasants out of backwater rural areas into the more advanced economies of the cities. For most of the past two centuries, the shift toward fewer farmers has generally been assumed to be a kind of progress. The substitution of high-powered diesel tractors for slow-moving women and men with hoes, or of large mechanized industrial farms for clusters of small "old fashioned" farms, is typically seen as the way to a more abundant and affordable food supply. Our urban-centered society has even come to view rural life, especially in the form of small family-owned businesses, as backwards or boring, fit only for people who wear overalls and go to bed early—far from the sophistication and dynamism of the city.

Urban life does offer a wide array of opportunities, attractions, and hopes—some of them falsely created by urban-oriented commercial media—that many farm families decide to pursue willingly. But city life often turns out to be a disappointment, as displaced farmers find themselves lodged in crowded slums, where unemployment and ill-health are the norm and where they are worse off than they were back home. Much evidence suggests that farmers aren't so much being lured to the city as they are being driven off their farms by a variety of structural changes in the way the global food chain operates. Bob Long, a rancher in McPherson County, Nebraska, stated in a recent *New York Times* article that passing the farm onto his son would be nothing less than "child abuse."

As long as cities are under the pressure of population growth (a situation expected to continue at least for the next three or four decades), there will always be pressure for a large share of humanity to subsist in the countryside. Even in highly urbanized North America and Europe, roughly 25 percent of the population—275 million people—still reside in rural areas. Meanwhile, for the 3 billion Africans, Asians, and Latin Americans who remain in the countryside—and who will be there for the foreseeable future—the marginalization of farmers has set up a vicious cycle of low educational achievement, rising infant mortality, and deepening mental distress.

Hired Hands on Their Own Land

In the 18th and 19th centuries, farmers weren't so trapped. Most weren't wealthy, but they generally enjoyed stable incomes and strong community ties. Diversified farms yielded a range of raw and processed goods that the farmer could typically sell in a local market. Production costs tended to be much lower than now, as many of the needed inputs were home-grown: the farmer planted seed that he or she had saved from the previous year, the farm's cows or pigs provided fertilizer, and the diversity of crops—usually a large range of grains, tubers, vegetables, herbs, flowers, and fruits for home use as well as for sale—effectively functioned as pest control.

Things have changed, especially in the past half-century, according to Iowa State agricultural economist Mike Duffy. "The end of World War II was a watershed period," he says. "The widespread introduction of chemical fertilizers and synthetic pesticides, produced as part of the war effort, set in motion dramatic changes in how we farm—and a dramatic decline in the number of farmers." In the post-war period, along with increasing mechanization, there was an increasing tendency to "outsource" pieces of the work that the farmers had previously done themselves—from producing their own fertilizer to cleaning and packaging their harvest. That outsourcing, which may have seemed like a welcome convenience at the time, eventually boomeranged: at first it enabled the farmer to increase output, and thus profits, but when all the other farmers were doing it too, crop prices began to fall.

Before long, the processing and packaging businesses were adding more "value" to the purchased product than the farmer, and it was those businesses that became the

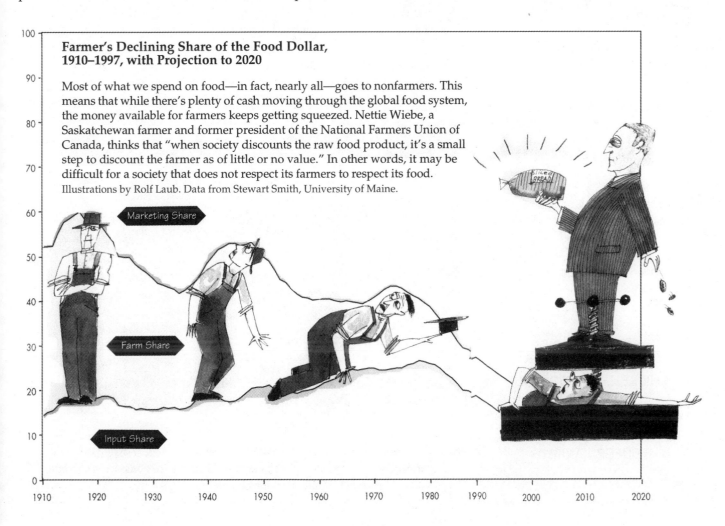

Farmer's Declining Share of the Food Dollar, 1910–1997, with Projection to 2020

Most of what we spend on food—in fact, nearly all—goes to nonfarmers. This means that while there's plenty of cash moving through the global food system, the money available for farmers keeps getting squeezed. Nettie Wiebe, a Saskatchewan farmer and former president of the National Farmers Union of Canada, thinks that "when society discounts the raw food product, it's a small step to discount the farmer as of little or no value." In other words, it may be difficult for a society that does not respect its farmers to respect its food.
Illustrations by Rolf Laub. Data from Stewart Smith, University of Maine.

Marketing Share

Farm Share

Input Share

ConAgra: *Vertical Integration, Horizontal Concentration, Global Omnipresence*

Three conglomerates (ConAgra/DuPont, Cargill/Monsanto, and Novartis/ADM) dominate virtually every link in the North American (and increasingly, the global) food chain. Here's a simplified diagram of one conglomerate.

KEY: ⬇ Vertical integration of production links, from seed to supermarket ◄► Concentration within a link

INPUTS
Distribution of farm chemicals, machinery, fertilizer, and seed

◄► 3 companies dominate North American farm machinery sector
6 companies control 63% of global pesticide market
4 companies control 69% of North American seed corn market
3 companies control 71% of Canadian nitrogen fertilizer capacity
ConAgra distributes all of these inputs, and is in a joint venture with DuPont to distribute DuPont's transgenic high-oil corn seed.

FARMS

◄► The farm sector is rapidly consolidating in the industrial world, as farms "get big or get out." Many go under contract with **ConAgra** and other conglomerates; others just go under. In the past 50 years, the number of farmers has declined by 86% in Germany, 85% in France, 85% in Japan, 64% in the U.S., 59% in South Korea, and 59% in the U.K.

GRAIN COLLECTION

◄► A proposed merger of Cargill and Continental Grain will control half of the global grain trade; **ConAgra** has about one-quarter.

GRAIN MILLING

◄► **ConAgra** and 3 other companies account for 62% of the North American market.

PRODUCTION OF BEEF, PORK, TURKEY, CHICKEN, AND SEAFOOD

◄► **ConAgra** ranks 3rd in cattle feeding and 5th in broiler production.

◄► **ConAgra** Poultry, Tyson Foods, Perdue, and 3 other companies control 60% of U.S. chicken production

PROCESSING OF BEEF, PORK, TURKEY, CHICKEN, AND SEAFOOD

◄► IBP, **ConAgra,** Cargill, and Farmland control 80% of U.S. beef packing

◄► Smithfield, **ConAgra**, and 3 other companies control 75% of U.S. pork packing

SUPERMARKETS

◄► **ConAgra** divisions own Wesson oil, Butterball turkeys, Swift Premium meats, Peter Pan peanut butter, Healthy Choice diet foods, Hunt's tomato sauce, and about 75 other major brands.

dominant players in the food industry. Instead of farmers outsourcing to contractors, it became a matter of large food processors buying raw materials from farmers, on the processors' terms. Today, most of the money is in the work the farmer no longer does—or even controls. In the United States, the share of the consumer's food dollar that trickles back to the farmer has plunged from nearly 40 cents in 1910 to just above 7 cents in 1997, while the shares going to input (machinery, agrochemicals, and seeds) and marketing (processing, shipping, brokerage, advertising, and retailing) firms have continued to expand. (See graph "Farmer's Declining Share of the Food Dollar") The typical U.S. wheat farmer, for instance, gets just 6 cents of the dollar spent on a loaf of bread—so when you buy that loaf, you're paying about as much for the wrapper as for the wheat.

Ironically, then, as U.S. farms became more mechanized and more "productive," a self-destructive feedback loop was set in motion: over-supply and declining crop prices cut into farmers' profits, fueling a demand for more technology aimed at making up for shrinking margins by increasing volume still more. Output increased dramatically, but expenses (for tractors, combines, fertilizer, and seed) also ballooned—while the commodity prices stagnated or declined. Even as they were looking more and more modernized, the farmers were becoming less and less the masters of their own domain.

On the typical Iowa farm, the farmer's profit margin has dropped from 35 percent in 1950 to 9 percent today. In order to generate the same income, this farm would need to be roughly four times as large today as in 1950—or the farmer would need to get a night job. And that's precisely what we've seen in most industrialized nations: fewer farmers on bigger tracts of land producing a greater share of the total food supply. The farmer with declining margins buys out his neighbor and expands or risks being cannibalized himself.

There is an alternative to this huge scaling up, which is to buck the trend and bring some of the input-supplying and post-harvest processing—and the related profits—back onto the farm. But more self-sufficient farming would be highly unpopular with the industries that now make lucrative profits from inputs and processing. And since these industries have much more political clout than the farmers do, there is little support for rescuing farmers from their increasingly servile condition—and the idea has been largely forgotten. Farmers continue to get the message that the only way to succeed is to get big.

The traditional explanation for this constant pressure to "get big or get out" has been that it improves the efficiency of the food system—bigger farms replace smaller farms, because the bigger farms operate at lower costs. In some respects, this is quite true. Scaling up may allow a farmer to spread a tractor's cost over greater acreage, for example. Greater size also means greater leverage in purchasing inputs or negotiating loan rates—increasingly important as satellite-guided combines and other equipment make farming more and more capital-intensive. But these economies of scale typically level off. Data for a wide range of crops produced in the United States show that the lowest production costs are generally achieved on farms that are much smaller than the typical farm now is. But large farms can tolerate lower margins, so while they may not *produce* at lower cost, they can afford to *sell* their crops at lower cost, if forced to do so—as indeed they are by the food processors who buy from them. In short, to the extent that a giant farm has a financial benefit over a small one, it's a benefit that goes only to the processor—not to the farmer, the farm community, or the environment.

This shift of the food dollar away from farmers is compounded by intense concentration in every link of the food chain—from seeds and herbicides to farm finance and retailing. In Canada, for example, just three companies control over 70 percent of fertilizer sales, five banks provide the vast majority of agricultural credit, two companies control over 70 percent of beef packing, and five companies dominate food retailing. The merger of Philip Morris and Nabisco will create an empire that collects nearly 10 cents of every dollar a U.S. consumer spends on food, according to a company spokesperson. Such high concentration can be deadly for the bottom line, allowing agribusiness firms to extract higher prices for the products farmers buy from them, while offering lower prices for the crop they buy from the farmers.

An even more worrisome form of concentration, according to Bill Heffernan, a rural sociologist at the University of Missouri, is the emergence of several clusters of firms that—through mergers, takeovers, and alliances with other links in the food chain—now possess "a seamless and fully vertically integrated control of the food system from gene to supermarket shelf." (See diagram "ConAgra") Consider the recent partnership between Monsanto and Cargill, which controls seeds, fertilizers, pesticides, farm finance, grain collection, grain processing, livestock feed processing, livestock production, and slaughtering, as well as some well-known processed food brands. From the standpoint of a company like Cargill, such alliances yield tremendous control over costs and can therefore be extremely profitable.

But suppose you're the farmer. Want to buy seed to grow corn? If Cargill is the only buyer of corn in a hundred mile radius, and Cargill is only buying a particular Monsanto corn variety for its mills or elevators or feedlots, then if you don't plant Monsanto's seed you won't have a market for your corn. Need a loan to buy the seed? Go to Cargill-owned Bank of Ellsworth, but be sure to let them know which seed you'll be buying. Also mention that you'll be buying Cargill's Saskferco brand fertilizer. OK, but once the corn is grown, you don't like the idea of having to sell to Cargill at the prices it dictates? Well, maybe you'll feed the corn to your pigs, then, and sell them to the highest bidder. No problem—Cargill's Excel Corporation buys pigs, too. OK, you're moving to the city, and renouncing the farm life! No more home-made grits for breakfast, you're buying corn flakes. Well, good news: Cargill Foods supplies corn flour to the top cereal makers. You'll notice, though, that all the big brands of corn flakes seem to have pretty much the same hefty price per ounce. After all, they're all made by the agricultural oligopoly.

As these vertical food conglomerates consolidate, Heffernan warns, "there is little room left in the global food system for independent farmers"—the farmers being increasingly left with "take it or leave it" contracts from the remaining conglomerates. In the last two decades, for example, the share of American agricultural output produced under contract has more than tripled, from 10 percent to 35 percent—and this doesn't include the contracts that farmers must sign to plant genetically engineered seed. Such centralized control of the food system, in which farmers are in effect reduced to hired hands on their own land, reminds Heffernan of the Soviet-style state farms, but with the Big Brother role now being played by agribusiness executives. It is also reminiscent of the "company store" which once dominated small American mining or factory towns, except that if you move out of town now, the store is still with you. The company store has gone global.

With the conglomerates who own the food dollar also owning the political clout, it's no surprise that agricultural policies—including subsidies, tax breaks, and environmental legislation at both the national and international levels—do not generally favor the farms. For example, the conglomerates command growing influence over both private and public agricultural research priorities, which might explain why the U.S. Department of Agriculture (USDA), an agency ostensibly beholden to farmers, would help to develop the seed-sterilizing Terminator technology—a biotechnology that offers farmers only greater dependence on seed companies. In some cases the influence is indirect, as manifested in government funding decisions, while in others it is more blatant. When Novartis provided $25 million to fund a research partnership with the plant biology department of the University of California at Berkeley, one of the conditions was that Novartis has the first right of refusal for any patentable inventions. Under those circumstances, of course, the UC officials—mindful of where their funding comes from—have strong incentives to give more attention to technologies like the Terminator seed, which shifts profit away from the farmer, than to technologies that directly benefit the farmer or the public at large.

Even policies that are touted to be in the best interest of farmers, like liberalized trade in agricultural products, are increasingly shaped by non-farmers. Food traders, processors, and distributors, for example, were some of the principal architects of recent revisions to the General Agreement on Trade and Tariffs (GATT)—the World Trade Organization's predecessor—that paved the way for greater trade flows in agricultural commodities. Be-fore these revisions, many countries had mechanisms for assuring that their farmers wouldn't be driven out of their own domestic markets by predatory global traders. The traders, however, were able to do away with those protections.

The ability of agribusiness to slide around the planet, buying at the lowest possible price and selling at the highest, has tended to tighten the squeeze already put in place by economic marginalization, throwing every farmer on the planet into direct competition with every other farmer. A recent UN Food and Agriculture Organization assessment of the experience of 16 developing nations in implementing the latest phase of the GATT concluded that "a common reported concern was with a general trend towards the concentration of farms," a process that tends to further marginalize small producers and exacerbate rural poverty and unemployment. The sad irony, according to Thomas Reardon, of Michigan State University, is that while small farmers in all reaches of the world are increasingly affected by cheap, heavily subsidized imports of foods from outside of their traditional rural markets, they are nonetheless often excluded from opportunities to participate in food exports themselves. To keep down transaction costs and to keep processing standardized, exporters and other downstream players prefer to buy from a few large producers, as opposed to many small producers.

As the global food system becomes increasingly dominated by a handful of vertically integrated, international corporations, the servitude of the farmer points to a broader society-wide servitude that OPEC-like food cartels could impose, through their control over food prices and food quality. Agricultural economists have already noted that the widening gap between retail food prices and farm prices in the 1990s was due almost exclusively to exploitation of market power, and not to extra services provided by processors and retailers. It's questionable whether we should pay as much for a bread wrapper as we do for the nutrients it contains. But beyond this, there's a more fundamental question. Farmers are professionals, with extensive knowledge of their local soils, weather, native plants, sources of fertilizer or mulch, native pollinators, ecology, and community. If we are to have a world where the land is no longer managed by such professionals, but is instead managed by distant corporate bureaucracies interested in extracting maximum output at minimum cost, what kind of food will we have, and at what price?

Agrarian Services

No question, large industrial farms can produce lots of food. Indeed, they're designed to maximize quantity. But when the farmer becomes little more than the lowest-cost producer of raw materials, more than his own welfare will suffer. Though the farm sector has lost power and profit, it is still the one link in the agrifood chain accounting for the largest share of agriculture's public goods—including half the world's jobs, many of its most vital communities, and many of its most diverse landscapes. And in providing many of these goods, small farms clearly have the advantage.

Local economic and social stability: Over half a century ago, William Goldschmidt, an anthropologist working at the USDA, tried to assess how farm structure and size affect the health of rural communities. In California's San Joaquin Valley, a region then considered to be at the cutting edge of agricultural industrialization, he identified two small towns that were alike in all basic economic and geographic dimensions, including value of agricultural production, except in farm size. Comparing the two, he found an inverse correlation between the sizes of the farms and the well-being of the communities they were a part of.

The small-farm community, Dinuba, supported about 20 percent more people, and at a considerably higher level of living—including lower poverty rates, lower levels of economic and social class distinctions, and a lower crime rate—than the large-farm community of Arvin. The majority of Dinuba's residents were independent entrepreneurs, whereas fewer than 20 percent of Arvin's residents were—most of the others being agricultural laborers. Dinuba had twice as many business establishments as Arvin, and did 61 percent more retail business. It had more schools, parks, newspapers, civic organizations, and churches, as well as better physical infrastructure—paved streets, sidewalks, garbage disposal, sewage disposal and other public services. Dinuba also had more institutions for democratic decision making, and a much broader participation by its citizens. Political scientists have long recognized that a broad base of independent entrepreneurs and property owners is one of the keys to a healthy democracy.

The distinctions between Dinuba and Arvin suggest that industrial agriculture may be limited in what it can do for a community. Fewer (and less meaningful) jobs, less local spending, and a hemorrhagic flow of profits to absentee landowners and distant suppliers means that industrial farms can actually be a net drain on the local economy. That hypothesis has been corroborated by Dick Levins, an agricultural economist at the University of Minnesota. Levins studied the economic receipts from Swift County, Iowa, a typical Midwestern corn and soybean community, and found that although total farm sales are near an all-time high, farm income there has been dismally low—and that many of those who were once the financial stalwarts of the community are now deeply in debt. "Most of the U.S. Corn Belt, like Swift County, is a colony, owned and operated by people who don't live there and for the benefit of those who don't live there," says Levin. In fact, most of the land in Swift County is rented, much of it from absentee landlords.

This new calculus of farming may be eliminating the traditional role of small farms in anchoring rural economies—the kind of tradition, for example, that we saw in the emphasis given to the support of small farms by Japan, South Korea, and Taiwan following World War II. That emphasis, which brought radical land reforms and targeted investment in rural areas, is widely cited as having been a major stimulus to the dramatic economic boom those countries enjoyed.

Not surprisingly, when the economic prospects of small farms decline, the social fabric of rural communities begins to tear. In the United States, farming families are more than twice as likely as others to live in poverty. They have less education and lower rates of medical protection, along with higher rates of infant mortality, alcoholism, child abuse, spousal abuse, and mental stress. Across Europe, a similar pattern is evident. And in sub-Saharan Africa, sociologist Deborah Bryceson of the Netherlands-based African Studies Centre has studied the dislocation of small farmers and found that "as de-agrarianization proceeds, signs of social dysfunction associated with urban areas [including petty crime and breakdowns of family ties] are surfacing in villages."

People without meaningful work often become frustrated, but farmers may be a special case. "More so than other occupations, farming represents a way of life and defines who you are," says Mike Rosemann, a psychologist who runs a farmer counseling network in Iowa. "Losing the family farm, or the prospect of losing the family farm, can generate tremendous guilt and anxiety, as if one has failed to protect the heritage that his ancestors worked to hold onto." One measure of the despair has been a worldwide surge in the number of farmers committing suicide. In 1998, over 300 cotton farmers in Andhra Pradesh, India, took their lives by swallowing pesticides that they had gone into debt to purchase but that had nonetheless failed to save their crops. In Britain, farm workers are two-and-a-half times more likely to commit suicide than the rest of the population. In the United States, official statistics say farmers are now five times as likely to commit suicide as to die from farm accidents, which have been traditionally the most frequent cause of unnatural death for them. The true number may be even higher, as suicide hotlines report that they often receive calls from farmers who want to know which sorts of accidents (Falling into the blades of a combine? Getting shot while hunting?) are least likely to be investigated by insurance companies that don't pay claims for suicides.

Whether from despair or from anger, farmers seem increasingly ready to rise up, sometimes violently, against government, wealthy landholders, or agribusiness giants. In recent years we've witnessed the Zapatista revolution in Chiapas, the seizing of white-owned farms by landless blacks in Zimbabwe, and the attacks of European farmers on warehouses storing genetically engineered seed. In the book *Harvest of Rage*, journalist Joel Dyer links the 1995 Oklahoma City bombing that killed nearly 200 people—

In the Developing World, an Even Deeper Farm Crisis

"One would have to multiply the threats facing family farmers in the United States or Europe five, ten, or twenty times to get a sense of the handicaps of peasant farmers in less developed nations," says Deborah Bryceson, a senior research fellow at the African Studies Centre in the Netherlands. Those handicaps include insufficient access to credit and financing, lack of roads and other infrastructure in rural areas, insecure land tenure, and land shortages where population is dense.

Three forces stand out as particularly challenging to these peasant farmers:

Structural adjustment requirements, imposed on indebted nations by international lending institutions, have led to privatization of "public commodity procurement boards" that were responsible for providing public protections for rural economies. "The newly privatized entities are under no obligation to service marginal rural areas," says Rafael Mariano, chairman of a Filipino farmers' union. Under the new rules, state protections against such practices as dumping of cheap imported goods (with which local farmers can't compete) were abandoned at the same time that state provision of health care, education, and other social services was being reduced.

Trade liberalization policies associated with structural adjustment have reduced the ability of nations to protect their agricultural economies even if they want to. For example, the World Trade Organization's Agreement on Agriculture will forbid domestic price support mechanisms and tariffs on imported goods—some of the primary means by which a country can shield its own farmers from overproduction and foreign competition.

The growing emphasis on agricultural grades and standards—the standardizing of crops and products so they can be processed and marketed more "efficiently"—has tended to favor large producers, and to marginalize smaller ones. Food manufacturers and supermarkets have emerged as the dominant entities in the global agri-food chain, and with their focus on brand consistency, ingredient uniformity, and high volume, smaller producers often are unable to deliver—or aren't even invited to bid.

Despite these daunting conditions, many peasant farmers tend to hold on long after it has become clear that they can't compete. One reason, says Peter Rosset of the Institute for Food and Development Policy, is that "even when it gets really bad, they will cling to agriculture because of the fact that it at least offers some degree of food security—that you can feed yourself." But with the pressures now mounting, particularly as export crop production swallows more land, even that fallback is lost.

as well as the rise of radical right and antigovernment militias in the U.S. heartland—to a spreading despair and anger stemming from the ongoing farm crisis. Thomas Homer-Dixon, director of the Project on Environment, Population, and Security at the University of Toronto, regards farmer dislocation, and the resulting rural unemployment and poverty, as one of the major security threats for the coming decades. Such dislocation is responsible for roughly half of the growth of urban populations across the Third World, and such growth often occurs in volatile shantytowns that are already straining to meet the basic needs of their residents. "What was an extremely traumatic transition for Europe and North America from a rural society to an urban one is now proceeding at two to three times that speed in developing nations," says Homer-Dixon. And, these nations have considerably less industrialization to absorb the labor. Such an accelerated transition poses enormous adjustment challenges for India and China, where perhaps a billion and a half people still make their living from the land.

Ecological stability: In the Andean highlands, a single farm may include as many as 30 to 40 distinct varieties of potato (along with numerous other native plants), each having slightly different optimal soil, water, light, and temperature regimes, which the farmer—given enough time—can manage. (In comparison, in the United States, just four closely related varieties account for about 99 percent of all the potatoes produced.) But, according to Karl Zimmerer, a University of Wisconsin sociologist, declining farm incomes in the Andes force more and more growers into migrant labor forces for part of the year, with serious effects on farm ecology. As time becomes constrained, the farmer manages the system more homogenously—cutting back on the number of traditional varieties (a small home garden of favorite culinary varieties may be the last refuge of diversity), and scaling up production of a few commercial varieties. Much of the traditional crop diversity is lost.

Complex farm systems require a highly sophisticated and intimate knowledge of the land—something small-scale, full-time farmers are more able to provide. Two or three different crops that have different root depths, for example, can often be planted on the same piece of land, or crops requiring different drainage can be planted in close proximity on a tract that has variegated topography. But these kinds of cultivation can't be done with heavy tractors moving at high speed. Highly site-specific and management-intensive cultivation demands ingenuity and awareness of local ecology, and can't be achieved by heavy equipment and heavy applications of agrochemicals. That isn't to say that being small is always sufficient to ensure ecologically sound food production, because economic adversity can drive small farms, as well as big ones, to compromise sustainable food production by transmogrifying the craft of land stewardship into the crude labor of commodity production. But a large-scale, highly mechanized farm is simply not equipped to preserve landscape complexity. Instead, its normal modus is to use blunt management tools, like crops that have been genetically engineered to churn out insecticides, which obviate the need to scout the field to see if spraying is necessary at all.

In the U.S. Midwest, as farm size has increased, cropping systems have gotten more simplified. Since 1972, the number of counties with more than 55 percent of their acreage planted in corn and soybeans has nearly tripled, from 97 to 267. As farms scaled up, the great simplicity of managing the corn-soybean rotation—an 800 acre farm, for instance, may require no more than a couple of weeks planting in the spring and a few weeks harvesting in the fall—became its big selling point. The various arms of the agricultural economy in the region, from extension services to grain elevators to seed suppliers, began to solidify around this corn-soybean rotation, reinforcing the farmers' movement away from other crops. Fewer and fewer farmers kept livestock, as beef and hog production became "economical" only in other parts of the country where it was becoming more concentrated. Giving up livestock meant eliminating clover, pasture mixtures, and a key source of fertilizer in the Midwest, while creating tremendous manure concentrations in other places.

But the corn and soybean rotation—one monoculture followed by another—is extremely inefficient or "leaky" in its use of applied fertilizer, since low levels of biodiversity tend to leave a range of vacant niches in the field, including different root depths and different nutrient preferences. Moreover, the Midwest's shift to monoculture has subjected the country to a double hit of nitrogen pollution, since not only does the removal and concentration of livestock tend to dump inordinate amounts of feces in the places (such as Utah and North Carolina) where the livestock operations are now located, but the monocultures that remain in the Midwest have much poorer nitrogen retention than they would if their cropping were more complex. (The addition of just a winter rye crop to the corn-soy rotation has been shown to reduce nitrogen runoff by nearly 50 percent.) And maybe this disaster-in-the-making should really be regarded as a triple hit, because in addition to contaminating Midwestern water supplies, the runoff ends up in the Gulf of Mexico, where the nitrogen feeds massive algae blooms. When the algae die, they are decomposed by bacteria, whose respiration depletes the water's oxygen—suffocating fish, shellfish,

and all other life that doesn't escape. This process periodically leaves 20,000 square kilometers of water off the coast of Louisiana biologically dead. Thus the act of simplifying the ecology of a field in Iowa can contribute to severe pollution in Utah, North Carolina, Louisiana, *and* Iowa.

The world's agricultural biodiversity—the ultimate insurance policy against climate variations, pest outbreaks, and other unforeseen threats to food security—depends largely on the millions of small farmers who use this diversity in their local growing environments. But the marginalization of farmers who have developed or inherited complex farming systems over generations means more than just the loss of specific crop varieties and the knowledge of how they best grow. "We forever lose the best available knowledge and experience of place, including what to do with marginal lands not suited for industrial production," says Steve Gleissman, an agroecologist at the University of California at Santa Cruz. The 12 million hogs produced by Smithfield Foods Inc., the largest hog producer and processor in the world and a pioneer in vertical integration, are nearly genetically identical and raised under identical conditions—regardless of whether they are in a Smithfield feedlot in Virginia or Mexico.

As farmers become increasingly integrated into the agribusiness food chain, they have fewer and fewer controls over the totality of the production process—shifting more and more to the role of "technology applicators," as opposed to managers making informed and independent decisions. Recent USDA surveys of contract poultry farmers in the United States found that in seeking outside advice on their operations, these farmers now turn first to bankers and then to the corporations that hold their contracts. If the contracting corporation is also the same company that is selling the farm its seed and fertilizer, as is often the case, there's a strong likelihood that the company's procedures will be followed. That corporation, as a global enterprise with no compelling local ties, is also less likely to be concerned about the pollution and resource degradation created by those procedures, at least compared with a farmer who is rooted in that community. Grower contracts generally disavow any environmental liability.

And then there is the ecological fallout unique to large-scale, industrial agriculture. Colossal confined animal feeding operations (CAFOs)—those "other places" where livestock are concentrated when they are no longer present on Midwestern soy/corn farms—constitute perhaps the most egregious example of agriculture that has, like a garbage barge in a goldfish pond, overwhelmed the scale at which an ecosystem can cope. CAFOs are increasingly the norm in livestock production, because, like crop monocultures, they allow the production of huge populations of animals which can be slaughtered and marketed at rock-bottom costs. But the disconnection between the livestock and the land used to produce their feed means that such CAFOs generate gargantuan amounts of waste, which the surrounding soil cannot possibly absorb. (One farm in Utah will raise over five million hogs in a year, producing as much waste each day as the city of Los Angeles.) The waste is generally stored in large lagoons, which are prone to leak and even spill over during heavy storms. From North Carolina to South Korea, the overwhelming stench of these lagoons—a combination of hydrogen sulfide, ammonia, and methane gas that smells like rotten eggs—renders miles of surrounding land uninhabitable.

A different form of ecological disruption results from the conditions under which these animals are raised. Because massive numbers of closely confined livestock are highly susceptible to infection, and because a steady diet of antibiotics can modestly boost animal growth, overuse of antibiotics has become the norm in industrial animal production. In recent months, both the Centers for Disease Control and Prevention in the United States and the World Health Organization have identified such industrial feeding operations as principal causes of the growing antibiotic resistance in food-borne bacteria like *salmonella* and *campylobacter*. And as decisionmaking in the food chain grows ever more concentrated—confined behind fewer corporate doors—there may be other food safety issues that you won't even hear about, particularly in the burgeoning field of genetically modified organisms (GMOs). In reaction to growing public concern over GMOs, a coalition that ingenuously calls itself the "Alliance for Better Foods"—actually made up of large food retailers, food processors, biotech companies and corporate-financed farm organizations—has launched a $50 million public "educational" campaign, in addition to giving over $676,000 to U.S. lawmakers and political parties in 1999, to head off the mandatory labeling of such foods.

Perhaps most surprising, to people who have only casually followed the debate about small-farm values versus factory-farm "efficiency," is the fact that a wide body of evidence shows that small farms are actually more productive than large ones—by as much as 200 to 1,000 percent greater output per unit of area. How does this jive with the often-mentioned productivity advantages of large-scale mechanized operations? The answer is simply that those big-farm advantages are always calculated on the basis of how much of *one crop* the land will yield per acre. The greater productivity of a smaller, more complex farm, however, is calculated on the basis of how much food *overall* is produced per acre. The smaller farm can grow several crops utilizing different root depths, plant heights, or nutrients, on the same piece of land simultaneously. It is this "polyculture" that offers the small farm's productivity advantage.

To illustrate the difference between these two kinds of measurement, consider a large Midwestern corn farm. That farm may produce more corn per acre than a small farm in which the corn is grown as part of a polyculture that also includes beans, squash, potato, and "weeds"

that serve as fodder. But in overall output, the polycrop—under close supervision by a knowledgeable farmer—produces much more food overall, whether you measure in weight, volume, bushels, calories, or dollars.

The inverse relationship between farm size and output can be attributed to the more efficient use of land, water, and other agricultural resources that small operations afford, including the efficiencies of intercropping various plants in the same field, planting multiple times during the year, targeting irrigation, and integrating crops and livestock. So in terms of converting inputs into outputs, society would be better off with small-scale farmers. And as population continues to grow in many nations, and the agricultural resources per person continue to shrink, a small farm structure for agriculture may be central to meeting future food needs.

Rebuilding Foodsheds

Look at the range of pressures squeezing farmers, and it's not hard to understand the growing desperation. The situation has become explosive, and if stabilizing the erosion of farm culture and ecology is now critical not just to farmers but to everyone who eats, there's still a very challenging question as to what strategy can work. The agribusiness giants are deeply entrenched now, and scattered protests could have as little effect on them as a mosquito bite on a tractor. The prospects for farmers gaining political strength on their own seem dim, as their numbers—at least in the industrial countries—continue to shrink.

A much greater hope for change may lie in a joining of forces between farmers and the much larger numbers of other segments of society that now see the dangers, to their own particular interests, of continued restructuring of the countryside. There are a couple of prominent models for such coalitions, in the constituencies that have joined forces to fight the Mississippi River Barge Capacity and Hidrovía Barge Capacity projects being pushed forward in the name of global soybean productivity.

The American group has brought together at least the following riverbedfellows:

- National environmental groups, including the Sierra Club and National Audubon Society, which are alarmed at the prospect of a public commons being damaged for the profit of a small commercial interest group;
- Farmers and farmer advocacy organizations, concerned about the inordinate power being wielded by the agribusiness oligopoly;
- Taxpayer groups outraged at the prospect of a corporate welfare payout that will drain more than $1 billion from public coffers;
- Hunters and fishermen worried about the loss of habitat;

- Biologists, ecologists, and birders concerned about the numerous threatened species of birds, fish, amphibians, and plants;
- Local-empowerment groups concerned about the impacts of economic globalization on communities;
- Agricultural economists concerned that the project will further entrench farmers in a dependence on the export of low-cost, bulk commodities, thereby missing valuable opportunities to keep money in the community through local milling, canning, baking, and processing.

A parallel coalition of environmental groups and farmer advocates has formed in the Southern hemisphere to resist the Hidrovía expansion. There too, the river campaign is part of a larger campaign to challenge the hegemony of industrial agriculture. For example, a coalition has formed around the Landless Workers Movement, a grassroots organization in Brazil that helps landless laborers to organize occupations of idle land belonging to wealthy landlords. This coalition includes 57 farm advocacy organizations based in 23 nations. It has also brought together environmental groups in Latin America concerned about the related ventures of logging and cattle

ranching favored by large landlords; the mayors of rural towns who appreciate the boost that farmers can give to local economies; and organizations working on social welfare in Brazil's cities, who see land occupation as an alternative to shantytowns.

The Mississippi and Hidrovía projects, huge as they are, still constitute only two of the hundreds of agro-industrial developments being challenged around the world. But the coalitions that have formed around them represent the kind of focused response that seems most likely to slow the juggernaut, in part because the solutions these coalitions propose are not vague or quixotic expressions of idealism, but are site-specific and practical. In the case of the alliance forming around the Mississippi River project, the coalition's work has included questioning the assumptions of the Corps of Engineers analysis, lobbying for stronger antitrust examination of agribusiness monopolies, and calling for modification of existing U.S. farm subsidies, which go disproportionately to large farmers. Environmental groups are working to re-establish a balance between use of the Mississippi as a barge mover and as an intact watershed. Sympathetic agricultural extensionists are promoting alternatives to the standard corn-soybean rotation, including certified organic crop production, which can simultaneously bring down

Past and Future: Connecting the Dots

Given the direction and speed of prevailing trends, how far can the decline in farmers go? The lead editorial in the September 13, 1999 issue of *Feedstuffs*, an agribusiness trade journal, notes that "Based on the best estimates of analysts, economists and other sources interviewed for this publication, American agriculture must now quickly consolidate all farmers and livestock producers into about 50 production systems… each with its own brands," in order to maintain competitiveness. Ostensibly, other nations will have to do the same in order to keep up.

To put that in perspective, consider that in traditional agriculture, each farm is an independent production system. In this map of Ireland's farms circa 1930, each dot represents 100 farms, so the country as a whole had many thousands of independent production systems. But if the *Feedstuffs* prognosis were to come to pass, this map would be reduced to a single dot. And even an identically keyed map of the much larger United States would show the country's agriculture reduced to just one dot.

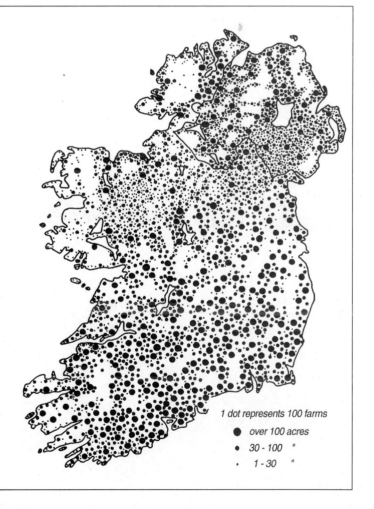

1 dot represents 100 farms

● over 100 acres
• 30 - 100   "
· 1 - 30   "

input costs and garner a premium for the final product, and reduce nitrogen pollution.

The United States and Brazil may have made costly mistakes in giving agribusiness such power to reshape the rivers and land to its own use. But the strategy of interlinked coalitions may be mobilizing in time to save much of the world's agricultural health before it is too late. Dave Brubaker, head of the Spira/GRACE Project on Industrial Animal Production at the Johns Hopkins University School of Public Health, sees these diverse coalitions as "the beginning of a revolution in the way we look at the food system, tying in food production with social welfare, human health, and the environment." Brubaker's project brings together public health officials focused on antibiotic overuse and water contamination resulting from hog waste; farmers and local communities who oppose the spread of new factory farms or want to close down existing ones; and a phalanx of natural allies with related campaigns, including animal rights activists, labor unions, religious groups, consumer rights activists, and environmental groups.

"As the circle of interested parties is drawn wider, the alliance ultimately shortens the distance between farmer and consumer," observes Mark Ritchie, president of the Institute for Agriculture and Trade Policy, a research and advocacy group often at the center of these partnerships. This closer proximity may prove critical to the ultimate sustainability of our food supply, since socially and ecologically sound buying habits are not just the passive *result* of changes in the way food is produced, but can actually be the most powerful *drivers of* these changes. The explosion of farmers' markets, community-supported agriculture, and other direct buying arrangements between farmers and consumers points to the growing numbers of nonfarmers who have already shifted their role in the food chain from that of choosing from the tens of thousands of food brands offered by a few dozen companies to bypassing such brands altogether. And, since many of the additives and processing steps that take up the bulk of the food dollar are simply the inevitable consequence of the ever-increasing time commercial food now spends in global transit and storage, this shortening of distance between grower and consumer will not only benefit the culture and ecology of farm communities. It will also give us access to much fresher, more flavorful, and more nutritious food. Luckily, as any food marketer can tell you, these characteristics aren't a hard sell.

Brian Halweil is a staff researcher at the Worldwatch Institute.

All the Wild Rivers

In the United States, the number of dams being torn down now outnumbers those being built. Elsewhere as well, people are having second thoughts about the great 20th-century enthusiasm for controlling and harnessing rivers.

by Curtis Runyan

In 1966, Floyd Dominy, the commissioner of the U.S. Bureau of Reclamation, gave a speech lambasting environmentalists for their opposition to damming up the Grand Canyon national park. If the dams were not built, he told the audience, the Colorado River would be "useless to anyone." Dominy, head of the agency that led the charge in the United States' rush to dam up its rivers, concluded: "I've seen all the wild rivers I ever want to see."

Thirty years later, in 1998, Bruce Babbitt, the U.S. Secretary of the Interior, traveled across the country to several rivers on a "Sledgehammer tour"—not to break ground on new construction, but to tear four dams down. "America overshot the mark in our dam building frenzy," he said in a speech to the Ecological Society of America. "For most of this century, politicians have eagerly rushed in, amidst cheering crowds, to claim credit for the construction of 75,000 dams all across America. Think about that number. That means we have been building, on average, one large dam a day, every single day, since the Declaration of Independence. Many of these dams have become monuments, expected to last forever. You could say forever just got a lot shorter."

One of the world leaders in building new dams, the United States, is now leading the world in tearing them down. The country is now decommissioning more large dams than it builds each year, and has removed at least 465 of them, according to a study by American Rivers, Friends of the Earth, and Trout Unlimited. France and other countries are following suit. "It's striking how, in just two or three decades, the U.S. has gone from building dams to not building dams to taking some of them down," wrote Marc Reisner, author of *Cadillac Desert*, in the Earth Day 2000 edition of *Time* magazine. "What we're just beginning to understand is how water development has, like nuclear energy, amounted to a Faustian bargain between civilization and the natural world."

Ecologically, rivers are under siege. They are being drained, diverted, polluted, and blocked at a rate that has degraded freshwater ecosystems worldwide. With more than half of the world's rivers stopped up by at least one large dam (over 15 meters high), dams have played a significant role in destabilizing riverine ecology. For example, at least one fifth of the world's freshwater fish are now endangered or extinct. In addition, reservoirs behind dams have flooded vast amounts of the world's most fertile agricultural and forest land. Reservoirs also trap the sediment loads of rivers, reduce the supply of nutrients flowing downstream, release water at cooler temperatures, and disrupt healthy river ecosystems.

The ill-effects of dams are not confined to river valleys. Half of the world's dams were built to irrigate the farmland that now provides about 12 to 16 percent of the human food supply. However, channeling water to irrigate basins without good drainage has led to extensive salinization and waterlogging of soils. Bad drainage and poorly planned irrigation—including groundwater pumping—have reduced or ended the productivity of nearly one-fifth of the world's irrigated land.

The impact of dam building on communities has also been substantial. An estimated 40 to 80 million people have been physically displaced by the construction of dams. They have been flooded out, forced to move. One of the world's most massive engineering projects, the Three Gorges Dam in China, if completed could force the relocation of nearly 2 million people. Most frequently, the people affected are not those who receive the irrigation, electricity, or other benefits provided by dams. In fact, those who are resettled have rarely ever seen their livelihoods restored. "The poor, other vulnerable groups, and future generations are likely to bear a disproportionate share the social and environmental costs of large dam projects without gaining a commensurate share of the economic benefits," finds the World Commission on Dams (WCD), an independent, collaborative body consisting of dam construction industry representatives, anti-dam activists, and government officials, among others. The commission released its landmark report in November 2000, providing one of the first global surveys of dams with input from both supporters and critics.

The U.S. effort to consider dam removal or breaching is only part of a worldwide shift in thinking about dams. In almost every country in the world, the number of new dams being built is plummeting. Ninety-one large dams were built in the 1970s in Brazil, for instance. The number built dropped to 60 in the 1980s, and to 28 in the first six years of the 1990s (see graph above). Even where dams continue to be built, public acceptance is waning, says Owen Lammers of the Glen Canyon Action Network, an ambitious U.S. activist group pushing to tear down the massive Glen Canyon Dam in Arizona. "The number of dams being constructed is going down," says Lammers, "while the number facing resistance and severe criticism is going up."

Evoking the almost religious fervor with which dams have been built in the past, Prime Minister Jawaharlal Nehru called the massive concrete and earthen structures being put up around his country "the temples of modern India." But after a half-century of being regarded as technological marvels, many of these structures are being reinspected and rejected as boondoggles.

Still, the number of dams and dam projects that have been stopped or removed is only a tiny fraction of those that have

Number of New Large Dams Constructed

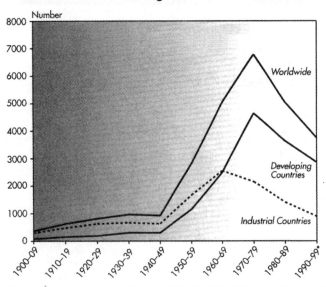

*1997–99 projected

Source: International Commission on Large Dams, *World Register of Dams, 1998*

Dams are categorized as large if they measure over 15 meters or have a reservoir of more than 3 million cubic meters. Data exclude more than 90 percent of the large dams in China. Data for 1997–99 are projected from 1990–96 dam commissioning rates.

been built in the past half century. And projects that face strong opposition may also be getting strong support from urban residents, large-scale farmers, or other groups that stand to benefit from a dam's construction. The Sardar Sarovar Dam on India's Narmada River, for example, is at the center of the country's debate over how development should occur. The Narmada Bachao Andolan (NBA), the local movement opposed to the damming, has rallied international attention against the project, which includes plans to construct 3,200 dams on the river. But despite the opposition, in October 2000 the Indian Supreme Court lifted its four-year stay on the project.

When the global rush to build dams reached its peak in the 1970s, on average two or three large dams were commissioned around the world every day. International lenders, governments, development agencies all felt they had found in dams a solution to many of the world's development dilemmas. Dams have played an important role in addressing hunger, drought, and lack of access to clean water and electricity. They generate 19 percent of the world's electricity supply, provide water for 30 to 40 percent of the world's irrigated land, and in some places help to reduce floods. But the benefits of controlling unruly waterways—building dams and creating reservoirs with the aim of halting floods, expanding irrigation, providing drinking water, and supplying hydroelectric energy—have always been assumed to overwhelmingly outweigh the costs, even though little was known about what these costs were. However, as researchers conduct more studies on the effects of dams and as more of the local people who are affected are consulted, the assumption that the benefits outweigh the costs has become less certain.

Now that more than 45,000 large dams (over 15 meters high) have been built around the world, a growing body of research

indicates that their costs may be higher than many ever imagined. The World Commission on Dams report finds that "In too many cases an unacceptable and often unnecessary price has been paid [to secure the benefits of dams], especially in social and environmental terms, by people displaced, by communities downstream, by taxpayers, and by the natural environment." Irrigation schemes haven't supplied projected revenues, hydropower dams have not met electricity-generation projections, drinking water supplies have been costly and often unreliable, and reservoirs have lost their usefulness as they fill with sediment. Recent studies have shown that the organic debris washed into reservoirs releases large amounts of greenhouse gases, raising questions as to whether hydroelectric dams really do produce clean, renewable energy. "Considering the enormous capital invested in large dams, it is surprising that substantive evaluations of project performance are few in number, narrow in scope, and poorly integrated," finds the report.

Even the World Bank, the world's largest international funder of dam projects (the Bank has invested $75 billion in 538 dams), has begun to have second thoughts. "Our involvement in large dams has been decreasing and is focusing more on financing dam rehabilitation and safety and much less on financing new dams," said World Bank President James Wolfensohn in November 2000. To put the Bank's shift fully in perspective, however, it might be noted that while opposition to dams has been increasingly effective, the most common reason for the dramatic drop in the growth of dams is simply that many countries are already at capacity—there are fewer and fewer safe or unprotected places left to build the structures.

The jury is still out on how the World Bank, which helped instigate the dam commission, will respond to the findings of the WDC report. Wolfensohn recently told an audience of Indian business reporters that "It is unfortunate that the World Bank could not understand the depth of the water crisis in Gujarat and had to pull out of the Narmada project," which is fiercely opposed by the NBA.

"We note and appreciate that the World Commission on Dams report vindicates many concerns raised by NGO campaigns," announced an international coalition of more than 100 nongovernmental anti-dam activist groups in November 2000. In many ways, the World Commission on Dams report provides an up-front review of adverse impacts that most dam projects are never subjected to. The activists contend that if the planning process proposed by the WCD had been followed in the past, many dams would never have been built. The report concludes that dam projects should require the consent of affected communities, participatory decision-making, examination of alternatives to dams, requirements to "sustain aquatic ecosystems," and mechanisms to ensure proper reimbursement to affected communities. The coalition of activists has called for suspension of all large dam projects until countries follow the report's recommendations for equitable, accountable, and participatory decision-making.

The debate over dams has come a long way since Dominy's call 30 years ago to silence all the rivers. And while the thinking about dams has expanded since then, so has the number of dams that choke the world's rivers. It is time to take the lessons learned from constructing more than 45,000 large dams around the world, and to incorporate them into our thinking about future planning for our rivers. *For more information see: The World Commission on Dams, www.dams.org*

Curtis Runyan is associate editor of WORLD WATCH.

From *World Watch*, January/February 2001, pp. 31–33. © 2001 by Worldwatch Institute, www.worldwatch.org.

Growing more Food with less Water

by Sandra Postel

Six thousand years ago farmers in Mesopotamia dug a ditch to divert water from the Euphrates River. With that successful effort to satisfy their thirsty crops, they went on to form the world's first irrigation-based civilization. This story of the ancient Sumerians is well known. What is not so well known is that Sumeria was one of the earliest civilizations to crumble in part because of the consequences of irrigation.

Sumerian farmers harvested plentiful wheat and barley crops for some 2,000 years thanks to the extra water brought in from the river, but the soil eventually succumbed to salinization—the toxic buildup of salts and other impurities left behind when water evaporates. Many historians argue that the poisoned soil, which could not support sufficient food production, figured prominently in the society's decline.

Far more people depend on irrigation in the modern world than did in ancient Sumeria. About 40 percent of the world's food now grows in irrigated soils, which make up 18 percent of global cropland [see "Top 10 Irrigators Worldwide"]. Farmers who irrigate can typically reap two or three harvests every year *and* get higher crop yields. As a result, the spread of irrigation has been a key factor behind the near tripling of global grain production since 1950. Done correctly, irrigation will continue to play a leading role in feeding the world, but as history shows, dependence on irrigated agriculture also entails significant risks.

Today irrigation accounts for two thirds of water use worldwide and as much as 90 percent in many developing countries. Meeting the crop demands projected for 2025, when the planet's population is expected to reach eight billion, could require an additional 192 cubic miles of water—a volume nearly equivalent to the annual flow of the Nile 10 times over. No one yet knows how to supply that much additional water in a way that protects supplies for future use.

Severe water scarcity presents the single biggest threat to future food production. Even now many freshwater sources—underground aquifers and rivers—are stressed beyond their limits.

As much as 8 percent of food crops grows on farms that use groundwater faster than the aquifers are replenished, and many large rivers are so heavily diverted that they don't reach the sea for much of the year. As the number of urban dwellers climbs to five billion by 2025, farmers will have to compete even more aggressively with cities and industry for shrinking resources.

If the world hopes to feed its burgeoning population, irrigation must become less wasteful and more widespread

Despite these challenges, agricultural specialists are counting on irrigated land to produce most of the additional food that will be needed worldwide. Better management of soil and water, along with creative cropping patterns, can boost production from cropland that is watered only by rainfall, but the heaviest burden will fall on irrigated land. To fulfill its potential, irrigated agriculture requires a thorough redesign organized around two primary goals: cut water demands of mainstream agriculture and bring low-cost irrigation to poor farmers.

Fortunately, a great deal of room exists for improving the productivity of water used in agriculture. A first line of attack is to increase irrigation efficiency. At present, most farmers irrigate their crops by flooding their fields or channeling the water down parallel furrows, relying on gravity to move the water across the land. The plants absorb only a small fraction of the water; the rest drains into rivers or aquifers, or evaporates. In many locations this practice not only wastes and pollutes water but also degrades the land through erosion, waterlogging and salinization. More efficient and environmentally sound technologies exist that could reduce water demand on farms by up to 50 percent.

Drip systems rank high among irrigation technologies with significant untapped potential. Unlike flooding techniques, drip systems enable farmers to deliver water directly to the plants' roots drop by drop, nearly eliminating waste. The water travels at low pressure through a network of perforated plastic tubing installed on or below the surface of the soil, and it emerges through small holes at a slow but steady pace. Because the plants enjoy an ideal moisture environment, drip irrigation usually offers the added bonus of higher crop yields. Studies in India, Israel, Jordan, Spain and the U.S. have shown time and again that drip irrigation reduces water use by 30 to 70 percent and increases crop yield by 20 to 90 percent compared with flooding methods.

Technologies exist that could reduce water demand on farms by up to 50 percent

Sprinklers can perform almost as well as drip methods when they are designed properly. Traditional high-pressure irrigation sprinklers spray water high into the air to cover as large a land area as possible. The problem is that the more time the water spends in the air, the more of it evaporates and blows off course before reaching the plants. In contrast, new low-energy sprinklers deliver water in small doses through nozzles positioned just above the ground. Numerous farmers in Texas who have installed such sprinklers have found that their plants absorb 90 to 95 percent of the water that leaves the sprinkler nozzle.

Despite these impressive payoffs, sprinklers service only 10 to 15 percent of the world's irrigated fields, and drip systems account for just over 1 percent. The higher costs of these technologies (relative to simple flooding methods) have been a barrier to their spread, but so has the prevalence of national water policies that discourage rather than foster efficient water use. Many governments have set very low prices for publicly supplied irrigation, leaving farmers with little motivation to invest in ways to conserve water or to improve efficiency. Most authorities have also failed to regulate groundwater pumping, even in regions where aquifers are overtapped. Farmers might be inclined to conserve their own water supplies if they could profit from selling the surplus, but a number of countries prohibit or discourage this practice.

Efforts aside from irrigation technologies can also help reduce agricultural demand for water. Much potential lies in scheduling the timing of irrigation to more precisely match plants' water needs. Measurements of climate factors such as temperature and precipitation can be fed into a computer that calculates how much water a typical plant is consuming. Farmers can use this figure to determine, quite accurately, when and how much to irrigate their particular crops throughout the growing season. A 1995 survey conducted by the University of California at Berkeley found that, on average, farmers in California who used this tool reduced water use by 13 percent and achieved an 8 percent increase in yield—a big gain in water productivity.

An obvious way to get more benefit out of water is to use it more than once. Some communities use recycled wastewater. Treated wastewater accounts for 30 percent of Israel's agricultural water supply, for instance, and this share is expected to climb to 80 percent by 2025. Developing new crop varieties offers potential as well. In the quest for higher yields, scientists have already exploited many of the most fruitful agronomic options for growing more food with the same amount of water. The hybrid wheat and rice varieties that spawned the green revolution, for example, were bred to allocate more of the plants' energy—and thus their water uptake—into edible grain. The widespread adoption of high-yielding and early-maturing rice varieties has led to a roughly threefold increase in the amount of rice harvested per unit of water consumed—a tremendous achievement. No strategy in sight—neither conventional breeding techniques nor genetic engineering—could repeat those gains on such a grand scale, but modest improvements are likely.

Yet another way to do more with less water is to reconfigure our diets. The typical North American diet, with its large share of animal products, requires twice as much water to produce as the less meat-intensive diets common in many Asian and some European countries. Eating lower on the food chain could allow the same volume of water to feed two Americans instead of one, with no loss in overall nutrition.

Reducing the water demands of mainstream agriculture is critical, but irrigation will never reach its potential to alleviate rural hunger and poverty without additional efforts. Among the world's approximately 800 million undernourished people are millions of poor farm families who could benefit dramatically from access to irrigation water or to technologies that enable them to use local water more productively.

Top 10 Irrigators Worldwide

SOURCE: UN FAO AGROSTAT database, 1998

Most of these people live in Asia and Africa, where long dry seasons make crop production difficult or impossible without irrigation. For them, conventional irrigation technologies are too expensive for their small plots, which typically encompass fewer than five acres. Even the least expensive motorized pumps that are made for tapping groundwater cost about $350, far out of reach for farmers earning barely that much in a year. Where affordable irrigation technologies have been made available, however, they have proved remarkably successful.

I traveled to Bangladesh in 1998 to see one of these successes firsthand. Torrential rains drench Bangladesh during the monsoon months, but the country receives very little precipitation the rest of the year. Many fields lie fallow during the dry season, even though groundwater lies less than 20 feet below the surface. Over the past 17 years a foot-operated device called a treadle pump has transformed much of this land into productive, year-round farms.

To an affluent Westerner, this pump resembles a StairMaster exercise machine and is operated in much the same way. The user pedals up and down on two long bamboo poles, or treadles, which in turn activate two steel cylinders. Suction pulls groundwater into the cylinders and then dispenses it into a channel in the field. Families I spoke with said they often treadled four to six hours a day to irrigate their rice paddies and vegetable plots. But the hard work paid off: not only were they no longer hungry during the dry season, but they had surplus vegetables to take to market.

Costing less than $35, the treadle pump has increased the average net income for these farmers—which is often as little as a dollar a day—by $100 a year. To date, Bangladeshi farmers have purchased some 1.2 million treadle pumps, raising the productivity of more than 600,000 acres of farmland. Manufactured and marketed locally, the pumps are injecting at least an additional $350 million a year into the Bangladeshi economy.

In other impoverished and water-scarce regions, poor farmers are reaping the benefits of newly designed low-cost drip and sprinkler systems. Beginning with a $5 bucket kit for home gardens, a spectrum of drip systems keyed to different income levels and farm sizes is now enabling farmers with limited access to water to irrigate their land efficiently. In 1998 I spoke with farmers in the lower Himalayas of northern India, where crops are grown on terraces and irrigated with a scarce communal water supply. They expected to double their planted area with the increased efficiency brought about by affordable drip systems.

Bringing these low-cost irrigation technologies into more widespread use requires the creation of local, private-sector supply chains—including manufacturers, retailers and installers—as well as special innovations in marketing. The treadle pump has succeeded in Bangladesh in part because local businesses manufactured and sold the product and marketing specialists reached out to poor farmers with creative techniques, including an open-air movie and village demonstrations. The challenge is great, but so is the potential payoff. Paul Polak, a pioneer in the field of low-cost irrigation and president of International Development Enterprises in Lakewood, Colo., believes a realistic goal for the next 15 years is to reduce the hunger and poverty of 150 million of the world's poorest rural people through the spread of affordable small-farm irrigation techniques. Such an accomplishment would boost net income among the rural poor by an estimated $3 billion a year.

Over the next quarter of a century the number of people living in water-stressed countries will climb from 500 million to three billion. New technologies can help farmers around the world supply food for the growing population while simultaneously protecting rivers, lakes and aquifers. But broader societal changes—including slower population growth and reduced consumption—will also be necessary. Beginning with Sumeria, history warns against complacency when it comes to our agricultural foundation. With so many threats to the sustainability and productivity of our modern irrigation base now evident, it is a lesson worth heeding.

Further Information

SALT AND SILT IN ANCIENT MESOPOTAMIAN AGRICULTURE. Thorkild Jacobsen and Robert M. Adams in *Science*, Vol. 128, pages 1251–1258; November 21, 1958.

PILLAR OF SAND: CAN THE IRRIGATION MIRACLE LAST? Sandra Postel. W. W. Norton, 1999.

GROUNDWATER IN RURAL DEVELOPMENT. Stephen Foster et al. Technical paper No. 463. World Bank, Washington, D.C., 2000.

Irrigation and land-use databases are maintained by the United Nations Food and Agriculture Organization at http://apps.fao.org

SANDRA POSTEL directs the Global Water Policy Project in Amherst, Mass., and is a visiting senior lecturer in environmental studies at Mount Holyoke College. She is also a senior fellow of the Worldwatch Institute, where she served as vice president for research from 1988 to 1994.

OCEANS
Are on the Critical List

*The primary threats to the planet's seas—
overfishing, habitat degradation, pollution,
introduction of alien species, and climate
change—are largely human-induced.*

BY ANNE PLATT MCGINN

OCEANS FUNCTION as a source of food and fuel, a means of trade and commerce, and a base for cities and tourism. Worldwide, people obtain much of their animal protein from fish. Ocean-based deposits meet one-fourth of the world's annual oil and gas needs, and more than half of world trade travels by ship. More important than these economic figures, however, is the fact that humans depend on oceans for life itself. Harboring a greater variety of animal body types (phyla) than terrestrial systems and supplying more than half of the planet's ecological goods and services, the oceans play a commanding role in the Earth's balance of life.

Due to their large physical volume and density, oceans absorb, store, and transport vast quantities of heat, water, and nutrients. The oceans store about 1,000 times more heat than the atmosphere does, for example. Through processes such as evaporation and photosynthesis, marine systems and species help regulate the climate, maintain a livable atmosphere, convert solar energy into food, and break down natural wastes. The value of these "free" services far surpasses that of

ocean-based industries. Coral reefs alone, for instance, are estimated to be worth $375,000,000,000 annually by providing fish, medicines, tourism revenues, and coastal protection for more than 100 countries.

Despite the importance of healthy oceans to our economy and well-being, we have pushed the world's oceans perilously close to—and in some cases past—their natural limits. The warning signs are clear. The share of overexploited marine fish species jumped from almost none in 1950 to 35% in 1996, with an additional 25% nearing full exploitation. More than half of the world's coastlines and 60% of the coral reefs are threatened by human activities, including intensive coastal development, pollution, and overfishing.

In January, 1998, as the United Nations was launching the Year of the Ocean, more than 1,600 marine scientists, fishery biologists, conservationists, and oceanographers from across the globe issued a joint statement entitled "Troubled Waters." They agreed that the most pressing threats to ocean health are human-induced, in-

cluding species overexploitation, habitat degradation, pollution, introduction of alien species, and climate change. The impacts of these five threats are exacerbated by poorly planned commercial activities and coastal population growth.

Yet, many people still consider the oceans as not only inexhaustible, but immune to human interference. Because scientists just recently have begun to piece together how ocean systems work, society has yet to appreciate—much less protect—the wealth of oceans in its entirety. Indeed, current courses of action are rapidly undermining this wealth.

Nearly 1,000,000,000 people, predominantly in Asia, rely on fish for at least 30% of their animal protein supply. Most of these fish come from oceans, but with increasing frequency they are cultured on farms rather than captured in the wild. Aquaculture, based on the traditional Asian practice of raising fish in ponds, constitutes one of the fastest-growing sectors in world food production.

In addition to harvesting food from the sea, people have traditionally relied on oceans as a transportation route. Sea trade currently is dominated by multinational companies that are more influenced by the rise and fall of stock prices than by the tides and trade winds. Modern fishing trawlers, oil tankers, aircraft carriers, and container ships follow a path set by electronic beams, satellites, and computers.

Society derives a substantial portion of energy and fuel from the sea—a trend that was virtually unthinkable a century ago. In an age of falling trade barriers and mounting pressures on land-based resources, new ocean-based industries such as tidal and thermal energy production promise to become even more vital to the workings of the world economy. Having increased sixfold between 1955 and 1995, the volume of international trade is expected to triple again by 2020, according to the U.S. National Oceanographic and Atmospheric Administration, and 90% of it is expected to move by ocean.

In contrast to familiar fishing grounds and sea passageways, the depths of the ocean were long believed to be a vast wasteland that was inhospitable, if not completely devoid of life. Since the first deployment of submersibles in the 1930s and more advanced underwater acoustics and pressure chambers in the 1960s, scientific and commercial exploration has helped illuminate life in the deep sea and the geological history of the ancient ocean. Mining for sand, gravel, coral, and minerals (including sulfur and, most recently, petroleum) has taken place in shallow waters and continental shelves for decades, although offshore mining is severely restricted in some national waters.

Isolated, but highly concentrated, deep-sea deposits of manganese, gold, nickel, and copper, first discovered in the late 1970s, continue to tempt investors. These valuable nodules have proved technologically difficult and expensive to extract, given the extreme pressures and depths of their location. An international compromise on the deep seabed mining provisions of the Law of the Sea in 1994 has opened the way to some mining in international waters, but it appears unlikely to lead to much as long as mineral prices remain low, demand is largely met from the land, and the cost of underwater operations remains prohibitively high.

Perhaps more valuable than the mineral wealth in oceans are still-undiscovered living resources—new forms of life, potential medicines, and genetic material. For example, in 1997, medical researchers stumbled across a compound in dogfish that stops the spread of cancer by cutting off the blood supply to tumors. The promise of life-saving cures from marine species is gradually becoming a commercial reality for bioprospectors and pharmaceutical companies as anti-inflammatory and cancer drugs have been discovered and other leads are being pursued.

Tinkering with the ocean for the sake of shortsighted commercial development, whether for mineral wealth or medicine, warrants close scrutiny, however. Given how little we know—a mere 1.5% of the deep sea has been explored, let alone adequately inventoried—any development could be potentially irreversible in these unique environments. Although seabed mining is subjected to some degree of international oversight, prospecting for living biological resources is completely unregulated.

During the past 100 years, scientists who work both underwater and among marine fossils found high in mountains have shown that the tree of life has its evolutionary roots in the sea. For about 3,200,000,000 years, all life on Earth was marine. A complex and diverse food web slowly evolved from a fortuitous mix there of single-celled algae, bacteria, and several million trips around the sun. Life remained sea-bound until 245,000,000 years ago, when the atmosphere became oxygen-rich.

Thanks to several billion years' worth of trial and error, the oceans today are home to a variety of species that have no descendants on land. Thirty-two out of 33 animal life forms are represented in marine habitats. (Only insects are missing.) Fifteen of these are exclusively marine phyla, including those of comb jellies, peanut worms, and starfish. Five phyla, including that of sponges, live predominantly in salt water. Although, on an individual basis, marine species account for just nine percent of the 1,800,000 species described for the entire planet, there may be as many as 10,000,000 additional species in the sea that as yet have not been classified.

In addition to hosting a vast array of biological diversity, the marine environment performs such vital functions as oxygen production, nutrient recycling, storm protection, and climate regulation—services that often are taken for granted.

Marine biological activity is concentrated along the world's coastlines (where sunlit surface waters receive nutrients and sediments from land-based runoff, river deltas, and rainfall) and in upwelling systems (where cold, nutrient-rich, deep-water currents run up against continental margins). It provides 25% of the planet's primary biological productivity and an estimated 80–90% of the global commercial fish catch. It is estimated that coastal environments account for 38% of the goods and services provided by the Earth's ecosystems, while open oceans contribute an additional 25%. The value of all marine goods and services is estimated at 21 trillion dollars annually, 70% more than terrestrial systems come to.

Oceans are vital to both the chemical and biological balance of life. The same mechanism that created the present atmosphere—photosynthesis—continues to feed the marine food chain. Phytoplankton—tiny microscopic plants—take carbon dioxide (CO_2) from the atmosphere and convert it into oxygen and simple sugars, a form of carbon that can be consumed by marine animals. Other types of phytoplankton process nitrogen and sulfur, thereby helping the oceans function as a biological pump.

Although most organic carbon is consumed in the marine food web and eventually returned to the atmosphere via respiration, the unused balance rains down to the deep waters that make up the bulk of the ocean, where it is stored temporarily. Over the course of millions of years, these deposits have accumulated to the point that most of the world's organic carbon, approximately 15,000,000 gigatons (a gigaton equals 1,000,000,000 tons), is sequestered in marine sediments, compared with 4,000 gigatons in land-based reserves. On an annual basis, about one-third of the world's carbon emissions—around two gigatons—is taken up by oceans, an amount roughly equal to the uptake by land-based resources. If deforestation continues to diminish the ability of forests to absorb carbon dioxide, oceans are expected to play a more important role in regulating the planet's CO_2 budget in the future as human-induced emissions keep rising.

Perhaps no other example so vividly illustrates the connections between the oceans and the atmosphere than El Niño. Named after the Christ child because it usually appears in December, the El Niño Southern Oscillation takes place when trade winds and ocean surface currents in the eastern and central Pacific Ocean reverse direction. Scientists do not know what triggers the shift, but the aftermath is clear: Warm surface waters essentially pile up in the eastern Pacific and block deep, cold waters from upwelling, while a low pressure system hovers over South America, collecting heat and moisture that would otherwise be distributed at sea. This produces severe weather conditions in many parts of the world—increased precipitation, heavy flooding, drought, fire, and deep freezes—which, in turn, have enormous economic impact. During the 1997–98 El Niño, for instance, Argentina lost more than $3,000,000,000 in agricultural products due to these ocean-climate reactions, and Peru reported a 90% drop in anchovy harvests compared with the previous year.

A sea of problems

As noted earlier, the primary threats to oceans are largely human-induced and synergistic. Fishing, for example, has drastically altered the marine food web and underwater habitat areas. Meanwhile, the ocean's front line of defense—the coastal zone—is crumbling from years of degradation and fragmentation, while its waters have been treated as a waste receptacle for generations. The combination of overexploitation, the loss of buffer areas, and a rising tide of pollution has suffocated marine life and the livelihoods based on it in some areas. Upsetting the marine ecosystem in these ways has, in turn, given the upper hand to invasive species and changes in climate.

Overfishing poses a serious biological threat to ocean health. The resulting reductions in the genetic diversity of the spawning populations make it more difficult for the species to adapt to future environmental changes. The orange roughy, for instance, may have been fished down to the point where future recovery is impossible. Moreover, declines in one species can alter predator-prey relations and leave ecosystems vulnerable to invasive species. The overharvesting of triggerfish and pufferfish for souvenirs on coral reefs in the Caribbean has sapped the health of the entire reef chain. As these fish declined, populations of their prey—sea urchins—exploded, damaging the coral by grazing on the protective layers of algae and hurting the local reef-diving industry.

These trends have enormous social consequences as well. The welfare of more than 200,000,000 people around the world who depend on fishing for their income and food security is severely threatened. As the fish disappear, so do the coastal communities that depend on fishing for their way of life. Subsistence and small-scale fishers, who catch nearly half of the world's fish, suffer the greatest losses as they cannot afford to compete with large-scale vessels or changing technology. Furthermore, the health of more than 1,000,000,000 poor consumers who depend on minimal quantities of fish to constitute their diets is at risk as an ever-growing share of fish—83% by value—continues to be exported to industrial countries each year.

Despite a steadily growing human appetite for fish, large quantities are wasted each year because the fish are undersized or a nonmarketable sex or species, or because a fisher does not have a permit to catch them and must therefore throw them out. The United Nations' Food and Agricultural Organization estimates that discards of fish alone—not counting marine mammals, seabirds, and turtles—total 20,000,000 tons, equivalent to one-fourth of the annual marine catch. Many of these fish do not survive the process of getting entangled in gear, being brought on board, and then tossed back to sea.

Another threat to habitat areas stems from trawling, with nets and chains dragged across vast areas of mud, rocks, gravel, and sand, essentially sweeping everything in the vicinity. By recent estimates, all the ocean's continental shelves are trawled by fishers at least once every two years, with some areas hit several times a season. Considered a major cause of habitat degradation, trawling disturbs bottom-dwelling communities as well as localized species diversity and food supplies.

The conditions that make coastal areas so productive for fish—proximity to nutrient flows and tidal mixing and their place at the crossroads between land and water—also make them vulnerable to human assault. Today, nearly 40% of the world lives within 60 miles of a coastline. As more people move to coastal areas and further stress the seams between land and sea, coastal ecosystems are losing ground.

Human activities on land cause a large portion of offshore contamination. An estimated 44% of marine pollution comes from land-based pathways, flowing down rivers into tidal estuaries, where it bleeds out to sea; an additional 33% is airborne pollution that is carried by winds and deposited far off shore. From nutrient-rich sediments, fertilizers, and human waste to toxic heavy metals and synthetic chemicals, the outfall from human society ends up circulating in the fluid and turbulent seas.

Excessive nutrient loading has left some coastal systems looking visibly sick. Seen from an airplane, the surface waters of Manila Bay in the Philippines resemble green soup due to dense carpets of algae. Nitrogen and phosphorus are necessary for life and, in limited quantities, can help boost plant productivity, but too much of a good thing can be bad. Excessive nutrients build up and create conditions that are conducive to outbreaks of dense algae blooms, also known as "red tides." The blooms block sunlight, absorb dissolved oxygen, and disrupt food-web dynamics. Large portions of the Gulf of Mexico are now considered a biological "dead zone" due to algal blooms.

The frequency and severity of red tides has increased in the past couple of decades. Some experts link the recent outbreaks to increasing loads of nitrogen and phosphorus from nutrient-rich wastewater and agricultural runoff in poorly flushed waters.

Organochlorines, a fairly recent addition to the marine environment, are proving to have pernicious effects. Synthetic organic compounds such as chlordane, DDT, and PCBs are used for everything from electrical wiring to pesticides. Indeed, one reason they are so difficult to control is that they are ubiquitous. The organic form of tin (tributyltin), for example, is used in most of the world's marine paints to keep barnacles, seaweed, and other organisms from clinging to ships. Once the paint is dissolved in the water, it accumulates in mollusks, scallops, and rock crabs, which are consumed by fish and marine mammals.

As part of a larger group of chemicals known collectively as persistent organic pollutants (POPs), these compounds are difficult to control because they do not degrade easily. Highly volatile in warm temperatures, POPs tend to circulate toward colder environments where the conditions are more stable, such as the Arctic Circle. Moreover, they do not dissolve in water, but are lipid-soluble, meaning that they accumulate in the fat tissues of fish that are then consumed by predators at a more concentrated level.

POPs have been implicated in a wide range of animal and human health problems—from suppression of immune systems, leading to higher risk of illness and infection, to disruption of the endocrine system, which is linked to birth defects and infertility. Their continued use in many parts of the world poses a threat to marine life and fish consumers everywhere.

Because marine species are extremely sensitive to fluctuations in temperature, changes in climate and atmospheric conditions pose high risks to them. Recent evidence shows that the thinning ozone layer above Antarctica has allowed more ultraviolet-B radiation to penetrate the waters. This has affected photosynthesis and the growth of phytoplankton and macroalgae. The effects are not limited to the base of the food chain. By striking aquatic species during their most vulnerable stages of life and reducing their food supply at the same time, increases in UV-B could have devastating impacts on world fisheries production.

Because higher temperatures cause water to expand, a warming world may trigger more frequent and damaging storms. Ironically, the coastal barriers, seawalls, jetties, and levees that are designed to protect human settlements from storm surges likely exacerbate the problem of coastal erosion and instability, as they create deeper inshore troughs that boost wave intensity and sustain winds.

Depending on the rate and extent of warming, global sea levels may rise as much as three feet by 2100—up to five times as much as during the last century. Such a rise would flood most of New York City, including the entire subway system and all three major airports. Economic damages and losses could cost the global economy as much as $970,000,000,000 in 2100, according to the Organisation for Economic Co-operation and Development. The human costs would be unimaginable, especially in the low-lying, densely populated river deltas of Bangladesh, China, Egypt, and Nigeria.

These damages could be just the tip of the iceberg. Warmer temperatures would likely accelerate polar ice cap melting and could boost this rising wave by several feet. Just four years after a large portion of Antarctica melted, another large ice sheet fell off into the Southern Sea in February, 1998, rekindling fears that global warming could ignite a massive thaw that would flood coastal areas worldwide. Because oceans play such a vital role in regulating the Earth's climate and maintaining a healthy planet, minor changes in ocean circulation or in its temperature or chemical balance could have repercussions many orders of magnitude larger than the sum of human-induced wounds.

While understanding past climatic fluctuations and predicting future developments are an ongoing challenge for scientists, there is clear and growing evidence of the overuse—indeed abuse—that many marine ecosystems and species are suffering from direct human actions. The situation is probably much worse, for many sources of danger are still unknown or poorly monitored. The need to take preventive and decisive action on behalf of oceans is more important than ever.

Saving the oceans

Scientists' calls for precaution and protective measures are largely ignored by policymakers, who focus on enhancing commerce, trade, and market supply and look to extract as much from the sea as possible, with little regard for the effects on marine species or habitats. Overcoming the interest groups that favor the status quo will require engaging all potential stakeholders and reformulating the governance equation to incorporate the stewardship obligations that come with the privilege of use.

Fortunately for the planet, a new sea ethic is emerging. From tighter dumping regulations to recent international agreements, policymakers have made initial progress toward the goal of cleaning up humans' act. Still, much more is needed in the way of public education to build political support for marine conservation.

To boost ongoing efforts, two key principles are important. First, any dividing up of the waters should be based on equity, fairness, and need as determined by dependence on the resource and the best available scientific knowledge, not simply on economic might and political pressure. In a similar vein, resource users should be responsible for their actions, with decision-making and accountability shared by stakeholders and government officials. Second, given the uncertainty in scientific knowledge and management capabilities, it is necessary to err on the side of caution and take a precautionary approach.

Replanting mangroves and constructing artificial reefs are two concrete steps that help some fish stocks rebound quickly while letting people witness firsthand the results of their labors. Once they see the immediate payoff of their work, they are more likely to stay involved in longer-term protection efforts, such as marine sanctuaries, which involve removing an area from use entirely.

Marine protected areas are an important tool to help marine scientists and resource planners incorporate an ecologically based approach to oceans protection. By limiting accessibility and easing pressures on the resource, these areas allow stocks to rebound and profits to return. Globally, more than 1,300 marine and coastal sites have some form of protection, but most lack effective on-the-ground management.

Meanwhile, efforts to establish marine refuges and parks lag far behind similar efforts on land. The World Heritage Convention, which identifies and protects areas of special significance to mankind, identifies just 31 sites that include either a marine or a coastal component, out of a total of 522. John Waugh, Senior Program Officer of the World Conservation Union-U.S., and others argue that the World Heritage List could be extended to a number of marine hotspots and should include representative areas of the continental shelf, the deep sea, and the open ocean. Setting these and other areas aside as off-limits to commercial development can help advance scientific understanding of marine systems and provide refuge for threatened species.

To address the need for better data, coral reef scientists have enlisted the help of recreational scuba divers. Sport divers who volunteer to collect data are given basic training to identify and survey fish and coral species and conduct rudimentary site assessments. The data then are compiled and put into a global inventory that policymakers use to monitor trends and to target intervention. More efforts like these— that engage the help of concerned individuals and volunteers—could help overcome funding and data deficiencies and build greater public awareness of the problems plaguing the world's oceans.

Promoting sustainable ocean use also means shifting demand away from environmentally damaging products and extraction techniques. To this end, market forces, such as charging consumers more for particular fish and introducing industry codes of conduct, can be helpful. In April, 1996, the World Wide Fund for Nature teamed up with one of the world's largest manufacturers of seafood products, Anglo-Dutch Unilever, to create economic incentives for sustainable fishing. Implemented through an independent Marine Stewardship Council, fisheries products that are harvested in a sustainable manner will qualify for an ecolabel. Similar efforts could help convince industries to curb wasteful practices and generate greater consumer awareness of the need to choose products carefully.

Away from public oversight, companies engaged in shipping, oil and gas extraction, deep-sea mining, bioprospecting, and tidal and thermal energy represent a coalition of special interests whose activities help determine the fate of the oceans. It is crucial to get representatives of these industries engaged in implementing a new ocean charter that supports sustainable use.

Their practices not only affect the health of oceans, they help decide the pace of a transition toward a more sustainable energy economy, which, in turn, affects the balance between climate and oceans.

Making trade data and industry information publicly available is an important way to build industry credibility and ensure some degree of public oversight. While regulations are an important component of environmental protection, pressure from consumers, watchdog groups, and conscientious business leaders can help develop voluntary codes of action and standard industry practices that can move industrial sectors toward cleaner and greener operations. Economic incentives targeted to particular industries, such as low-interest loans for thermal projects, can aid companies in making a quicker transition to sustainable practices.

The fact that oceans are so central to the global economy and to human and planetary health may be the strongest motivation for protective action. Although the range of assaults and threats to ocean health are broad, the benefits that oceans provide are invaluable and shared by all. These huge bodies of water represent an enormous opportunity to forge a new system of cooperative, international governance based on shared resources and common interests. Achieving these far-reaching goals, however, begins with the technically simple, but politically daunting, task of overcoming several thousand years' worth of ingrained behavior. It requires seeing oceans not as an economic frontier for exploitation, but as a scientific frontier for exploration and a biological frontier for careful use.

For generations, oceans have drawn people to their shores for a glimpse of the horizon, a sense of scale, and awe at nature's might. Today, oceans offer careful observers a different kind of awe—a warning that humans' impacts on the Earth are exceeding natural bounds and in danger of disrupting life. Protection efforts already lag far behind what is needed. How humans choose to react will determine the future of the planet. Oceans are not simply one more system under pressure—they are critical to man's survival. As Carl Safina writes in *The Song for the Blue Ocean*, "we need the oceans more than they need us."

Anne Platt McGinn *is a senior researcher, Worldwatch Institute, Washington, D.C.*

FEELING THE HEAT
LIFE IN THE GREENHOUSE

Except for nuclear war or a collision with an asteroid, no force has more potential to damage our planet's web of life than global warming. It's a "serious" issue, the White House admits, but nonetheless George W. Bush has decided to abandon the 1997 Kyoto treaty to combat climate change—an agreement the U.S. signed but the new President believes is fatally flawed. His dismissal last week of almost nine years of international negotiations sparked protests around the world and a face-to-face disagreement with German Chancellor Gerhard Schröder. Our special report examines the signs of global warming that are already apparent, the possible consequences for our future, what we can do about the threat and why we have failed to take action so far.

By MICHAEL D. LEMONICK

There is no such thing as normal weather. The average daytime high temperature for New York City this week should be 57°F, but on any given day the mercury will almost certainly fall short of that mark or overshoot it, perhaps by a lot. Manhattan thermometers can reach 65° in January every so often and plunge to 50° in July. And seasons are rarely normal. Winter snowfall and summer heat waves beat the average some years and fail to reach it in others. It's tough to pick out overall changes in climate in the face of these natural fluctuations. An unusually warm year, for example, or even three in a row don't necessarily signal a general trend.

Yet the earth's climate does change. Ice ages have frosted the planet for tens of thousands of years at a stretch, and periods of warmth have pushed the tropics well into what is now the temperate zone. But given the normal year-to-year variations, the only reliable signal that such changes may be in the works is a long-term shift in worldwide temperature.

And that is precisely what's happening. A decade ago, the idea that the planet was warming up as a result of human activity was largely theoretical. We knew that since the Industrial Revolution began in the 18th century, factories and power plants and automobiles and farms have been loading the atmosphere with heat-trapping gases, including carbon dioxide and methane. But evidence that the climate was actually getting hotter was still murky.

Not anymore. As an authoritative report issued a few weeks ago by the United Nations-sponsored Intergovernmental Panel on Climate Change makes plain, the trend toward a warmer world has unquestionably begun. Worldwide temperatures have climbed more than 1°F over the past

MAKING THE CASE THAT OUR CLIMATE IS CHANGING

From melting glaciers to rising oceans, the signs are everywhere. Global warming can't be blamed for any particular heat wave, drought or deluge, but scientists say a hotter world will make such extreme weather more frequent—and deadly.

EXHIBIT A

Thinning Ice

ANTARCTICA, home to these Adélie penguins, is heating up. The annual melt season has increased up to three weeks in 20 years.

MOUNT KILIMANJARO has lost 75% of its ice cap since 1912. The ice on Africa's tallest peak could vanish entirely within 15 years.

LAKE BAIKAL in eastern Siberia now feezes for the winter 1.1 days later than it did a century ago.

VENEZUELAN mountaintops had six glaciers in 1972. Today only two remain.

EXHIBIT B

Hotter Times

TEMPERATURES SIZZLED from Kansas to New England last May.

CROPS WITHERED and Dallas temperatures topped 100°F for 29 day-sstraight in a Texas hot spell that struck during the summer of 1998.

INDIA'S WORST heat shock in 50 years killed more than 2,500 people in May 1998.

CHERRY BLOSSOMS in Washington bloom seven days earlier in the spring than they did in 1970.

EXHIBIT C

Wild Weather

HEAVY RAINS in England and Wales made last fall Britain's wettest three-month period on record.

FIRES due to dry conditions and record-breaking heat consumed 20% of Samos Island, Greece, last July.

FLOODS along the Ohio River in March 1997 caused 30 deaths and at least $500 million in property damage.

HURRICAN FLOYD brought flooding rains and 130-m.p.h. winds through the Atlantic seabord in September1999, killing 77 people and leaving thousands homeless.

EXHIBIT D

Nature's Pain

PACIFIC SALMON populations fell sharply in 1997 and 1998, when local ocean temperatures rose 6°F.

POLAR BEARS in Hudson Bay are having fewer cubs, possibly as a result of earlier spring ice breakup.CORAL REEFS suffer from the loss of algae that color and nourish them. The process, called bleaching, is caused by warmer oceans.DISEASES like dengue fever are expanding their reach northward in the U.S.

BUTTERFLIES are relocating to higher latitudes. The Edith's Checkerspot butterfly of western North America has moved almost 60 miles north in 100 years.

EXHIBIT E

Rising Sea Levels

CAPE HATTERAS Lighthouse was 1,500 ft. from the North Carolina shoreline when it was built in 1870. By the late 1980s teh ocean had crept to within 160 ft., and the lighthouse had to be moved to avoid collapse.

JAPANESE FORTIFICATIONS were built on Kosrae Island in the southwest Pacific Ocean during World War II to guard against U.S. Marines' invading the beach. Today the fortifications are awash at high tide.

FLORIDA FARMLAND up to 1,000 ft. inland from Biscayne Bay is being infiltrated by salt water, rendering the land too toxic for crops. Salt water is also nibbling at the edges of farms on Maryland's Eastern Shore.

BRAZILIAN SHORELINE in the region of Recife receded more than 6 ft. a year from 1915 to 1950 and more than 8 ft. a year from 1985 to 1995.

century, and the 1990s were the hottest decade on record. After analyzing data going back at least two decades on everything from air and ocean temperatures to the spread and retreat of wildlife, the IPCC asserts that this slow but steady warming has had an impact on no fewer than 420 physical processes and animal and plant species on all continents.

Glaciers, including the legendary snows of Kilimanjaro, are disappearing from mountaintops around the globe. Coral reefs are dying off as the seas get too warm for comfort. Drought is the norm in parts of Asia and Africa. El Niño events, which trigger devastating weather in the eastern Pacific, are more frequent. The Arctic permafrost is starting to melt. Lakes and rivers in colder climates are freezing later and thawing earlier each year. Plants and animals are shifting their ranges poleward and to higher altitudes, and migration patterns for animals as diverse as polar bears, butterflies and beluga whales are being disrupted.

Faced with these hard facts, scientists no longer doubt that global warming is happening, and almost nobody questions the fact that humans are at least partly responsible. Nor are the changes over. Already, humans have increased the concentration of carbon dioxide, the most abundant heat-trapping gas in the atmosphere, to 30% above pre-industrial levels—and each year the

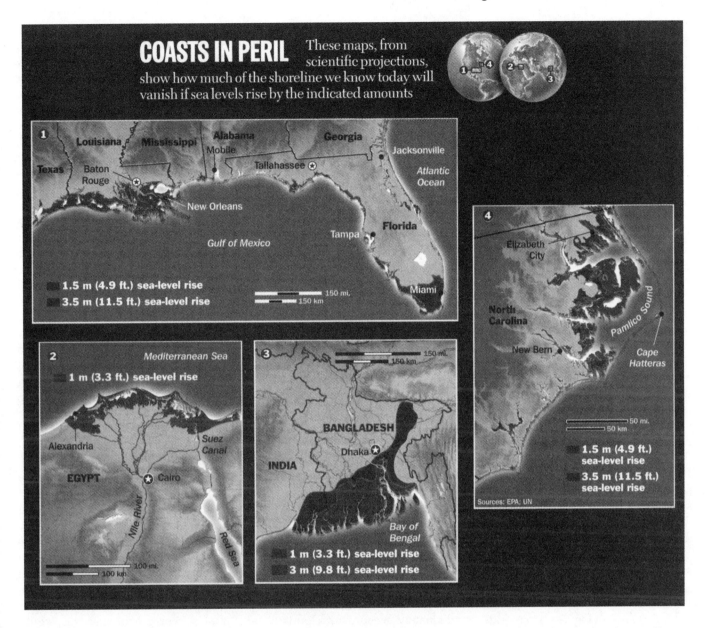

COASTS IN PERIL These maps, from scientific projections, show how much of the shoreline we know today will vanish if sea levels rise by the indicated amounts

1.
- 1.5 m (4.9 ft.) sea-level rise
- 3.5 m (11.5 ft.) sea-level rise
150 mi.
150 km

Louisiana Mississippi Alabama Mobile Georgia
Texas Baton Rouge Tallahassee Jacksonville
New Orleans Atlantic Ocean
Gulf of Mexico Florida Tampa Miami

2. Mediterranean Sea
- 1 m (3.3 ft.) sea-level rise
Alexandria Suez Canal
EGYPT Cairo
Nile River Red Sea
100 mi.
100 km

3. 150 mi. 150 km
BANGLADESH
INDIA Dhaka
Bay of Bengal
- 1 m (3.3 ft.) sea-level rise
- 3 m (9.8 ft.) sea-level rise

4. Elizabeth City
North Carolina
New Bern Pamlico Sound Cape Hatteras
50 mi.
50 km
- 1.5 m (4.9 ft.) sea-level rise
- 3.5 m (11.5 ft.) sea-level rise
Sources: EPA; UN

rate of increase gets faster. The obvious conclusion: temperatures will keep going up.

Unfortunately, they may be rising faster and heading higher than anyone expected. By 2100, says the IPCC, average temperatures will increase between 2.5°F and 10.4°F—more than 50% higher than predictions of just a half-decade ago. That may not seem like much, but consider that it took only a 9°F shift to end the last ice age. Even at the low end, the changes could be problematic enough, with storms getting more frequent and intense, droughts more pronounced,

coastal areas ever more severely eroded by rising seas, rainfall scarcer on agricultural land and ecosystems thrown out of balance.

But if the rise is significantly larger, the result could be disastrous. With seas rising as much as 3 ft., enormous areas of densely populated land—coastal Florida, much of Louisiana, the Nile Delta, the Maldives, Bangladesh—would become uninhabitable. Entire climatic zones might shift dramatically, making central Canada look more like central Illinois, Georgia more like Guatemala. Agriculture would be

thrown into turmoil. Hundreds of millions of people would have to migrate out of unlivable regions.

Public health could suffer. Rising seas would contaminate water supplies with salt. Higher levels of urban ozone, the result of stronger sunlight and warmer temperatures, could worsen respiratory illnesses. More frequent hot spells could lead to a rise in heat-related deaths. Warmer temperatures could widen the range of disease-carrying rodents and bugs, such as mosquitoes and ticks, increasing the incidence of dengue fever, malaria, encephalitis,

Lyme disease and other afflictions. Worst of all, this increase in temperatures is happening at a pace that outstrips anything the earth has seen in the past 100 million years. Humans will have a hard enough time adjusting, especially in poorer countries, but for wildlife, the changes could be devastating.

Like any other area of science, the case for human-induced global warming has uncertainties—and like many pro-business lobbyists, President Bush has proclaimed those uncertainties a reason to study the problem further rather than act. But while the evidence is circumstantial, it is powerful, thanks to the IPCC's painstaking research. The U.N.-sponsored group was organized in the late 1980s. Its mission: to sift through climate-related studies from a dozen different fields and integrate them into a coherent picture. "It isn't just the work of a few green people," says Sir John Houghton, one of the early leaders who at the time ran the British Meteorological Office. "The IPCC scientists come from a wide range of backgrounds and countries."

Measuring the warming that has already taken place is relatively simple; the trick is unraveling the causes and projecting what will happen over the next century. To do that, IPCC scientists fed a wide range of scenarios involving varying estimates of population and economic growth, changes in technology and other factors into computers. That process gave them about 35 estimates, ranging from 6 billion to 35 billion tons, of how much excess carbon dioxide will enter the atmosphere.

Then they loaded those estimates into the even larger, more powerful computer programs that attempt to model the planet's climate. Because no one climate model is considered definitive, they used seven different versions, which yielded 235 independent predictions of global temperature increase. That's where the range of 2.5°F to 10.4°F (1.4°C to 5.8°C) comes from.

The computer models were criticized in the past largely because the climate is so complex that the limited hardware and software of even a half-decade ago couldn't do an adequate simulation. Today's climate models, however, are able to take into account the heat-trapping effects not just of CO_2 but also of other greenhouse gases, including methane. They can also factor in natural variations in the sun's energy and the effect of substances like dust from volcanic eruptions and particulate matter spewed from smokestacks.

That is one reason the latest IPCC predictions for temperature increase are higher than they were five years ago. Back in the mid-1990s, climate models didn't include the effects of the El Chichon and Mount Pinatubo volcanic eruptions, which threw enough dust into the air to block out some sunlight and slow down the rate of warming. That effect has dissipated, and the heating should start to accelerate. Moreover, the IPCC noted, many countries have begun to reduce their emissions of sulfur dioxide in order to fight acid rain. But sulfur dioxide particles, too, reflect sunlight; without this shield, temperatures should go up even faster.

The models still aren't perfect. One major flaw, agree critics and champions alike, is that they don't adequately account for clouds. In a warmer world, more water will evaporate from the oceans and presumably form more clouds. If they are billowy cumulus clouds, they will tend to shade the planet and slow down warming; if they are high, feathery cirrus clouds, they will trap even more heat.

Research by M.I.T. atmospheric scientist Richard Lindzen suggests that warming will tend to make cirrus clouds go away. Another critic, John Christy of the University of Alabama in Huntsville, says that while the models reproduce the current climate in a general way, they fail to get right the amount of warming at different levels in the atmosphere. Neither Lindzen nor Christy (both IPCC authors) doubts, however, that humans are influencing the climate. But they question how much—and how high temperatures will go. Both scientists are distressed that only the most extreme scenarios, based on huge population growth and the maximum use of dirty fuels like coal, have made headlines.

It won't take the greatest extremes of warming to make life uncomfortable for large numbers of people. Even slightly higher temperatures in regions that are already drought- or flood-prone would exacerbate those conditions. In temperate zones, warmth and increased CO_2 would make some crops flourish—at first. But beyond 3° of warming, says Bill Easterling, a professor of geography and agronomy at Penn State and a lead author of the IPCC report, "there would be a dramatic turning point. U.S. crop yields would start to decline rapidly." In the tropics, where crops are already at the limit of their temperature range, the decrease would start right away.

Even if temperatures rise only moderately, some scientists fear, the climate would reach a "tipping point"—a point at which even a tiny additional increase would throw the system into violent change. If peat bogs and Arctic permafrost warm enough to start releasing the methane stored within them, for example, that potent greenhouse gas would suddenly accelerate the heat-trapping process.

By contrast, if melting ice caps dilute the salt content of the sea, major ocean currents like the Gulf Stream could slow or even stop, and so would their warming effects on northern regions. More snowfall reflecting more sunlight back into space could actually cause a net cooling. Global warming could, paradoxically, throw the planet into another ice age.

Even if such a tipping point doesn't materialize, the more drastic effects of global warming might be only postponed rather than avoided. The IPCC's calculations end with the year 2100, but the warming won't.

World Bank chief scientist, Robert Watson, currently serving as IPCC chair, points out that the CO_2 entering the atmosphere today will be there for a century. Says Watson: "If we stabilize (CO_2 emissions) now, the concentration will continue to go up for hundreds of years. Temperatures will rise over that time."

That could be truly catastrophic. The ongoing disruption of ecosystems and weather patterns would be bad enough. But if temperatures reach the IPCC's worst-case levels and stay there for as long as 1,000 years, says Michael Oppenheimer, chief scientist at Environmental Defense, vast ice sheets in Greenland and Antarctica could melt, raising sea level more than 30 ft. Florida would be history, and every city on the U.S. Eastern seaboard would be inundated.

In the short run, there's not much chance of halting global warming, not even if every nation in the world ratifies the Kyoto Protocol tomorrow. The treaty doesn't require reductions in carbon dioxide emissions until 2008. By that time, a great deal of damage will already have been done. But we can slow things down. If action today can keep the climate from eventually reaching an unstable tipping point or can finally begin to reverse the warming trend a century from now, the effort would hardly be futile. Humanity embarked unknowingly on the dangerous experiment of tinkering with the climate of our planet. Now that we know what we're doing, it would be utterly foolish to continue.

Reported by David Bjerklie,
Robert H. Boyle and
Andrea Dorfman/New York and
Dick Thompson/Washington

UNIT 6

Pollution: The Hazards of Growth

Unit Selections

Key Points to Consider

- Why has the U.S. federal government refused to acknowledge carbon dioxide as a pollutant and therefore subject to federal regulation? Is present and pending legislation sufficient to stem the flow of new greenhouse gases from U.S. industries?

- Why is groundwater pollution so difficult to trace and to monitor? What mechanisms might be employed to reduce the contributions of agriculture and industry to the contamination of the world's important freshwater supply?

- Describe the differences between "point source" and "nonpoint source" pollution. What are some of the differences in how these pollution sources can be managed? Is government regulation the only or the best answer to the water pollution problem?

- What are some of the most significant improvements in environmental quality made during the last 30 years in the United States? Do you think the U.S. environment is better or worse than it was 30 years ago?

 **Links: www.dushkin.com/online/**
These sites are annotated in the World Wide Web pages.

IISDnet
http://www.iisd.org/default.asp
Persistant Organic Pollutants (POP)
http://irptc.unep.ch/pops/
School of Labor and Industrial Relations (SLIR): Hot Links
http://www.lir.msu.edu/hotlinks/
Space Research Institute
http://arc.iki.rssi.ru/Welcome.html
Worldwatch Institute
http://www.worldwatch.org

Of all the massive technological changes that have combined to create our modern industrial society, perhaps none has been as significant for the environment as the chemical revolution. The largest single threat to environmental stability is the proliferation of chemical compounds for a nearly infinite variety of purposes, including the universal use of organic chemicals (fossil fuels) as the prime source of the world's energy systems. The problem is not just that thousands of new chemical compounds are being discovered or created each year, but that their long-term environmental effects are often not known until an environmental disaster involving humans or other living organisms occurs. The problem is exacerbated by the time lag that exists between the recognition of potentially harmful chemical contamination and the cleanup activities that are ultimately required.

A critical part of the process of dealing with chemical pollutants is the identification of toxic and hazardous materials, a problem that is intensified by the myriad ways in which a vast number of such materials, natural and man-made, can enter environmental systems. Governmental legislation and controls are important in correcting the damages produced by toxic and hazardous materials such as DDT, PCBs, or CFCs; in limiting fossil

fuel burning; or in preventing the spread of living organic hazards such as pests and disease-causing agents. Unfortunately, as evidenced by most of the articles in this unit, we are losing the battle against harmful substances regardless of legislation, and chemical pollution of the environment is probably getting worse rather than better.

The first article in this unit deals with one of the newest legislative approaches to several forms of pollution resulting from the chemical revolution: the increases in sulfur dioxide, nitrogen oxides, and mercury in soil, water, and air as a consequence of industrial activities, transportation, and heavy applications of artificial fertilizer. In "Three Pollutants and an Emission," author Dallas Burtraw, a senior fellow at Resources for the Future, describes how the U.S. Congress may be poised to enact the first major pollution legislation in more than a decade—but is still failing to recognize carbon dioxide as a pollutant, calling it an "emission" instead. In the second article in this unit, an emphasis is also placed on organic or biological pollution related to what may be humanity's most important environmental problem: the quality of the global supply of freshwater. In "Groundwater Shock: The Polluting of the World's Major Freshwater Stores," Payal Sampat notes that the vast majority (97 percent) of the world's freshwater supply lies not in the visible surface systems of lakes and streams but in underground aquifers. This precious reserve, used for virtually every purpose from drinking to irrigating crops, is becoming polluted by surface processes related to agricultural, commercial, industrial, domestic, transportation, and other human activities. Payal notes that while much of the world worries about what is happening in the atmosphere (global warming), what happens below our feet may ultimately be of as much concern.

The section's third selection continues with the theme of water pollution and the interrelationship between polluter and pollutant. In "Water Quality: The Issues," author Mary H. Cooper notes the tremendous strides that have been made in cleaning up the nation's water supply since the passage of the Clean Water Act of 1972. But she also notes that nearly 40 percent of the inland and coastal waters in the United States are unfit for fishing, swimming, or drinking. A basic part of the problem, she claims, is that the clean water legislation dealt only with pollution from point-sources: factories, power plants, wastewater treatment facilities, and so on. But much of the water pollution anywhere is attributable to nonpoint sources, such as runoff from crop and animal agriculture, urban areas, roads, and forests. By their nature, nonpoint sources are much more difficult to control and Cooper advocates the development of nonregulatory approaches to mitigating this form of water pollution.

Finally, the concluding article in the section, "Statehouse to Greenhouse" offers a breath of optimism. An optimistic note is struck by academic Barry G. Rabe in this, the unit's final selection. Rabe notes the strides in environmental regulatory legislation made by states in the United States, particularly in terms of the emission of greenhouse gasses. It is not unusual, Rabe notes, for states to produce regulations in advance of federal action on the same problems, and he states that states are doing so may serve as a predictive device to suggest federal action will come soon.

The pollution problem might appear nearly impossible to solve. Yet as the last article notes, solutions exist: massive cleanup campaigns to remove existing harmful chemicals from the environment and to severely restrict their future use; strict regulation of the production, distribution, use, and disposal of potentially hazardous chemicals; the development of sound biological techniques to replace existing uses of chemicals for such purposes as pest control; the adoption of energy and material resource conservation policies; and the use of more conservative and protective agricultural and construction practices. We now possess the knowledge and the tools to ensure that environmental cleanup is carried through. It will not be an easy task, and it will be terribly expensive. It will also demand a new way of thinking about humankind's role in the environmental systems upon which all life forms depend. If we do not complete the task, however, the support capacity of the environment may be damaged or diminished beyond our capacities to repair it. The consequences would be fatal for all who inhabit this planet.

Three Pollutants and an Emission

A Playbill for the Multipollutant Legislative Debate

By Dallas Burtraw

For the first time since 1990, Congress may be poised to enact major clean air legislation. Proposals now before Congress would affect electricity generation and large industrial facilities only and, as such, fall short of a full Clean Air Act reauthorization. But in other ways, the proposals are dramatic. They address multiple pollutants, broadly applying market-based approaches and aggregate emission caps, and would mandate deep emission cuts.

All the Congressional proposals and that of the Bush administration would address emissions of three pollutants—sulfur dioxide, nitrogen oxides, and mercury—from large stationary sources. The affected plants account for two-thirds of total U.S. emissions of sulfur dioxide, a quarter of nitrogen oxides, and a third of mercury. The proposals differ in whether they address carbon dioxide, even though the sources in question give off 40 percent of all U.S. emissions of carbon dioxide, a key greenhouse gas. Although carbon dioxide regulation is destined one day to play the leading role in clean air policy and thereby shape controls on the other three pollutants, for now it is backstage. Paradoxically, however, even from backstage carbon dioxide plays a critical role in the proposals' prospects for enactment.

The stage was set in spring of last year, when President Bush used previous clean air legislation to distinguish the three conventional pollutants from carbon dioxide. Because carbon dioxide was not identified as a pollutant in the Clean Air Act, the president termed it an "emission," not a "pollutant" and thus attempted to keep carbon dioxide out of the debate. The president's Clean Skies Initiative, announced this past Valentine's Day, excludes mandatory controls on carbon dioxide until at least 2012. Instead, the proposal embraces the goal of simply maintaining the modest rate of reduction in the intensity of carbon emissions per dollar of gross domestic product that was achieved in the last decade. Champions in Congress are divided mostly along partisan lines. Some proposals, including most notably that of Senator James Jeffords (I-VT), have kept carbon dioxide in the script.

All the proposals would make important—and costly—cuts in emissions of sulfur dioxide, nitrogen oxides, and mercury. But costs of the proposals that include carbon dioxide go far higher—so high that they would pretty well swamp the costs of controlling the other pollutants and thereby determine the direction of investments and technological choices for the future. Including carbon dioxide in the upcoming legislation would lead industry to begin to change its choice of fuel for power production. But if carbon dioxide emissions continue unregulated, coal—which now accounts for about 51 percent of electricity production—will continue to fuel power plants at about the same level. As recent studies by the Energy Information Administration and Environmental Protection Agency show, three-pollutant legislation that excluded carbon dioxide would necessitate retrofitting coal-fired facilities with ambitious controls for the conventional pollutants, but would have little effect on coal use and carbon dioxide emissions.

Legislation that also mandates big cuts in carbon dioxide emissions, however, would require many plants to switch from coal to natural gas. The switch to natural gas, which burns more cleanly than coal, would also mean sizable cuts in the other pollutants.

Will the Bush administration's proposal to leave out carbon dioxide and target only the three conventional pollutants resolve the future of carbon dioxide regulation, or will the issue be revisited? Power producers who invest hundreds of millions of dollars to comply with cuts mandated by three-pollutant legislation will not expect to see major new carbon dioxide controls just around the corner. Passing three-pollutant legislation would thus seem implicitly to commit the nation to a long-term delay in carbon dioxide policy and to an energy policy with a large coal component for many years. Indeed, for a future Congress to consider regulating carbon dioxide right on the heels of passing three-pollutant legislation could be perceived by power producers as highly unfair. Anticipating that reaction, environmental advocates may have little incentive to compromise over some pretty dicey issues involved with controls on the conventional pollutants.

A key goal of the proposed multipollutant legislation is to improve on the current regulatory framework by giving industry greater regulatory predictability and certainty about compliance requirements. Paradoxically, however, achieving such certainty, as well as taking into account the interests of industry, may require giving carbon dioxide cuts at least a small role in the multipollutant strategy.

Reducing Multipollutant Levels

Debate over the legislative proposals involves three main questions. The first is what level of control to impose for each of the three pollutants. The stakes here are high, for the leading proposals would impose large costs—several billion dollars—and reap large benefits as well. But the costs are spread roughly equally over the three pollutants, while the benefits are not.

Reducing sulfur dioxide emissions will likely yield the lion's share of quantifiable benefits. Sulfur dioxide as a gas is recognized as a potent health threat, but it is more noteworthy in economic terms for its role in forming secondary particulates and acidifying soil and water.

In the public policy world, sulfur dioxide hit marquee status with passage of the 1990 Clean Air Act Amendments, which mandated a 50 percent cut in aggregate emissions from electricity generation. At the time, Paul Portney, of Resources for the Future (RFF)—the only analyst bold enough to offer an estimate of the benefits and costs of the amendments—predicted that benefits of sulfur dioxide cuts would about justify the costs. Subsequent analysis by the EPA and RFF suggests that two unexpected developments caused benefits to exceed costs by far. First, since 1990, a cascade of new research—most recently, a study published in the *Journal of the American Medical Association* early in March—has linked secondary particulates from sulfur dioxide emissions with premature mortality, raising the quantifiable economic benefits of sulfur dioxide cuts many times over what was expected. Second, cutting sulfur dioxide proved less costly than anticipated.

In 1990, although the primary benefits were ultimately found to stem from improvements in health, arguably the main justification for the sulfur dioxide cuts was acid rain. Since 1990, sulfur deposition in sensitive ecological areas has fallen dramatically, but recovery has been slower than expected, leading many advocates in the northeast to call for further sulfur dioxide cuts of as much as 75 percent. Meeting that target will mean more than simply switching to low-sulfur coal at more facilities. Compliance will require the widespread installation of scrubbers—and will impose large capital costs on industry. Even so, from a benefit-cost perspective, the proposed sulfur dioxide cut is an easy pill to swallow.

Nitrogen Oxides

Pending multipollutant proposals would reduce nitrogen oxides from electricity generation to some 50–75 percent of current baseline emissions. The cuts have multifaceted and important benefits but are harder to justify on a cost-benefit basis than the cuts in sulfur dioxide.

Nitrogen oxides have effects similar to those of sulfur dioxide. They contribute to secondary particulates but are usually thought to be less potent in affecting health. They are also instrumental in acidification, but again most analyses attribute to them a lesser role. Both pollutants impair visibility. Nitrogen oxides also contribute to forming ground-level ozone, another pollutant identified in the Clean Air Act, but health scientists and economists assign lower benefits to reducing ozone than to reducing particulates.

The 1990 Clean Air Act Amendments first imposed modest cuts in nitrogen oxides emissions at power plants. In complying with the 1990 law, plants have already made the least expensive cuts. Dramatic further reductions will require widespread installation of post-combustion controls (mostly, selective catalytic reduction) and will be quite costly.

In sum, viewed from the basis of each ton reduced, the benefits of cutting nitrogen oxides appear to be equal to (or perhaps a little less than) those of cutting sulfur dioxide. But the cost per ton of the nitrogen oxide reductions is greater than the cost per ton of cutting sulfur dioxide. Thus, benefit-cost considerations would argue for greater relative reductions in sulfur dioxide—at least as long as they continue to yield the biggest bang for the buck. But current proposals would reduce the emissions of both by equal percentages.

Mercury Levels

Proposed mercury cuts are even larger, in percentage terms, than cuts in the other two pollutants. Though highly toxic and with potentially profound ecological and human health effects, mercury is emitted in much smaller quantities. And the benefits of reducing mercury emissions are hard to quantify. The Clean Air Act would normally require maximum achievable emission reduction for mercury as a hazardous air pollutant, leading to more than 90 percent removal, but the 1990 amendments gave mercury a temporary special exception. Mercury may, however, come under strict control in the next few years within the existing regulatory process unless exempted by legislation or blocked by litigation. Whichever ending is played out—implementation or exemption—will determine the baseline against which the multipollutant legislative proposals are measured. Environmental advocates seem to expect a baseline that reflects tightening controls on mercury. Realistically, however, long delays in implementing mercury controls on electricity generation are likely unless multipollutant legislation is passed.

Cost-effectiveness is an important issue. Byron Swift of the Environmental Law Institute notes that cutting use of mercury from various small sources through pollution prevention—for example, by placing a declining cap on the use of mercury in products and process—would be far less costly than controlling mercury emitted in power production. That would argue for pursuing other, less costly options before pushing far in the power sector.

A key question in the multipollutant legislative debate is what role to give to limiting carbon dioxide emissions.

Furthermore, installing post-combustion controls for sulfur dioxide and nitrogen oxides would yield sizable ancillary reductions in mercury. Selective catalytic reduction controls for nitrogen oxides oxidize much of the available mercury, and sulfur dioxide scrubbers capture the oxidized mercury. Although technical experience is limited and prediction is thus uncertain, simultaneous controls for sulfur dioxide and nitrogen oxides probably cut more than half of mercury emissions. Making greater cuts would require installing new technologies, such as activated carbon injection, that are as yet unproven in widescale application. Their cost is especially high when viewed in marginal cost terms, because the incremental reductions (those over and above the ancillary benefits of sulfur dioxide and nitrogen oxides controls) are small.

In sum, moderate mercury emissions cuts are available as part of a multipollutant strategy that targets sulfur dioxide and nitrogen oxides comprehensively, and these cuts are reflected in the administration's proposal. Making further modest cuts would encourage development of new technologies, but the dramatic reduction targets for mercury controls embedded in Jeffords' legislative proposals come at a high cost.

New Source Review

A second question in the debate over multipollutant legislation is whether to revise requirements for the New Source Review (NSR) program, originally established by the 1977 Clean Air Act and now under scrutiny by the Bush administration. The program requires sources that add or modify equipment that could generate new emissions to use new technology to cut emissions. New source performance standards characterize allowable emission rates for either modified or new sources anywhere in the nation. Sources that increase emissions in areas of the country that do not meet federal air quality standards are held to even stricter emission levels and must cut emissions in other facilities to offset new emissions. To some,

NSR appears to be a bargaining chip in the debate; to others it is the central issue.

The NSR program's biggest problem is uncertainty. A firm that modifies a source does not know whether it will trigger NSR requirements and what those requirements might be until a determination is made by the EPA. The technologies that are acceptable in areas that do not meet federal standards are subject to change.

Differences in emission standards between facilities under NSR and those not under NSR lead to inefficient spending on pollution control. In addition, firms under NSR may try to avoid expense and uncertainty by delaying investments to modernize facilities. And new sources in areas that do not require emission offsets increase emissions in the aggregate.

Advocates vigorously defend NSR, arguing that it has led to dramatic emission cuts and that it identifies new technologies and innovations and brings them to application. But a key goal of the legislative debate must be to streamline and increase the predictability of the NSR process as it applies to electricity generation.

Design

During the mid-1990s Washington sharply revised its view of market-based approaches to environmental regulation. After years of suspicion, the default question is now: why not use market-based approaches? All the pending multipollutant proposals would regulate sulfur dioxide and nitrogen oxides within an emission allowance trading program based on the trading program established in the 1990 Clean Air Act Amendments. But how the trading programs are designed and whether mercury is included are major issues still in play.

The existing regional cap and trading program for nitrogen oxides operates during the five-month summer ozone season in 11 northeastern states. It will be expanded to a 19-state program in 2004. Current three-pollutant proposals would make the program national and year-round. Recent research suggests that operating the program year-round in the 19-state region would improve its cost-effectiveness dramatically. But expanding it nationwide is harder to justify because population density is much less, and consequently the aggregate benefits of reducing exposure to pollution are much less. It would also lead to a divergence in the distribution of benefits and costs across the nation that could be its undoing in Congress.

It remains to be seen whether mercury would be regulated through trading or through inflexible technology standards. Textbook treatment of pollution implies that the marginal costs of control among facilities converge as emission reduction targets approach 100 percent, because essentially every facility has to eliminate the emission. But Alex Farrell of Carnegie Mellon University has demonstrated that the marginal costs of mercury control among facilities fan out widely as controls approach 100 percent because of differences in coal types and in the other pollution controls that are in place. That argues for a trading program, which would be a departure from the usual controls on hazardous air pollutants.

One other design issue is a sleeper, but its importance will surface soon enough. How tradable emission allowances are allocated portends potentially large transfers of wealth and affects total costs as well. Because of expanded competition in the electricity industry, the value of the emission allowances may far outstrip the cost of compliance for power producers when allowances are allocated at zero cost. And prices for consumers may go higher than is justified by producer costs. When a similar scenario unfolded as industry complied with the 1987 Montreal Protocol to protect stratospheric ozone, Congress captured the "windfall profits" by taxing them. The same thing may happen here.

Looking Ahead

Compromise would enhance the cost-effectiveness of the multipollutant legislation and, at least from the view in the balcony seats, increase its chance of passage. Sulfur dioxide reductions of the size being proposed have apparent justification, but the nitrogen oxide reductions proposed for the entire nation are hard to justify economically and in any case face political opposition in the western states. A compelling alternative would be to transform the seasonal program in the 19 eastern states into an annual one and to impose somewhat less stringent controls in the other states. The case for maximum reductions in mercury emissions is the weakest. A better strategy would be to claim the mercury reductions that are achieved ancillary to imposed reductions on sulfur dioxides and nitrogen oxides as a victory. In fact, this approach may provide a justification for the cost-effectiveness of nationwide nitrogen oxide controls that is missing if those controls come on top of controls on mercury (and vice versa). One could also question whether the NSR program needs to exist at all if stringent caps are placed on emissions. But unless those caps decline continuously over time, the NSR program will be essential from the perspective of environmentalists.

The administration's proposal embodies many of these suggestions with respect to levels of control (72 percent reduction for sulfur dioxide and 67 percent reduction for nitrogen oxides). It goes most of the way to meeting the Jeffords proposal with respect to sulfur dioxide and nitrogen oxides (75 percent reductions for both pollutants), but calls for mercury reductions that would be mostly ancillary to these (69 percent) rather than the maximum achievable (90 percent). The big difference between the proposals is timing. The Jeffords proposal would achieve these reductions a decade sooner than does the administration plan. Given the maturity of control technologies, the shorter timeline is practical and preferable.

The most important consideration, however, is what role is given to limiting carbon dioxide emissions. Large cuts do not appear plausible, but zero reductions do nothing to resolve the most important environmental uncertainty facing the electricity industry. A policy of modest cuts, beginning now and implemented through market-based policies, would stabilize the setting for new investment, provide incentive for innovation in technologies and institutions, and provide the opportunity to learn about the costs of climate change policy.

Every negotiation is shaped by expectations about what will happen if it breaks down. If the multipollutant debate collapses, what is expected is a return to the status quo—the so-called emission baseline. But like so much in this debate, the baseline itself is uncertain—so much so that the parties may find it more useful to compromise over legislation than to continue the familiar battles over a shifting baseline. Environmental advocates feel that regulatory inertia favors continuing reductions in conventional emissions. Regulatory calendars stretching to 2015 foresee the implementation of new particulate standards, mercury controls, and regional haze rules that will affect the various emissions individually. But advocates of multipollutant legislation have to feel their possible advances are at risk under the Bush administration.

The bellwether may be how the administration completes its ongoing review of NSR. If it backs away from proposed changes that would give power plants more freedom to expand or modernize without coming under NSR requirements, industry may feel it has the most to gain from legislation. If the administration review leads to substantial revisions or elimination of NSR, the baseline will be changed. Either fur may fly in Congress, or the stage may be set for an important compromise around multipollutant legislation. Or maybe both.

Dallas Burtraw is a senior fellow at Resources for the Future.

From the *Brookings Review*, Spring 2002, pp. 14-17, p. 48. © 2002 by the Brookings Institution Press, Washington, DC. Reprinted by permission.

Groundwater *Shock*

The Polluting of the World's Major Freshwater Stores

Scientists have shown that the world deep beneath our feet is essential to the life above. Ancient myths depicted the Underworld as a place of damnation and death. Now, the spreading contamination of major aquifers threatens to turn the myth into a tragic reality.

by Payal Sampat

The Mississippi River occupies a mythic place in the American imagination, in part because it is so huge. At any given moment, on average, about 2,100 billion liters of water are flowing across the Big Muddy's broad bottom. If you were to dive about 35 feet down and lie on that bottom, you might feel a sense of awe that the whole river was on top of you. But in one very important sense, you'd be completely wrong. At any point in time, only 1 percent of the water in the Mississippi River system is in the part of the river that flows downstream to the Gulf of Mexico. The other 99 percent lies beneath the bottom, locked in massive strata of rock and sand.

This is a distinction of enormous consequence. The availability of clean water has come to be recognized as perhaps the most critical of all human security issues facing the world in the next quarter-century—and what is happening to water buried under the bottoms of rivers, or under our feet, is vastly different from what happens to the "surface" water of rivers, lakes, and streams. New research finds that contrary to popular belief, it is groundwater that is most dangerously threatened. Moreover, the Mississippi is not unique in its ratio of surface to underground water; worldwide, 97 percent of the planet's liquid freshwater is stored in aquifers.

In the early centuries of civilization, surface water was the only source we needed to know about. Human population was less than a tenth of one percent the size it is now; settlements were on river banks; and the water was relatively clean. We still think of surface water as being the main resource. So it's easy to think that the problem of contamination is mainly one of surface water: it is polluted rivers and streams that threaten health in times of flood, and that have made waterborne diseases a major killer of humankind. But in the past century, as population has almost quadrupled and rivers have become more depleted and polluted, our dependence on pumping groundwater has soared—and as it has, we've made a terrible discovery. Contrary to the popular impression that at least the waters from our springs and wells are pure, we're uncovering a pattern of pervasive pollution there too. And in these sources, unlike rivers, the pollution is generally irreversible.

This is largely the work of another hidden factor: the rate of groundwater renewal is very slow in comparison with that of surface water. It's true that some aquifers recharge fairly quickly, but the average recycling time for groundwater is 1,400 years, as opposed to only 20 days for river water. So when we pump out groundwater, we're effectively removing it from aquifers for generations to come. It may evaporate and return to the atmosphere quickly enough, but the resulting rainfall (most of which falls back into the oceans) may take centuries to recharge the aquifers once they've been depleted. And because water in aquifers moves through the Earth with glacial slowness, its pollutants continue to accumulate. Unlike rivers, which flush themselves into the oceans, aquifers become sinks for pollutants, decade after decade—thus further diminishing the amount of clean water they can yield for human use.

Perhaps the largest misconception being exploded by the spreading water crisis is the assumption that the ground we stand on—and what lies beneath it—is solid, unchanging, and inert. Just as the advent of climate change has awakened us to the fact that the air over our heads is an arena of enormous forces in the midst of titanic shifts, the water crisis has revealed that, slow-moving though it may be, groundwater is part of a system of powerful hydrological interactions—between earth, surface water, sky, and sea— that we ignore at our peril. A few years ago, reflecting on how human activity is beginning to affect climate, Columbia University scientist Wallace Broecker warned, "The climate system is an angry beast and we are poking it with sticks." A similar statement might now be made about the system under our feet. If we continue to drill holes into it—expecting it to swallow our waste and yield freshwater in return—we may be toying with an outcome no one could wish.

Valuing Groundwater

For most of human history, groundwater was tapped mainly in arid regions where surface water was in short supply. From Egypt to Iran, ancient Middle Eastern civilizations used periscope-like conduits to funnel spring water from mountain slopes to nearby towns—a technology that allowed settlement to spread out from the major rivers. Over the centuries, as populations and cropland expanded, innovative well-digging techniques evolved in China, India, and Europe. Water became such a valuable resource that some cultures developed elaborate mythologies imbuing underground water and its seekers with special powers. In medieval Europe, people called water witches or dowsers were believed to be able to detect groundwater using a forked stick and mystical insight.

In the second half of the 20th century, the soaring demand for water turned the dowsers' modern-day counterparts into a major industry. Today, major aquifers are tapped on every continent, and groundwater is the primary source of drinking water for more than 1.5 billion people worldwide (see table, *Groundwater as a Share of Drinking Water Use by Region*). The aquifer that lies beneath the Huang-Huai-Hai plain in eastern China alone supplies drinking water to nearly 160 million people. Asia as a whole relies on its groundwater for nearly one-third of its drinking water supply. Some of the largest cities in the developing world—Jakarta, Dhaka, Lima, and Mexico City, among them—depend on aquifers for almost all their wa-

ter. And in rural areas, where centralized water supply systems are undeveloped, groundwater is typically the sole source of water. More than 95 percent of the rural U.S. population depends on groundwater for drinking.

A principal reason for the explosive rise in groundwater use since 1950 has been a dramatic expansion in irrigated agriculture. In India, the leading country in total irrigated area and the world's third largest grain producer, the number of shallow tubewells used to draw groundwater surged from 3,000 in 1960 to 6 million in 1990. While India doubled the amount of its land irrigated by surface water between 1950 and 1985; it increased the area watered by aquifers 113-fold. Today, aquifers supply water to more than half of India's irrigated land. The United States, with the third highest irrigated area in the world, uses groundwater for 43 percent of its irrigated farmland. Worldwide, irrigation is by far the biggest drain on freshwater: it accounts for about 70 percent of the water we draw from rivers and wells each year.

Other industries have been expanding their water use even faster than agriculture—and generating much higher profits in the process. On average, a ton of water used in industry generates roughly $14,000 worth of output—about 70 times as much profit as the same amount of water used to grow grain. Thus, as the world has industrialized, substantial amounts of water have been shifted from farms to more lucrative factories. Industry's share of total consumption has reached 19 percent and is likely to continue rising rapidly. The amount of water available for drinking is thus constrained not only by a limited resource base, but by competition with other, more powerful users.

And as rivers and lakes are stretched to their limits—many of them dammed, dried up, or polluted—we're growing more and more dependent on groundwater for all these uses. In Taiwan, for example, the share of water supplied by groundwater almost doubled from 21 percent in 1983 to over 40 percent in 1991. And Bangladesh, which was once almost entirely river- and stream-dependent, dug over a million wells in the 1970s to substitute for its badly polluted surface-water supply. Today, almost 90 percent of its people use only groundwater for drinking.

Even as our dependence on groundwater increases, the availability of the resource is becoming more limited. On almost every continent, many major aquifers are being drained faster than their natural rate of recharge. Groundwater depletion is most severe in parts of India, China, the United States, North Africa, and the Middle East. Under certain geological conditions, groundwater overdraft can cause aquifer sediments to compact, permanently shrinking the aquifer's storage capacity. This loss can be quite considerable, and irreversible. The amount of water storage capacity lost because of aquifer compaction in California's Central Valley, for example, is equal to more than 40 percent of the combined storage capacity of all human-made reservoirs across the state.

As the competition among factories, farms, and households intensifies, it's easy to overlook the extent to which freshwater is also required for essential ecological services. It is not just rainfall, but groundwater welling up from beneath, that replenishes rivers, lakes, and streams. In a study of 54 streams in different parts of the country, the U.S. Geological Survey (USGS) found that groundwater is the source for more than half the flow, on average. The 492 billion gallons (1.86 cubic kilometers) of water aquifers add to U.S. surface water bodies each day is nearly equal to the daily flow of the Mississippi. Groundwater provides the base contribution for the Mississippi, the Niger, the Yangtze, and many more of the world's great rivers—some of which would otherwise not be flowing year-round. Wetlands, important habitat for birds, fish, and other wildlife, are often largely groundwater-fed, created in places where the water table overflows to the surface on a constant basis. And while providing surface bodies with enough water to keep them stable, aquifers also help prevent them from flooding: when it rains heavily, aquifers beneath rivers soak up the excess water, preventing the surface flow from rising too rapidly and overflowing onto neighboring fields and towns. In tropical Asia, where the hot season can last as long as 9 months, and where monsoon rains can be very intense, this dual hydrological service is of critical value.

Groundwater as a Share of Drinking Water Use by Region

Region	Share of Drinking Water from Groundwater	People Served
	(percent)	(millions)
Asia-Pacific	32	1,000 to 1,200
Europe	75	200 to 500
Latin America	29	150
United States	51	135
Australia	15	3
Africa	NA	NA
World		1,500 to 2,000

Sources: UNEP, OECD, FAO, U.S. EPA, Australian EPA.

Numerous studies have tracked the extent to which our increasing demand on water has made it a resource critical to a degree that even gold and oil have never been. It's the most valuable thing on Earth. Yet, ironically, it's the thing most consistently overlooked, and most widely used as a final resting place for our waste. And, of course, as contamination spreads, the supplies of usable water get tighter still.

Tracking the Hidden Crisis

In 1940, during the Second World War, the U.S. Department of the Army acquired 70 square kilometers of land around Weldon Spring and its neighboring towns near St. Louis, Missouri. Where farmhouses and barns had been, the Army established the world's largest TNT-producing facility. In this sprawling warren of plants, toluene (a compo-

nent of gasoline) was treated with nitric acid to produce more than a million tons of the explosive compound each day when production was at its peak.

Part of the manufacturing process involved purifying the TNT—washing off unwanted "nitroaromatic" compounds left behind by the chemical reaction between the toluene and nitric acid. Over the years, millions of gallons of this red-colored muck were generated. Some of it was treated at wastewater plants, but much of it ran off from the leaky treatment facilities into ditches and ravines, and soaked into the ground. In 1945, when the Army left the site, soldiers burned down the contaminated buildings but left the red-tinged soil and the rest of the site as they were. For decades, the site remained abandoned and unused.

Then, in 1980, the U.S. Environmental Protection Agency (EPA) launched its "Superfund" program, which required the cleaning up of several sites in the country that were contaminated with hazardous waste. Weldon Spring made it to the list of sites that were the highest priority for cleanup. The Army Corps of Engineers was assigned the task, but what the Corps workers found baffled them. They expected the soil and vegetation around the site to be contaminated with the nitroaromatic wastes that had been discarded there. When they tested the groundwater, however, they found that the chemicals were showing up in people's wells, in towns several miles from the site—a possibility that no one had anticipated, because the original pollution had been completely localized. Geologists determined that there was an enormous plume of contamination in the water below the TNT factory—a plume that over the previous 35 years had flowed through fissures in the limestone rock to other parts of the aquifer.

The Weldon Spring story may sound like an exceptional case of clumsy planning combined with a particularly vulnerable geological structure. But in fact there is nothing exceptional about it all. Across the United States, as well as in parts of Europe, Asia, and Latin America, human activities are sending massive quantities of chemicals and pollutants into groundwater. This isn't entirely new, of course; the subterranean world has always been a receptacle for whatever we need to dispose of—whether our sewage, our garbage, or our dead. But the enormous volumes of waste we now send underground, and the deadly mixes of chemicals involved, have created problems never before imagined.

What Weldon Spring shows is that we can't always anticipate where the pollution is going to turn up in our water, or how long it will be from the time it was deposited until it reappears. Because groundwater typically moves very slowly—at a speed of less than a foot a day, in some cases—damage done to aquifers may not show up for decades. In many parts of the world, we are only just beginning to discover contamination caused by practices of 30 or 40 years ago. Some of the most egregious cases of aquifer contamination now being unearthed date back to Cold War era nuclear testing and weapons-making, for example. And once it gets into groundwater, the pollution usually persists: the enormous volume, inaccessibility, and slow rate at which groundwater moves make aquifers virtually impossible to purify.

As this covert crisis unfolds, we are barely beginning to understand its dimensions. Few countries track the health of their aquifers—their enormous size and remoteness make them extremely expensive to monitor. As the new century begins, even hydrogeologists and health officials have only a hazy impression of the likely extent of groundwater damage in different parts of the world. Nonetheless, given the data we now have, it is possible to sketch a rough map of the regions affected, and the principal threats they face (see map, *Groundwater Contamination Hotspots* and table, *Some Major Threats to Groundwater*).

The Filter that Failed: Pesticides in Your Water

Pesticides are designed to kill. The first synthetic pesticides were introduced in the 1940s, but it took several decades of increasingly heavy use before it became apparent that these chemicals were injuring non-target organisms—including humans. One reason for the delay was that some groups of pesticides, such as organochlorines, usually have little effect until they bioaccumulate. Their concentration in living tissue increases as they move up the food chain. So eventually, the top predators—birds of prey, for example—may end up carrying a disproportionately high burden of the toxin. But bioaccumulation takes time, and it may take still more time before the effects are discovered. In cases where reproductive systems are affected, the aftermath of this chemical accumulation may not show up for a generation.

Even when the health concerns of some pesticides were recognized in the 1960s, it was easily assumed that the real dangers lay in the dispersal of these chemicals among animals and plants—not deep underground. It was assumed that very little pesticide would leach below the upper layers of soil, and that if it did, it would be degraded before it could get any deeper. Soil, after all, is known to be a natural filter, which purifies water as it trickles through. It was thought that industrial or agricultural chemicals, like such natural contaminants as rock dust, or leaf mold, would be filtered out as the water percolated through the soil.

But over the past 35 years, this seemingly safe assumption has proved mistaken. Cases of extensive pesticide contamination of groundwater have come to light in farming regions of the United States, Western Europe, Latin America, and South Asia. What we now know is that pesticides not only leach into aquifers, but sometimes remain there long after the chemical is no longer used. DDT, for instance, is still found in U.S. waters even though its use was banned 30 years ago. In the San Joaquin Valley of California, the soil fumigant DBCP (dibromochloropropane), which was used intensively in fruit orchards before it was banned in 1977, still lurks in the region's water supplies. Of 4,507 wells sampled by the USGS between 1971 and 1988, nearly a third had DBCP levels that were at least 10 times higher than allowed by the current drinking water standard.

In places where organochlorines are still widely used, the risks continue to mount. After half a century of spraying in the eastern Indian states of West Bengal and Bihar, for example, the Central Pollution Control Board found DDT in groundwater at levels as high as 4,500 micrograms per liter—several thousand times higher than what is considered a safe dose.

The amount of chemical that reaches groundwater depends on the amount used above ground, the geology of the region, and the characteristics of the pesticide itself. In some parts of the midwestern United States, for example, although pesticides are used intensively, the impermeable soils of the region make it difficult for the chemicals to percolate underground. The fissured aquifers of southern Arizona, Florida, Maine, and southern California, on the other hand, are very vulnerable to pollution—and these too are places where pesticides are applied in large quantities.

Pesticides are often found in combination, because most farms use a range of toxins to destroy different kinds of insects, fungi, and plant diseases. The USGS detected two or more pesticides in groundwater at nearly a quarter of the sites sampled in its National Water Quality Assessment between 1993 and 1995. In the Central Columbia Plateau aquifer, which extends over the states of Washington and Idaho, more than two-thirds of water samples contained multiple pesticides. Scientists aren't entirely sure what happens when these chemicals and their various metabolites come together. We don't even have standards for the many hundred *individual* pesticides in use—the EPA has drinking water standards for just 33 of these compounds—to say nothing of the infinite variety of toxic blends now trickling into the groundwater.

While the most direct impacts may be on the water we drink, there is also concern about what occurs when the pesticide-laden water below farmland is pumped back up for irrigation. One apparent consequence is a reduction in crop yields.

In 1990, the now-defunct U.S. Office of Technology Assessment reported that herbicides in shallow groundwater had the effect of "pruning" crop roots, thereby retarding plant growth.

From Green Revolution to Blue Baby: the Slow Creep of Nitrogen

Since the early 1950s, farmers all over the world have stepped up their use of nitrogen fertilizers. Global fertilizer use has grown ninefold in that time. But the larger doses of nutrients often can't be fully utilized by plants. A study conducted over a 140,000 square kilometer region of Northern China, for example, found that crops used on average only 40 percent of the nitrogen that was applied. An almost identical degree of waste was found in Sri Lanka. Much of the excess fertilizer dissolves in irrigation water, eventually trickling through the soil into underlying aquifers.

Joining the excess chemical fertilizer from farm crops is the organic waste generated by farm animals, and the sewage produced by cities. Livestock waste forms a particularly potent tributary to the stream of excess nutrients flowing into the environment, because of its enormous volume. In the United States, farm animals produce 130 times as much waste as the country's people do—with the result that millions of tons of cow and pig feces are washed into streams and rivers, and some of the nitrogen they carry ends up in groundwater. To this Augean burden can be added the innumerable leaks and overflows from urban sewage systems, the fertilizer runoff from suburban lawns, golf courses, and landscaping, and the nitrates leaking (along with other pollutants) from landfills.

There is very little historical information available about trends in the pollution of aquifers. But several studies show that nitrate concentrations have increased as fertilizer applications and population size have grown. In California's San Joaquin-Tulare Valley, for instance, nitrate levels in groundwater increased 2.5 times between the 1950s and 1980s—a period in which fertilizer inputs grew six-fold. Levels in Danish groundwater have nearly tripled since the 1940s. As with pesticides, the aftermath of this multi-sided assault of excess nutrients has only recently begun to become visible, in part because of the slow speed at which nitrate moves underground.

What happens when nitrates get into drinking water? Consumed in high concentrations—at levels above 10 milligrams (mg) per liter, but usually on the order of 100 mg/liter—they can cause infant methemoglobinemia, or so-called blue-baby syndrome. Because of their low gastric acidity, infant digestive systems convert nitrate to nitrite, which blocks the oxygen-carrying capacity of a baby's blood, causing suffocation and death. Since 1945, about 3,000 cases have been reported worldwide—nearly half of them in Hungary, where private wells have particularly high concentrations of nitrates. Ruminant livestock such as goats, sheep, and cows are vulnerable to methemoglobinemia in much the same way infants are, because their digestive systems also quickly convert nitrate to nitrite. Nitrates are also implicated in digestive tract cancers, although the epidemiological link is still uncertain.

In cropland, nitrate pollution of groundwater can have a paradoxical effect. Too much nitrate can weaken plants' immune systems, making them more vulnerable to pests and disease. So when nitrate-laden groundwater is used to irrigate crops that are also being fertilized, the net effect may be to reduce, rather than to increase production. This kind of over-fertilizing makes wheat more susceptible to wheat rust, for example, and it makes pear trees more vulnerable to fire blight.

In assembling studies of groundwater from around the world, we have found that nitrate pollution is pervasive—but has become particularly severe in the places where human population—and the demand for high food productivity—is most concentrated. In the northern Chinese counties of Beijing, Tianjin, Hebei, and Shandong, nitrate concentrations in groundwater exceeded 50 mg/liter in more than half of the locations studied. (The World Health Organization [WHO] drinking water guideline is 10 mg/liter.) In some places, the concentration had risen as high as 300 mg/liter. Since then, these levels may have increased, as fertilizer applications have escalated since the tests were carried out in 1995 and will likely increase even more as China's population (and demand for food) swells, and as more farmland is lost to urbanization, industrial development, nutrient depletion, and erosion.

Reports from other regions show similar results. The USGS found that about 15 percent of shallow groundwater sampled below agricultural and urban areas in the United States had nitrate concentrations higher than the 10 mg/liter guideline. In Sri Lanka, 79 percent of wells sampled by the British Geological Survey had nitrate levels that exceeded this guideline. Some 56 percent of wells tested in the Yucatan peninsula in Mexico had levels above 45 mg/liter. And the European Topic Centre on Inland Waters found that in Romania and Moldova, more than 35 percent of the sites sampled had nitrate concentrations higher than 50 mg/liter.

From Tank of Gas to Drinking Glass: the Pervasiveness of Petrochemicals

Drive through any part of the United States, and you'll probably pass more gas stations than schools or churches. As you pull into a station to fill up, it may not occur to you that you're parked over one of the most pervasive threats to ground-water: an underground storage tank (UST) for petroleum. Many of these tanks were installed two or three decades ago and, having been left in place long past their expected lifetimes, have rusted through in places—allowing a steady leakage of gasoline into the ground. Because they're underground, they're expensive to dig up and repair, so the leakage in some cases continues for years.

Petroleum and its associated chemicals—benzene, toluene, and gasoline additives such as MTBE—constitute the most common category of groundwater contaminant found in aquifers in the United States. Many of these chemicals are also known or suspected to be cancer-causing. In 1998, the EPA found that over 100,000 commercially owned petroleum USTs were leaking, of which close to 18,000 are known to have contaminated groundwater. In Texas, 223 of 254 counties report leaky USTs, resulting in a silent disaster that, according to the EPA, "has affected, or has the potential to affect, virtually every major and minor aquifer in the state." Household tanks, which store home heating oil, are a problem as well. Although the household tanks aren't subject to the same regulations and inspections as commercial ones, the EPA says they are "undoubtedly leaking." Outside the United States, the world's ubiquitous petroleum storage tanks are even less monitored, but spot tests suggest that the threat of leakage is omnipresent in the industrialized world. In 1993, petroleum giant Shell reported that a third of its 1,100 gas stations in the United Kingdom were known to have contaminated soil and

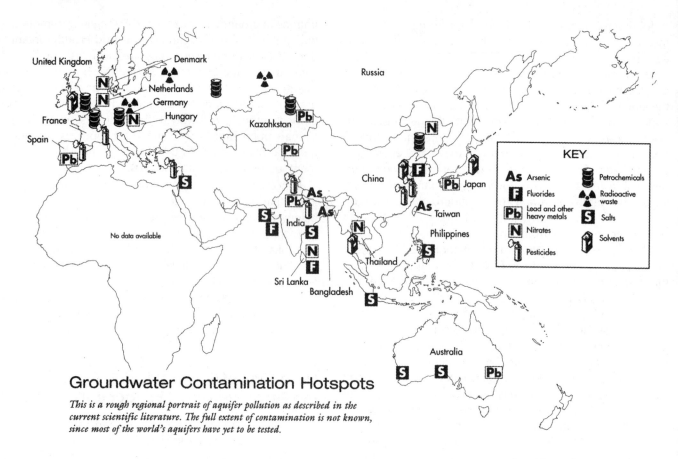

Groundwater Contamination Hotspots

This is a rough regional portrait of aquifer pollution as described in the current scientific literature. The full extent of contamination is not known, since most of the world's aquifers have yet to be tested.

(graphic continues on next page)

groundwater. Another example comes from the eastern Kazakh town of Semipalatinsk, where 6,460 tons of kerosene have collected in an aquifer under a military airport, seriously threatening the region's water supplies.

The widespread presence of petrochemicals in groundwater constitutes a kind of global malignancy, the danger of which has grown unobtrusively because there is such a great distance between cause and effect. An underground tank, for example, may take years to rust; it probably won't begin leaking until long after the people who bought it and installed it have left their jobs. Even after it begins to leak, it may take several more years before appreciable concentrations of chemicals appear in the aquifer—and it will likely be years beyond that before any health effects show up in the local population. By then, the trail may be decades old. So it's quite possible that any cancers occurring today as a result of leaking USTs might originate from tanks that were installed half a century ago. At that time, there were gas tanks sufficient to fuel 53 million cars in the world; today there are enough to fuel almost 10 times that number.

From Sediment to Solute: the Emerging Threat of Natural Contaminants

In the early 1990s, several villagers living near India's West Bengal border with Bangladesh began to complain of skin sores that wouldn't go away. A researcher at Calcutta's Jadavpur University, Dipanker Chakraborti, recognized the lesions immediately as early symptoms of chronic arsenic poisoning. In later stages, the disease can lead to gangrene, skin cancer, damage to vital organs, and eventually, death. In the months that followed, Chakraborti began to get letters from doctors and hospitals in Bangladesh, who were seeing streams of patients with similar symptoms. By 1995, it was clear that the country faced a crisis of untold proportions, and that the source of the poisoning was water from tubewells, from which 90 percent of the country gets its drinking water.

Experts estimate that today, arsenic in drinking water could threaten the health of 20 to 60 million Bangladeshis—up to half the country's population—and another 6 to 30 million people in West Bengal. As many as 1 million wells in the region may be contaminated with the heavy metal at levels between 5 and 100 times the WHO drinking water guidelines of 0.01 mg/liter.

How did the arsenic get into groundwater? Until the early 1970s, rivers and ponds supplied most of Bangladesh's drinking water. Concerned about the risks of water-borne disease, the WHO and international aid agencies launched a well-drilling program to tap groundwater instead. However, the agencies, not aware that soils of the Ganges aquifers are naturally rich in arsenic, didn't test the sediment before drilling tubewells. Because the effects of chronic arsenic poisoning can take up to 15 years to appear, the epidemic was not addressed until it was well under way.

Groundwater Contamination Hotspots
(continued from previous page)

Scientists are still debating what chemical reactions released the arsenic from the mineral matrix in which it is naturally bound up. Some theories implicate human activities. One hypothesis is that as water was pumped out of the wells, atmospheric oxygen entered the aquifer, oxidizing the iron pyrite sediments, and causing the arsenic to dissolve. An October 1999 article in the scientific journal *Nature* by geologists from the Indian Institute of Technology suggests that phosphates from fertilizer runoff and decaying organic matter may have played a role. The nutrient might have spurred the growth of soil microorganisms, which helped to loosen arsenic from sediments.

Salt is another naturally occurring groundwater pollutant that is introduced by human activity. Normally, water in coastal aquifers empties into the sea. But when too much water is pumped out of these aquifers, the process is reversed: seawater moves inland and enters the aquifer. Because of its high salt content, just 2 percent of seawater mixed with freshwater makes the water unusable for drinking or irrigation. And once salinized, a freshwater aquifer can remain contaminated for a very long time. Brackish aquifers often have to be abandoned because treatment can be very expensive.

In Manila, where water levels have fallen 50 to 80 meters because of overdraft, seawater has flowed as far as 5 kilometers into the Guadalupe aquifer that lies below the city. Saltwater has traveled several kilometers inland into aquifers beneath Jakarta and Madras, and in parts of the

U.S. state of Florida. Saltwater intrusion is also a serious problem on islands such as the Maldives and Cyprus, which are very dependent on aquifers for water supply.

Fluoride is another natural contaminant that threatens millions in parts of Asia. Aquifers in the drier regions of western India, northern China, and parts of Thailand and Sri Lanka are naturally rich in fluoride deposits. Fluoride is an essential nutrient for bone and dental health, but when consumed in high concentrations, it can lead to crippling damage to the neck and back, and to a range of dental problems. The WHO estimates that 70 million people in northern China, and 30 million in northwestern India are drinking water with high fluoride levels.

A Chemical Soup

With just over a million residents, Ludhiana is the largest city in Punjab, India's breadbasket state. It is also an important industrial town, known for its textile factories, electroplating industries, and metal foundries. Although the city is entirely dependent on groundwater, its wells are now so polluted with industrial and urban wastes that the water is no longer safe to drink. Samples show high levels of cyanide, cadmium, lead, and pesticides. "Ludhiana City's groundwater is just short of poison," laments a senior official at India's Central Ground Water Board.

Like Ludhiana's residents, more than a third of the planet's people live and work in densely settled cities, which occupy just 2 percent of the Earth's land area. With the labor force thus concentrated, factories and other centers of employment also group together around the same urban areas. Aquifers in these areas are beginning to mirror the increasing density and diversity of the human activity above them. Whereas the pollutants emanating from hog farms or copper mines may be quite predictable, the waste streams flowing into the water under cities contain a witch's brew of contaminants.

Ironically, a major factor in such contamination is that in most places people have learned to dispose of waste—to remove it from sight and smell—so effectively that it is easy to forget that the Earth is a closed ecological system in which nothing permanently disappears. The methods normally used to conceal garbage and other waste—landfills, septic tanks, and sewers—become the major conduits of chemical pollution of groundwater. In the United States, businesses drain almost 2 million kilograms of assorted chemicals into septic systems each year, contaminating the drinking water of 1.3 million people. In many parts of the developing world, factories still dump their liquid effluents onto the ground and wait for it to disappear. In the Bolivian city of Santa Cruz, for example, a shallow aquifer that is the city's main water source has had to soak up the brew of sulfates, nitrates, and chlorides dumped over it. And even protected landfills can be a potent source of aquifer pollution: the EPA found that a quarter of the landfills in the U.S. state of Maine, for example, had contaminated groundwater.

In industrial countries, waste that is too hazardous to land-fill is routinely buried in underground tanks. But as these caskets age, like gasoline tanks, they eventually spring leaks. In California's Silicon Valley, where electronics industries store assorted waste solvents in underground tanks, local groundwater authorities found that 85 percent of the tanks they inspected had leaks. Silicon Valley now has more Superfund sites—most of them affecting groundwater—than any other area its size in the country. And 60 percent of the United States' liquid hazardous waste—34 billion liters of solvents, heavy metals, and radioactive materials—is directly injected into the ground. Although the effluents are injected below the deepest source of drinking water, some of these wastes have entered aquifers used for water supplies in parts of Florida, Texas, Ohio, and Oklahoma.

Shenyang, China, and Jaipur, India, are among the scores of cities in the developing world that have had to seek out alternate supplies of water because their groundwater has become unusable. Santa Cruz has also struggled to find clean water. But as it has sunk deeper wells in pursuit of pure supplies, the effluent has traveled deeper into the aquifer to replace the water pumped out of it. In places where alternate supplies aren't easily available, utilities will have to resort to increasingly elaborate filtration set-ups to make the water safe for drinking. In heavily contaminated areas, hundreds of different filters may be necessary. At present, utilities in the U.S. Midwest spend $400 million each year to treat water for just one chemical—atrazine, the most commonly detected pesticide in U.S. groundwater. When chemicals are found in unpredictable mixtures, rather than discretely, providing safe water may become even more expensive.

One Body, Many Wounds

The various incidents of aquifer pollution described may seem isolated. A group of wells in northern China have nitrate problems; another lot in the United Kingdom are laced with benzene. In each place it might seem that the problem is local and can be contained. But put them together, and you begin to see a bigger picture emerging. Perhaps most worrisome is that we've discovered as much damage as we have, despite the very limited monitoring and testing of underground water. And because of the time-lags involved—and given our high levels of chemical use and waste generation in recent decades—what's still to come may bring even more surprises.

Some of the greatest shocks may be felt in places where chemical use and disposal has climbed in the last few decades, and where the most basic measures to shield groundwater have not been taken. In India, for example, the Central Pollution Control Board (CPCB) surveyed 22 major industrial zones and found that groundwater in every one of them was unfit for drinking. When asked about these findings, CPCB chairman D.K. Biswas remarked, "The result is frightening, and it is my belief that we will get more shocks in the future."

Jack Barbash, an environmental chemist at the U.S. Geological Survey, points out that we may not need to wait for expensive tests to alert us to what to expect in our groundwater. "If you want to know what you're likely to find in aquifers near Shanghai or Calcutta, just look at what's used above ground," he says. "If you've been applying DDT to a field for 20 years, for example, that's one of the chemicals you're likely to find in the underlying groundwater." The full consequences of today's chemical-dependent and waste-producing economies may not become apparent for another generation, but Barbash and other scientists are beginning to get a sense of just how serious those consequences are likely to be if present consumption and disposal practices continue.

Changing Course

Farmers in California's San Joaquin Valley began tapping the area's seemingly boundless groundwater store in the late-nineteenth century. By 1912, the aquifer was so depleted that the water table had fallen by as much as 400 feet in some places. But the farmers continued to tap the resource to keep up with demand for their produce. Over time, the dehydration of the aquifer caused its clay soil to shrink, and the ground began to sink—or as geologists put it, to "subside." In some parts of the valley, the ground has subsided as much as 29 feet—cracking foundations, canals, and aqueducts.

When the San Joaquin farmers could no longer pump enough groundwater to meet their irrigation demands, they began to bring in water from the northern part of the state via the California Aqueduct. The imported water seeped into the compacted aquifer, which was not able to hold all of the incoming flow. The water table then rose to an abnormally high level, dissolving salts and minerals in soils that had not been previously submerged. The salty groundwater, welling up from below, began to poison crop roots. In response, the farmers installed drains under irrigated fields—designed to capture the excess water and divert it to rivers and reservoirs in the valley so that it wouldn't evaporate and leave its salts in the soil.

But the farmers didn't realize that the rocks and soils of the region contained substantial amounts of the mineral selenium, which is toxic at high doses. Some of the selenium leached into the drainage water, which was routed to the region's wetlands. It wasn't until the mid-1980s that the aftermath of this solution became apparent: ecologists noticed that thousands of waterfowl in the nearby Kesterson Reservoir were dying of selenium poisoning.

Hydrological systems are not easy to outmaneuver, and the San Joaquin farmers' experience serves as a kind of cautionary tale. Each of their stopgap solutions temporarily took care of an immediate obstacle, but led to a longer-term problem more severe than the original one. "Human understanding has lagged one step behind the inflexible realities governing the aquifer system," observes USGS hydrologist Frank Chapelle.

Some Major Threats to Groundwater

Threat	Sources	Health and Ecosystem Effects at High Concentrations	Principal Regions Affected
Pesticides	Runoff from farms, backyards, golf courses; landfill leaks.	Organochlorines linked to reproductive and endocrine damage in wildlife; organophosphates and carbamates linked to nervous system damage and cancers.	United States, Eastern Europe, China, India.
Nitrates	Fertilizer runoff; manure from livestock operations; septic systems.	Restricts amount of oxygen reaching brain, which can cause death in infants ("blue-baby syndrome"); linked to digestive tract cancers. Causes algal blooms and eutrophication in surface waters.	Midwestern and mid-Atlantic United States, North China Plain, Western Europe, Northern India.
Petro-chemicals	Underground petroleum storage tanks	Benzene and other petrochemicals can be cancer-causing even at low exposure.	United States, United Kingdom, parts of former Soviet Union.
Chlorinated Solvents	Effluents from metals and plastics degreasing; fabric cleaning, electronics and aircraft manufacture.	Linked to reproductive disorders and some cancers.	Western United States, industrial zones in East Asia.
Arsenic	Naturally occurring; possibly exacerbated by over-pumping aquifers and by phosphorus from fertilizers.	Nervous system and liver damage; skin cancers.	Bangladesh, Eastern India, Nepal, Taiwan.
Other Heavy Metals	Mining waste and tailings; landfills; hazardous waste dumps.	Nervous system and kidney damage; metabolic disruption.	United States, Central America and northeastern South America, Eastern Europe.
Fluoride	Naturally occurring.	Dental problems; crippling spinal and bone damage.	Northern China, Western India; parts of Sri Lanka and Thailand.
Salts	Seawater intrusion; de-icing salt for roads.	Freshwater unusable for drinking or irrigation.	Coastal China and India, Gulf coasts of Mexico and Florida, Australia, Philippines.

Major sources: European Environmental Agency, USGS, British Geological Survey.

Around the world, human responses to aquifer pollution thus far have essentially reenacted the San Joaquin Valley farmers' well-meaning but inadequate approach. In many places, various authorities and industries have fought back the contamination leak by leak, or chemical by chemical—only to find that the individual fixes simply don't add up. As we line landfills to reduce leakage, for instance, tons of pesticide may be running off nearby farms and into aquifers. As we mend holes in underground gas tanks, acid from mines may be seeping into groundwater. Clearly, it's essential to control the damage we've already inflicted, and to protect communities and ecosystems from the poisoned fallout. But given what we already know—that damage done to aquifers is mostly irreversible, that it can take years before groundwater pollution reveals itself, that chemicals react synergistically, and often in unanticipated ways—its now clear that a patchwork response isn't going to be effective. Given how much damage this pollution inflicts on public health, the environment, and the economy once it gets into the water, it's critical that emphasis be shifted from filtering out toxins to not using them in the first place. Andrew Skinner, who heads the International Association of Hydrogeologists, puts it this way: "Prevention is the only credible strategy."

To do this requires looking not just at individual factories, gas stations, cornfields, and dry cleaning plants, but at the whole social, industrial, and agricultural systems of which these businesses are a part. The ecological untenability of these systems is what's really poisoning the world's water. It is the predominant system of high-input agriculture, for example, that not only shrinks biodiversity with its vast monocultures, but also overwhelms the land—and the underlying water—with its massive applications of agricultural chemicals. It's the system of car-dominated, geographically expanding cities that not only generates unsustainable amounts of climate-disrupting greenhouse gases and acid rain-causing air pollutants, but also overwhelms aquifers and soils with petrochemicals, heavy metals, and sewage. An adequate response will require a thorough overhaul of each of these systems.

Begin with industrial agriculture. Farm runoff is a leading cause of groundwater pollution in many parts of Europe, the United States, China, and India. Lessening its impact calls for adopting practices that sharply reduce this runoff—or, better still, that require far smaller inputs to begin with. In most places, current practices are excessively wasteful. In Colombia, for example, growers spray flowers with as much as 6,000 liters of pesticide per hectare. In Brazil, orchards get almost 10,000 liters per hectare. Experts at the U.N. Food and Agricultural Organization say that with modified application techniques, these chemicals could be applied at one-tenth those amounts and still be effective. But while using more efficient pesticide applications would constitute a major improvement, there is also the possibility of reorienting agriculture to use very little synthetic pesticide at all. Recent studies suggest that farms can maintain high yields while using little or no synthetic input. One decade-long investigation by the Rodale Institute in Pennsylvania, for example, compared traditional manure and legume-based cropping systems which used no synthetic fertilizer or pesticides, with a conventional, high-intensity system. All three fields were planted with maize and soybeans. The researchers found that the traditional systems retained more soil organic matter and nitrogen—indicators of soil fertility—and leached 60 percent less nitrate than the conventional system. Although organic fertilizer (like its synthetic counterpart) is typically a potent source of nitrate, the rotations of diverse legumes and grasses helped fix and retain nitrogen in the soil. Yields for the maize and soybean crops differed by less than 1 percent between the three cropping systems over the 10-year period.

In industrial settings, building "closed-loop" production and consumption systems can help slash the quantities of waste that factories and cities send to landfills, sewers, and dumps—thus protecting aquifers from leaking pollutants. In places as far-ranging as Tennessee, Fiji, Namibia, and Denmark, environmentally conscious investors have begun to build "industrial symbiosis" parks in which the unusable wastes from one firm become the input for another. An industrial park in Kalundborg, Denmark diverts more than 1.3 million tons of effluent from landfills and septic systems each year, while preventing some 135,000 tons of carbon and sulfur from leaking into the atmosphere. Households, too, can become a part of this systemic change by reusing and repairing products. In a campaign organized by the Global Action Plan for the Earth, an international nongovernmental organization, thoughtful consumption habits have enabled some 60,000 households in the United States and Europe to reduce their waste by 42 percent and their water use by 25 percent.

As it becomes clearer to decisionmakers that the most serious threats to human security are no longer those of military attack but of pervasive environmental and social decline, experts worry about the difficulty of mustering sufficient political will to bring about the kinds of systemic—and therefore revolutionary—changes in human life necessary to turn the tide in time. In confronting the now heavily documented assaults of climate change and biodiversity loss, leaders seem on one hand paralyzed by how bleak the big picture appears to be—and on the other hand too easily drawn into denial or delay by the seeming lack of immediate consequences of such delay. But protecting aquifers may provide a more immediate incentive for change, if only because it simply may not be possible to live with contaminated groundwater for as long as we could make do with a gradually more irritable climate or polluted air or impoverished wildlife. Although we've damaged portions of some aquifers to the point of no return, scientists believe that a large part of the resource still remains pure—for the moment. That's not likely to remain the case if we continue to depend on simply stepping up the present reactive tactics of cleaning up more of the chemical spills, replacing more of the leaking gasoline tanks, placing more

plastic liners under landfills, or issuing more fines to careless hog farms and copper mines. To save the water in time requires the same fundamental restructuring of the global economy as does the stabilizing of the climate and biosphere as a whole—the rapid transition from a resource-depleting, oil- and coal-fueled, high-input industrial and agricultural economy to one that is based on renewable energy, compact cities, and a very light human footprint. We've been slow to come to grips with this, but it may be our thirst that finally makes us act.

"Heaven is Under Our Feet"

Throughout human history, people have feared that the skies would be the source of great destruction. During the Cold War, industrial nations feared nuclear attack from above, and spent vast amounts of their wealth to avert it. Now some of that fear has shifted to the threats of atmospheric climate change: of increasing ultraviolet radiation through the ozone hole, and the rising intensity of global warming-driven hurricanes and typhoons. Yet, all the while, as the worldwide pollution of aquifers now reveals, we've been slowly poisoning ourselves from beneath. What lies under terra firma may, in fact, be of as much concern as what happens in the firmament above.

The ancient Greeks created an elaborate mythology about the Underworld, or Hades, which they described as a dismal, lifeless place completely lacking the abundant fertility of the world above. Science and human experience have taught us differently. Hydrologists now know that healthy aquifers are essential to the life above ground—that they play a vital role not just in providing water to drink, but in replenishing rivers and wetlands and, through their ultimate effects on rainfall and climate, in nurturing the life of the land and air as well. But ironically, our neglectful actions now threaten to make the Greek myth a reality after all. To avert that threat now will require taking to heart what the hydrologists have found. As Henry David Thoreau observed a century-and-a-half ago, "Heaven is under our feet, as well as over our heads."

A Few Key Sources

Francis H. Chapelle, *The Hidden Sea: Ground Water, Springs, and Wells* (Tucson, AZ: Geoscience Press, Inc., 1997).

U.N. Environment Programme, *Groundwater: A Threatened Resource* (Nairobi: 1996).

European Environmental Agency, *Groundwater Quality and Quantity in Europe* (Copenhagen: 1999).

U.S. Geological Survey, *The Quality of Our Nation's Waters—Nutrients and Pesticides* (Reston, VA: 1999).

British Geological Survey et al., *Characterisation and Assessment of Groundwater Quality Concerns in Asia-Pacific Region* (Oxfordshire, UK: 1996).

Payal Sampat is a staff researcher at the Worldwatch Institute.

From *World Watch*, January/February 2000, pp. 10-22. © 2000 by the Worldwatch Institute (www.worldwatch.org). Reprinted by permission.

Water Quality: THE ISSUES

BY MARY H. COOPER

The last three decades have seen dramatic improvements in the quality of America's waterways. Majestic cormorants and herons are again feeding in many rivers, signaling the return of the fish they prey on. Swimmers are taking the plunge into waters that once would have sickened them. Anglers are catching fish that are healthy enough to eat.

"Twenty-eight years ago, the Potomac River was too dirty to swim in, Lake Erie was dying and [Ohio's] Cuyahoga River was so polluted it burst into flames," said Carol M. Browner, administrator of the Environmental Protection Agency (EPA), the agency created in 1970 to implement federal environmental policy.

"Many rivers and beaches were little more than open sewers," Browner continued. "Enactment of the [1972] Clean Water Act dramatically improved the health of rivers, lakes and coastal waters. It stopped billions of pounds of pollution from fouling the water and doubled the number of waterways safe for fishing and swimming. Today, many rivers, lakes, and coasts are thriving centers of healthy communities."[1]

But there is still cause for concern about water pollution in the United States. Even after three decades of federal cleanup efforts, about 40 percent of the nation's streams, rivers, lakes and coastal waters are still too dirty for people to fish or swim in, the EPA reports.[2] That's down just 20 percentage points from the 60 percent recorded in 1972.

The culprits today are less likely to be factories and sewage treatment plants, which were the main focus of the first round of clean-water efforts. Those "point sources" were easy to identify and target because they generally discharged pollutants directly into the water. The Clean Water Act required industries and municipalities to filter out wastes before they reached the water.

Although the most egregious point-source pollution has been reduced, a more insidious form of pollution continues to dirty the nation's waterways—runoff from city streets, suburban construction sites and farms. Contaminated by fertilizers and animal waste, agricultural runoff contains nitrogen, phosphorus and other nutrients, which deprive waterways of the oxygen needed to support aquatic life. Runoff also contains toxins washed into storm sewers from city streets, which kill fish outright and threaten human health.

The extent of impairment to water quality caused by nutrient-polluted runoff is especially evident in estuaries, the bays and tidal rivers that empty into the oceans. Nutrients from as far away as Montana travel down the Missouri and Mississippi rivers, picking up additional pollutants along the way. Emptying into the Gulf of Mexico, they cause a lifeless "dead zone" that spreads over an area that at times equals the state of New Jersey. Similar problems plague the nation's largest estuary, the Chesapeake Bay, where a regional cleanup campaign is trying to stem the impact of nutrient pollution. (*See sidebar*, "Regional Plan Offers Hope for Chesapeake Bay.")

"It's clear that nutrient pollution remains the most daunting challenge in restoring the Chesapeake," says Michael Hirshfield, senior vice president for resource protection at the Chesapeake Bay Foundation, in Annapolis, Md. "We've known for 20 years that the biggest water-quality problem the bay faced was having far too much nitrogen and phosphorus coming in. We've made some modest gains, particularly in dealing with point sources. But when you look at what's happening on the land in the watershed surrounding the bay, it's clear that if we're really going to bring back underwater grasses and have adequate dissolved oxygen and clear water, pretty much every sector of the population and the economy is going to have to do a lot more to control nutrient pollution."

Cleaning up non-point source pollution—which cannot be traced to easily identifiable discharge points—poses the toughest challenge to policy-makers. And proposals for the next generation of programs to achieve the Clean Water Act's goal of making all the nation's waterways "fishable" and "swimmable" are highly controversial.

"Our regulatory systems are set up for fundamentally simple one-cause, one-effect, one-regulation kinds of situations, where you've got a sewage-treatment plant or a factory that's discharging into one little waterbody," Hirshfield explains. "Nutrient pollution, where the effects of lots of non-point sources are both distant from the sources and cumulative, means that allocating responsibility is a real challenge to our sector-by-sector environmental laws."

To speed the process of curbing runoff, President Clinton last year proposed a new set of "non-point source" regulations. The new rules would require states to set pollution limits—known as "total maximum daily loads," or TMDLs—for all the estimated 20,000 waterways that are polluted and establish a timetable for reducing pollution to meet those limits. The states would be responsible for monitoring and enforcing the regulations, and

America's Polluted Waters, 1998

Pollution of the nation's streams, rivers, coastlines, estuaries and lakes is largely caused by non-point pollution, such as agricultural runoff, and the impact of high population density. Waters are considered threatened or impaired if they do not meet one or more state water-quality standards. In some cases, areas not shown to suffer from pollution actually may be polluted but were not identified as impaired.

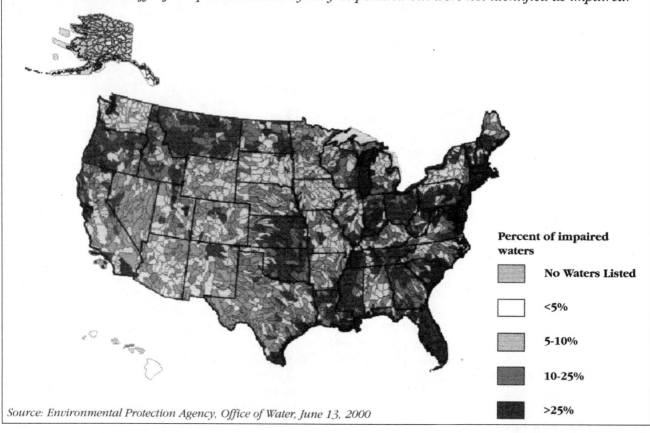

Percent of impaired waters

	No Waters Listed
	<5%
	5-10%
	10-25%
	>25%

Source: Environmental Protection Agency, Office of Water, June 13, 2000

EPA would get involved only if states are unable or unwilling to carry out the program.[3]

EPA officials say that the proposed regulations mark a departure from past clean-water efforts. "The TMDL program unquestionably represents a turning point in our nation's Clean Water Act programs," says J. Charles Fox, assistant administrator for EPA and director of the agency's Office of Water. "We've spent the better part of 30 years focusing on discrete point sources of pollution that have been heavily dependent on national and state-level regulations to effect pollution control. Today's challenges are much more diffuse, but the regulatory program is still going to have a very important role in the future as well."

A broad coalition of groups representing farmers, timber companies and builders says the regulations amount to bureaucratic excess. "We believe the agency has well overreached its regulatory authority," says Don Parrish, senior environmental policy specialist for the American Farm Bureau Federation, the country's leading farm lobby. "We don't believe that there's any federal implementation authority for TMDLs; we believe that authority is reserved for the states.

"The way I read the Clean Water Act," Parrish continues, "it gives EPA the authority to permit discharges. It does not give EPA the authority to permit people to operate. We still have a free-enterprise system here."

Because nutrient-laden runoff is the leading culprit in water pollution today, the new regulations would be directed largely toward curbing runoff at the source—such as farms, animal feedlots and timber operations. Farmers might be required to let more land lie fallow, for example, while feedlot operators might have to invest in new technology to store and treat manure to keep it out of waterways. Critics argue that the regulations would drive many operations out of business.

"It's going to be expensive to invest in these new technological fixes," Parrish says. "A lot of small operators are not going to be able to meet those requirements, and they're just going to load their animals on a trailer and take them to the local sale barn and go out of business."

In October, the Senate voted to delay implementation of the new TMDL rules until studies can be done to evaluate their costs and benefits.

As lawmakers take up the next phase of clean-water regulations, these are some of the issues they will consider:

Assessing U.S. Water Quality

About 40 percent of U.S. waters that were assessed in 1998 were not clean enough for fishing and swimming, including more than 290,000 miles of rivers and streams.* Leading pollutants in impaired waters include siltation, bacteria, nutrients and metals. Runoff from agricultural lands and urban areas is the primary source of these pollutants.

Quality of Rivers, Lakes and Estuaries

Waterbody Type	Total Size	Amount Assessed (% of total)	Good (% of assessed)	Good but Threatened (% of assessed)	Polluted (% of assessed)
Rivers (miles)	3,662,255	842,426 (23%)	463,441 (55%)	85,544 (10%)	291,264 (34%)
Lakes (acres)	41,593,748	17,390,370 (42%)	7,927,486 (46%)	1,565,175 (9%)	7,897,110 (45%)
Estuaries (sq. miles)	90,465	28,687 (32%)	13,439 (47%)	2,766 (10%)	12,482 (44%)

*About 32 percent of U.S. waters were assessed.
Note: Percentages may not add to 100% due to rounding.
Source: Environmental Protection Agency, "1998 National Water Quality Inventory Report to Congress," June 2000

Are proposed new rules to curb storm water and agricultural runoff the best way to clean up streams and rivers?

The federal regulatory system constructed by the Clean Water Act and its subsequent amendments marked a clear departure in environmental policy-making. Before the law, water, like land, fell under the jurisdiction of the states. The only exceptions were waterways that pass through more than one state, such as the Missouri River and the Great Lakes.

Congress changed that system because the states had clearly failed to stop the wholesale dumping of industrial pollutants and untreated sewage into the nation's waterways, especially with the economic boom and population increases that followed World War II. The Clean Water Act established a number of programs to clean up the water and shifted the primary responsibility for implementing them from the states to the federal EPA. In most cases EPA has delegated its authority to implement the law to state environmental protection agencies. In the 16 states that are either unable or unwilling to do the job, the federal agency assumes direct responsibility for enforcing its provisions.

Most observers agree that the regulatory approach has worked well to reduce point-source water pollution. It has allowed states that were not eager to offend polluting—but profitable and job-creating—industries to place the blame for the new rules squarely on the shoulders of the federal government. EPA issues the permits defining pollution thresholds to each polluter and is ultimately responsible for enforcing the law and imposing fines and other sanctions to repeat offenders. The states also have received federal funding under the law to help pay for upgrading municipal sewage-treatment plants, the other main point source of water pollution.

But the federal regulatory approach is far more controversial in the context of today's main threat to water quality, non-point source pollution, or runoff. Not only are the polluters harder to identify, they often are less able to bear the cost of mitigating the harm they cause to waterways, often hundreds of miles downstream. Many critics contend that neither EPA nor the states have collected enough scientifically valid data to justify requiring alleged polluters to shoulder the responsibility for fouled waterways.

"After nearly 30 years and $600 billion worth of hit-and-miss technologies, we still don't know what has been achieved or what still needs to be done," writes Richard A. Halpern, director of environmental affairs for the Hudson Institute's Center for Global Food Issues, a conservative think tank in Churchville, Va. "Also, the lack of real-world data leaves activists free to claim that 'our water continues to be poisoned,' and provides them a pretext for demanding an increasingly intrusive role in determining the national lifestyle."[4]

But supporters of the regulatory approach say it's neither necessary nor prudent to await the exact measurement of each and every tainted waterway to take steps to curb pollution. "I will be the first to admit that in many cases more science and more data would be better," says EPA's Fox. "But in the vast majority of cases we have enough information to make a rational decision today." He cites the validity of warnings in the early 1970s that excessive phosphorus levels were causing fish kills in the Great Lakes. "There were scientists at the time who protested that we didn't know exactly how much phosphorus was present. But the trends were unmistakable, and people in the late 1970s had the courage to make a decision to control phosphorus, and it clearly was the right decision."

Critics contend that federal regulations on runoff violate states' jurisdiction over land use. In September, the American Farm Bureau Federation filed a lawsuit in California to block the implementation of federal water-quality regulations that would restrict timber harvests in an effort to reduce runoff into Mendocino County's Garcia River.[5]

Point and Non-point Sources of Pollution

Factories and sewage-treatment plants (left side of river) were the main focus of U.S. clean-water efforts 30 years ago. But today the biggest remaining causes of water pollution are non-point sources, including runoff from city streets, suburban construction, agriculture and forestry (right side of river).

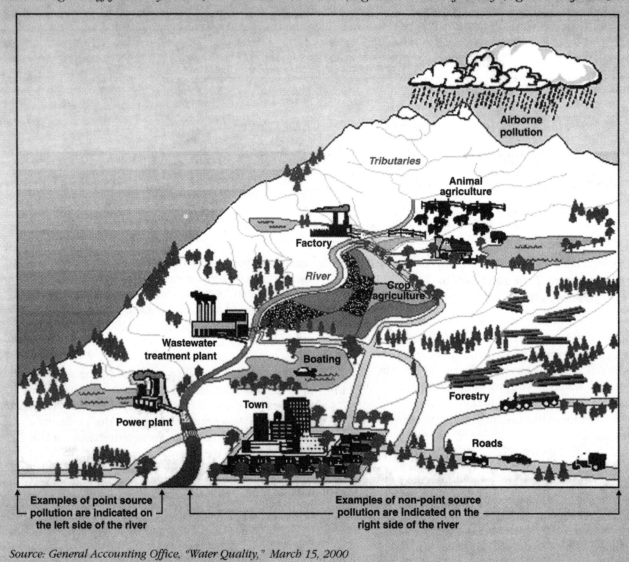

| Examples of point source pollution are indicated on the left side of the river | Examples of non-point source pollution are indicated on the right side of the river |

Source: General Accounting Office, "Water Quality," March 15, 2000

Removing trees and other vegetation from the land can increase runoff into nearby waterways, but not in an easily measured way, critics say. "As the statute defines total maximum daily loads (TMDLs), they imply a specific daily allocation of pollutant loads," Parrish says. "But you don't have daily loads with agricultural runoff; you only have agricultural runoff in the presence of snowmelt or a rainfall runoff event. And we have a real hard time applying what we believe is very specific, point-source statutory authority to a non-point source that is driven primarily by stormfall events."

Defenders of the regulatory nature of clean-water policies say there is room for accommodation in the proposed new rules. "I don't think anybody has ever said that every farm or every suburban development needs to get a permit," says Hirshfield of the Chesapeake Bay Foundation. He likens the clean-water rules to speed limits on the nation's roadways. "We don't put a machine in each one of our cars that prevents it from going over the speed limit; we post signs, and we have cops. The TMDL approach does not say that everybody needs to have a discharge permit; it says that everybody in a watershed who is contributing to the failure of a waterbody to meet its fishable, swimmable goals has to contribute to the solution."[6]

Can non-regulatory approaches achieve good water quality?

Not surprisingly, the strongest critics of EPA's policies offer a number of solutions to water-quality problems that do not in-

Regional Plan Offers Hope for Chesapeake Bay

Perhaps no body of water better illustrates the nature of today's threats to water quality—and the difficulty of neutralizing them—than the Chesapeake Bay. North America's largest estuary lies at the end of a vast watershed that encompasses six states—Delaware, Maryland, New York, Pennsylvania, Virginia and West Virginia—and the District of Columbia and includes much of the heavily populated East Coast corridor.

When John Smith sailed into the Chesapeake in 1607, he and the English settlers who accompanied him to the New World found a vast body of water that was teeming with oysters, crabs and many species of fish. After almost 400 years of intense development in the bay's watershed, however, industrial pollutants and human waste threatened to kill virtually all the bay's plant and animal species. Since passage of the Clean Air Act in 1970, most factories and sewage-treatment plants have installed technologies that filter out many of these pollutants. But like other waterways around the country, the bay is threatened today primarily by runoff filled with nutrients and chemicals from farms and city streets.

According to the Chesapeake Bay Foundation in Annapolis, Md., efforts to clean up the bay have made only modest improvements. In its annual report on the state of the bay, the group this year gave the bay a score of 28 on a scale of 100, which represents its pristine condition when the Europeans first arrived.[1] That's a 20 percent improvement since 1983, when the bay only scored 23.

Although foundation officials don't expect the estuary to ever return to its earlier condition, they hope to attain a score of 70 over the next two decades. The score is based on 13 factors that help determine the bay's overall health, including water clarity, the presence of underwater grasses and populations of oysters, striped bass and other aquatic species. Of these, only one—the presence of shad—improved last year, but the improvement from a score of 3 to 5 was hardly overwhelming.

The Chesapeake is not alone. "The Chesapeake may be the biggest, most thoroughly studied and best-known estuary in the country," says Michael Hirshfield, the foundation's senior vice president for resource protection. "But if you look at almost any estuary, whether it's Tampa Bay or Long Island Sound, nutrient pollution is a ubiquitous problem. One of the lessons of the last decade or so is that pretty much wherever you look near coastal waters you will see similar issues—point sources from sewage-treatment plants, livestock and agriculture, fallout from cars and power plants, runoff from suburban lawns and even dog poop."

Of all the causes of polluted runoff, the foundation singles out one as the main culprit—sprawl. "Chesapeake Bay restoration is being undermined by sloppy development," said William C. Baker, the foundation's president. "Millions of dollars are being spent to restore wetlands and underwater grasses, yet population growth and its ugly stepchild, sprawl, continue to threaten the bay."[2]

Congress has singled out estuaries for special treatment. On Oct. 25, lawmakers passed a law to establish a national estuary-protection program. It provides $275 million to restore more than a million acres of estuary habitat over the next decade. The new law also provides $40 million a year through fiscal 2005 for continued restoration of the Chesapeake Bay.

Meanwhile, environmental activists are guardedly optimistic about the impact of a new regional agreement to step up efforts to save the bay. By signing the new Chesapeake 2000 agreement in June, Maryland, Virginia, Pennsylvania and the District of Columbia agreed to preserve millions of acres of wetlands, which serve as a vital buffer between runoff and the bay, increase the bay's population of oysters, which filter pollutants from the water, and reduce the rate at which farmland and woodlands are lost to development.

"The Chesapeake Bay agreement and the participation of the three main bay states is pretty remarkable," Hirshfield says. "A drawback of this voluntary and consensus-based approach is that the bay cleanup is taking longer than it should have. We believe that you can get only so far with voluntary, consensus-based goals."

EPA officials say the regional approach shows promise for other waterways threatened by polluted runoff. "We learned in the mid-1980s that the base level of national regulations is not going to result in a clean Chesapeake Bay," says J. Charles Fox, EPA's assistant administrator for the Office of Water. "Defining how much more needed to be done was a job best done in partnership with the federal and state governments, and that's what happened. I think that model is going to be used a lot in the future in our national water programs."

---

[1] Chesapeake Bay Foundation, "The State of the Bay Report 2000."

[2] Baker spoke on Sept. 20, 2000, following the report's release. See Anita Huslin, "Bay Gets Failing Grade, but Progress Is Cited," *The Washington Post*, Sept. 21, 2000.

volve federal mandates. The Farm Bureau Federation, for example, says the federal regulatory approach is not appropriate for farmers or other non-point sources of water pollution and calls for a more collaborative effort involving farmers and surrounding communities at the state and local levels.

"What we've tried to do in the agricultural community and what we'd like to see done more specifically is take a closer look at these streams and rivers, and on a more collaborative basis," Parrish says. "We don't mind doing it with the state environmental protection agency or other interest groups within the states, but we would like to see a prioritization of where our water-quality problems are, then apply the resources to those

streams, rivers, lakes and estuaries in such a way that we can systematically address the worst cases first."

Funding for this approach, Parrish says, is available through existing farm bill programs that offer incentives for farmers to invest in clean-water technologies. "We believe that's a more targeted approach," he says. "Our approach would be one in which you actually find your water-quality problems and in a very systematic fashion get the community to buy into the effort and make the hard decisions at the local level to get them addressed."

Some current and former state regulators agree that a more flexible approach is needed. "There's no question about the fact that water-quality problems are site- and situation-specific problems that require site- and situation-specific solutions," says Becky Norton Dunlop, Virginia's secretary of natural resources during the administration of Republican Gov. George Allen and currently vice president for external relations at the conservative Heritage Foundation.

"People from the states understand the national goal to improve water quality should be able to set priorities for each river and move toward achieving clean water in ways that are fully embraced by the communities involved and based on good science," Dunlop continues. "I'm sure there are occasions when good science has been used by EPA, but what we've seen the agency do over the past eight years is use political science to make and implement national policy."

A few industries that are likely to be the focus of EPA's stepped-up effort to curb polluted runoff are taking steps on their own to clean up their operations—and ward off the heavy hand of impending federal regulations. Near the top of the list are so-called concentrated animal-feeding operations. Also known as factory farms, these facilities turn out huge numbers of hogs, cattle and chickens and increasingly are replacing smaller-scale family farms. The concentrations of animals have created vast amounts of manure and the inevitable horror stories. Last year, for example, a waste lagoon burst at a hog farm in Duplin County, N.C., spilling some 2 million gallons of hog manure into a tributary of the Northeast Cape Fear River.[7]

To avert future spills, the National Pork Producers Council has devised the livestock industry's most sweeping clean-water measure to date. The voluntary, industry-funded program employs third-party experts to help hog farmers assess their waste-management practices and invest in technologies to treat and store hog manure.

"Quite frankly, some of our livestock colleagues didn't appreciate what we were doing, but we realized that the states were taking action and setting up regulatory programs that we were concerned may not be affordable and practical for pork producers," says Deborah Atwood, an environmental consultant for the Des Moines, Iowa-based council. "So we decided that all pork producers, of all types and sizes, should have water permits."

Since the council introduced the $1.5-million program in 1997, Atwood says about 1,600 hog farmers have opted to participate. "The program has moved out of the pork industry now into the dairy, turkey, chicken and cattle industries," she says. "They're all realizing that it's very helpful to have a third set of eyes come to their operations and examine them for everything from odor to cleanliness to management. We've discovered that within six months over 67 percent of the problems have been fixed."

Another non-regulatory approach to mitigating water pollution is nutrient trading, a market-based approach that sets a goal for the total amount of nitrogen, phosphorus and other nutrients carried into waterways in runoff. The total amount of allowable pollution in a given waterway is then allocated among the farms, municipalities and other sources that choose to participate in the trading program. Those that can more easily reduce their pollution below their allowable limits earn credits, which they can then sell to polluters that can't afford to eliminate their discharges.

"With trading, whoever has the lowest cost for remediation would have an economic incentive to overcomply with the discharge limits, and in overcomplying they would generate credits," says Paul Faeth, director of the World Resources Institute's economics and population program in Washington, D.C., and the author of a recent study on nutrient trading.[8]

Faeth says the trading approach may greatly reduce the costs associated with further improvements in water quality. "If you go with traditional regulations, it can cost up to $24 for every pound of phosphorus kept out of the water," he says, "but if you go with the trading scheme you can get that down to as little as $2 a pound. We think that one of the key elements in terms of the political acceptance of further movement on water quality is to find ways of keeping those costs down."

Unlike some critics of EPA, however, Faeth says nutrient trading and other non-traditional methods of improving water quality are no substitute for regulations. "We need to move forward with regulation to improve water quality," he says. "But regulations should be coupled with flexibility."

Is tap water safe to drink?

Much of the nation's drinking water comes from lakes and rivers, some of the same waterways that fall under the protection of the Clean Water Act. But about half the nation's drinking water comes from groundwater, or deep underground reservoirs that generally do not mingle with surface waters.

To ensure the safety of the nation's drinking water, Congress passed the Safe Drinking Water Act in 1974. Unlike the Clean Water Act, which attempts to safeguard the cleanliness of all surface waters, the 1974 law generally is limited to regulating the quality of water as it comes out of the tap. In 1996, Congress amended the law to expand its scope to include assessments of water quality in all sources of drinking water.

"The bulk of our attention under the Safe Drinking Water Act is really focused on setting public-health standards for individual contaminants that we believe will protect public health," says EPA's Fox. "We then monitor and enforce the drinking-water treatment facilities and infrastructure to make sure that they can deliver that high-quality water at the tap."

Compared with many parts of the world, where outbreaks of cholera and other waterborne diseases kill about 10 million people each year, the United States has succeeded in safeguarding the purity of its drinking water—so much so that most people take the safety of their tap water for granted.[9]

But the growth of livestock operations and other activities that may pollute runoff into drinking-water supplies poses a health risk that occasionally challenges Americans' complacency. In 1993, for example, 400,000 people in Milwaukee, Wis., got sick, and 100 people died, after drinking municipal tap water that had been contaminated by *Cryptosporidium*, a parasite that is commonly found in livestock waste. Indeed, as many as 7 million Americans get sick from contaminated tap water each year.[10]

In addition to biological toxins, such as bacteria, toxic algae and metals, the water supply is threatened by chemical pollutants, including pesticides, used motor oil and other synthetic contaminants that may be deposited onto city streets and washed into waterways through storm sewers.

Even drugs, including antibiotics used to treat humans and to keep livestock healthy in crowded factory farms, can make their way into tap water.[11] Groundwater also may be contaminated when repeated depositions of toxins seep through soil and rock layers that separate groundwater from the surface.

Over the years, EPA has repeatedly added to the list of contaminants it regulates under the Safe Drinking Water Act. It now contains more than 80 contaminants and is likely to grow even longer as the adverse health effects of chemicals become apparent. The recent discovery in Los Angeles drinking water supplies of chromium 6, a cancer-causing byproduct of an otherwise harmless metal used in aircraft manufacturing and other industries, prompted calls for new standards to address the threat it poses to human health.[12]

Similarly, levels of arsenic, another carcinogen, were found to exceed new standards EPA proposed in May in 95 drinking-water systems in Colorado.[13] Indeed, the National Resources Defense Council (NRDC) estimates that as many as 56 million people in the 25 states that have thus far reported arsenic information to EPA currently are exposed to unsafe levels of arsenic in their tap water.[14]

In fact, scientists have recently determined that some of the chemicals used to purify tap water may be more dangerous than the contaminants themselves. Treating drinking water with chlorine has done much to improve water safety, but the disinfectant has been linked to serious health problems, bladder and colon cancer among them.

To reduce the risk of cancer linked to long-term drinking of chlorinated water, many communities are switching to ammonia. But that change is not without potentially harmful effects: Patients needing kidney dialysis and fish-tank owners will require special treatment of tap water to neutralize the ammonia.

These and other developments in the treatment and quality of drinking water have been brought to the attention of consumers since 1999, when a provision of the 1996 Safe Drinking Water Act Amendments went into effect requiring drinking-water utilities to provide "right-to-know" information to ratepaying consumers. Consumers now receive pamphlets describing the sources, treatment and quality of their tap water.

The law also increased the amount of funding authorized to help the states meet their water-quality responsibilities. But even that may not be enough to enable states to meet the law's stricter standards for drinking water. According to a recent survey by the U.S. General Accounting Office, "over 90 percent of the states predicted that their staffing levels would be less than adequate [to monitor water quality] in the future as a number of new program requirements and complex contaminant regulations take effect.[15]

"Generally speaking, this country has very high-quality drinking water," Fox says. "But we've also learned that we cannot take the quality of that water for granted, even in fairly sophisticated systems like the one Milwaukee had at the time of the *Cryptosporidium* outbreak. If we don't operate these systems very well, and if we aren't very conscious of what pollution can get into our source water, that can have very negative consequences on the quality of our drinking water."

Notes

1. Browner testified on Feb. 23, 2000, before the Senate Agriculture, Nutrition and Forestry Committee.

2. U.S. Environmental Protection Agency, "The Quality of Our Nation's Waters: A Summary of the National Water Quality Inventory: 1998 Report to Congress," June 2000.

3. See Charles Pope, "Clean Water: The Next Wave," *CQ Weekly*, March 18, 2000, pp. 585–586.

4. Richard A. Halpern, "1491 and All That," *American Outlook*, November/December 2000, p. 35.

5. The case, *Pronsolino v. Marcus*, is pending before the 9th U.S. Circuit Court of Appeals.

6. For background, see Mary H. Cooper, "Setting Environmental Priorities," *The CQ Researcher*, May 21, 1999, pp. 425–448.

7. "Spill Caused by Dike Failure," Associated Press Newswires, April 30, 1999.

8. Paul Faeth, "Fertile Ground: Nutrient Trading's Potential to Cost-Effectively Improve Water Quality," World Resources Institute, 2000.

9. Julie McCann, "On tap: The Story of Water: Solid Facts on a Liquid Asset," *National Post*, Oct. 1, 2000.

10. Natural Resources Defense Council, "Drinking Water," www.nrdc.org.

11. See Kathleen Fackelmann, "Drugs Found in Tap Water," *USA Today*, Nov. 8, 2000.

12. See "Chromium 6 Released for Years into L.A. River," The Associated Press, Oct. 1, 2000.

13. See Julia C. Martinez, "Arsenic Levels Too High," *The Denver Post*, Aug. 29, 2000.

14. Natural Resources Defense Council, "Drinking Water," www.nrdc.org.

15. General Accounting Office, "Drinking Water: Spending Constraints Could Affect States' Ability to Meet Increasing Program Requirements," Aug. 31, 2000, p. 4.

From *CQ Researcher*, November 24, 2000, pp. 955-964. © 2000 by Congressional Quarterly, Inc. Reprinted by permission.

Statehouse and Greenhouse

The States Are Taking the Lead on Climate Change

By Barry G. Rabe

Washington's role in greenhouse gas reduction remains as unclear today as it was in the late 1980s, when a convergence of research and steamy summers thrust the matter of climate change onto the national agenda. Although Vice President Al Gore personally negotiated key elements of the 1997 Kyoto Protocol, the Clinton-Gore administration never seriously pursued Senate ratification of the treaty. George W. Bush entered the White House last year voicing general support for addressing climate change, but he quickly withdrew the United States from direct involvement in ongoing post-Kyoto deliberations. It took him more than a year to offer any indication of what might follow. His Valentine's Day 2002 recommendations may buy him some political cover but are at best a fig leaf, likely to have little impact on greenhouse gas releases even if approved and implemented.

Ironically, while Washington has continued to stumble on the global warming issue, a number of states have launched constructive efforts to lower emissions of carbon dioxide, methane, and other greenhouse gases. Independently the states have passed more than three dozen laws—many during the past two years—establishing specific strategies. These strategies involve formal commitments in virtually every sector that can influence such heat-trapping gases, including electricity generation, air pollution regulation, transportation, forestry and natural resource preservation, and agriculture. The action at the state level has received remarkably little attention from environmentalists, journalists, or scholars, yet it includes elements of a new "policy architecture" for reducing effluents that the new Bush proposals ignore.

Policy for a Different Kind of Environmental Problem

International environmental agreements have a very mixed track record, particularly when it comes to implementation. In general, the more stakeholders, the greater the difficulty of monitoring and assuring compliance; the larger the economic dislocations, the greater the likely resistance to implementation. The Kyoto Protocol's scope of intended collective action is truly stunning, making it as complex an international agreement as has ever been negotiated in any sphere of public policy.

Protocol tinkering has continued, without input from the U.S. government, most notably in 2001 meetings in Germany and Morocco. Representatives of some 165 nations are trying to carve out rules of engagement for the much smaller subset of industrialized nations that would pledge to cut greenhouse gas emissions. Alas, American disengagement has provided a visible stage for widespread denunciation of the United States. At the same time, each round of international discussion results in more deal-cutting—on everything from more favorable measurement of "carbon sinks" to reduced targets for select nations—and exposes more potential loopholes. It also remains unclear whether, even if they ratify the agreement, many nations will take its implementation seriously.

Turning to the States: The Realm of the Possible

American states provide particularly fertile ground for policy innovation in climate change. First, many are quite large in terms of population, physical size, and resources devoted to environmental protection. They also spew a lot of harmful emissions. Indeed, if the American states were counted as sovereign nations, approximately half would rank among the top 60 national emitters of greenhouse gases around the globe. The annual carbon dioxide emissions of Texas, for example, exceed those of France. Indiana's exceed Indonesia's, and Georgia's exceed Venezuela's.

Second, states already have considerable jurisdiction over many spheres of environmental and energy policy with direct relevance to the climate change problem. State rules affect electricity rates, land use, waste management, and transportation. States also implement many federal environmental laws, issuing more than 90 percent of all environmental permits and conducting more than 75 percent of all enforcement actions.

Third, a growing body of scholarship suggests that far more innovation in American environmental and energy policy now emanates from the statehouses than from Congress. States now dominate policy formation in pollution prevention and cross-media regulatory integration, an exigency long neglected in Washington.

State governments, of course, vary markedly in their commitment to protecting the environment. The most active states in climate change policy are those along the Pacific and Atlantic coasts, but others too are redefining policy. The leading states tend not to be dominated by one political party; in fact, many of their most important initiatives have been enacted with bipartisan support, offering potential political lessons for federal policy. Often, affected industries have welcomed the opportunity to "claim credit" for early reductions and have not found adjustment costs particularly onerous.

Comprehensive Cuts: New Jersey

In 1998, then-governor of New Jersey, Christine Whitman (now chief of the Environmental Protection Agency) became the only North American political official obligated to implement the Kyoto Protocol. That year Whitman issued an executive order setting a goal of reducing the state's greenhouse gases by 3.5 percent below 1990 levels by 2005. If New Jersey succeeds, it would be in line to reach the larger cuts pledged under Kyoto by decade's end. To meet its commitment, New Jersey is using a comprehensive strategy coordinating every relevant state department or agency, from agriculture to transportation. Perhaps most noteworthy, the New Jersey Greenhouse Gas Registry facilitates intrastate crediting and trading of carbon dioxide emissions. The state expanded the registry in 1998 through a formal agreement with the Netherlands, sanctioning emission trading projects between an American state and a European nation. More recently, the state is trying to link more flexible approval of industrial permits with corporate pledges to reduce greenhouse gases.

Standards for New Plants: Oregon

Oregon has undertaken a decade-long search for ways of reducing its contributions to global warming. Its most important action derives from a series of laws signed by Democratic Governor John Kitzhaber establishing a formal standard for carbon dioxide releases from new electric power plants. Oregon has received many proposals for new facilities, particularly small- to moderate-sized plants that burn the most benign fossil fuel, natural gas. The new standard requires that carbon emissions from any new power plant proposed for operation in Oregon must be at least 17 percent below those of the most efficient natural gas-fired plant now operating in the United States. Proposed plants may meet this standard either by developing more efficient technologies or by purchasing carbon dioxide offsets through the Oregon Climate Trust, which is empowered to pursue carbon dioxide mitigation projects. Initial Climate Trust projects include solar rural electrification, geothermal heating, reforestation, and methane reuse from coal mines and sewage treatment plants.

Standards for Established Plants: Massachusetts

Whereas Oregon has concentrated on new energy facilities, Massachusetts has focused on some of its most established—and significant—greenhouse gas sources. In April 2001, Republican Governor Jane Swift issued regulations that establish carbon dioxide caps for six power plants that collectively produce 40 percent of the state's electricity. Each must reduce its carbon dioxide releases 10 percent below late 1990 levels by 2004-06. Options for attaining compliance include changing fuel or generating technologies, swapping carbon dioxide reduction credits with other plants in the state, or investing in off-system reductions. Several other states, including Illinois, New Hampshire, and New York, have been actively considering their own versions of this approach.

Looking Ahead

It is possible that the federal government will soon shift gears on climate change, developing a comprehensive and creative policy. It is also possible that the Chicago Cubs will reverse a near-century of futility and win the next World Series. Neither prospect seems plausible.

In the meantime the United States would be well advised to build on the best practices emerging at the state level. Congress and the executive could begin by requiring annual reporting on greenhouse gas releases and establishing formal metrics for banking and trading projects. They might also follow the lead of Massachusetts and its counterparts, expanding the "multipollutant" definition to include greenhouse gases and emulating the most promising state reduction strategies.

It might also be desirable to link the climate change question with the question of how to divide federal and state responsibilities for environmental and energy policy. Although virtually every study on the "future of environmental policy" since the first Earth Day in 1970 has recommended finding more rational ways to allocate these responsibilities, the old hodge-podge of intergovernmental duties and tensions persists.

The Clinton administration attempted some decentralization by offering to negotiate National Environmental Performance Partnership agreements with states. The goal was to spur more innovation and measurable performance improvements by offering states greater flexibility in regulatory compliance and in spending federal grant dollars. More than 40 states participated in some fashion, although it is not clear that they or the federal government have yet devoted much energy or creativity to this process.

The Bush administration could take this tool and invite states, individually or regionally, to outline a more state-based approach to climate change policy. It could challenge groups such as the National Governors Association, Environmental Council of the States, and National Association of State Energy Officials to weave a state-based strategy on greenhouse gas reductions into a larger compact on regulatory federalism. As long as performance measures were established—and goals attained—states could receive far more latitude than under traditional regulation. They could actively engage in interstate emissions trading, possibly making expanded use of interstate compacts or even state-to-nation trading as is now being examined between New Jersey and the Netherlands as well as a cluster of New England states and Maritime provinces.

Such a devolution of leadership would, in effect, reflect the policy reality of the past decade. Tapping into the experience of the states in setting realistic climate change policy—while recognizing its current limits—might finally move us beyond the fractious debate of the past decade.

Barry G. Rabe is professor of environmental policy in the School of Natural Resources and Environment and professor of public policy in the Gerald R. Ford School of Public Policy at the Univeristy of Michigan. He is writing a book on decentralized approaches to greenhouse gas reduction in the United States and Canada, supported by the Pew Center on Global Climate Change and the Canadian Embassy.

From the *Brookings Review,* Spring 2002, pp. 11-13. © 2002 by the Brookings Institution Press, Washington, DC. Reprinted by permission.

Environmental Information Retrieval

ON FINDING OUT MORE

There is probably more printed information on environmental issues, regulations, and concerns than on any other major topic. So much is available from such a wide and diverse group of sources, that the first effort at finding information seems an intimidating and even impossible task. Attempting to ferret out what agencies are responsible for what concerns, what organizations to contact for specific environmental information, and who is in charge of what becomes increasingly more difficult.

To list all of the governmental agencies private and public organizations, and journals devoted primarily to environmental issues is, of course, beyond the scope of this current volume. However, we feel that a short primer on environmental information retrieval should be included in order to serve as a springboard for further involvement; for it is through informed involvement that issues, such as those presented, will eventually be corrected.

I. SELECTED OFFICES WITHIN FEDERAL AGENCIES AND FEDERAL-STATE AGENCIES FOR ENVIRONMENTAL INFORMATION RETRIEVAL

Appalachian Regional Commission
1666 Connecticut Avenue, NW, Washington, DC 20235 (202) 884-7799
http://www.arc.gov

Council on Environmental Quality
Old Executive Office Bldg., Room 360, Washington, DC 20502 (202) 456-6224
http://www.whitehouse.gov/CEQ/

Delaware River Basin Commission
P.O. Box 7360, West Trenton, NJ 08628-0360 (609) 883-9500
http://www.state.nj.us/drbc/drbc.htm

Department of Agriculture
14th and Independence Avenue, SW, Washington, DC 20250 (202) 720-2791
http://www.usda.gov

Department of the Army (Corps of Engineers)
20 Massachusetts Ave., NW, Washington, DC 20314-1000 (202) 761-0660
http://www.usace.army.mil

Department of Commerce
14th and Constitution Ave. NW, Washington, DC 20230 (202) 482-2000
http://www.doc.gov

Department of Defense
Public Affairs, 1400 Defense Pentagon, Room 1E757, Washington, DC 20301-1400 (703) 697-5737
http://www.defenselink.mil/index.html

Department of Health and Human Services
200 Independence Avenue, SW, Washington, DC 20201 (202) 619-0257
http://www.os.dhhs.gov

Department of the Interior
1849 C Street, NW, Washington, DC 20240-0001 (202) 208-3100
http://www.doi.gov
- Bureau of Indian Affairs (202) 208-3711
- Bureau of Land Management (202) 452-5125
- National Park Service (202) 208-6843
- United States Fish and Wildlife Service (202) 208-4131

Department of State, Bureau of Oceans and International Environmental and Scientific Affairs
2201 C Street, NW, Washington, DC 20520 (202) 647-2492
http://www.state.gov/www/global/oes/index.html

Department of the Treasury, U.S. Customs Service
1300 Pennsylvania Avenue, NW, Washington, DC 20229 (202) 927-1000
http://www.customs.ustreas.gov

Environmental Protection Agency (EPA)
401 M Street, SW, Washington, DC 20460 (202) 260-2090
- Region 1, One Congress Street, John F. Kennedy Building, 11th Floor, Boston, MA 02203-0001 (617) 565-3420 (888) 373-7341
  http://www.epa.gov/region01/ (Connecticut, Maine, Massachusetts, New Hampshire, Rhode Island, Vermont)
- Region 2, 290 Broadway, New York, NY 10007-1866 (212) 637-3000
  http://www.epa.gov/region02/ (New Jersey, New York, Puerto Rico, Virgin Islands)
- Region 3, 1650 Arch Street, Philadelphia, PA 19100-2029 (215) 814-5000 (800) 438-2474
  http://www.epa.gov/region03/ (Delaware, District of Columbia, Maryland, Pennsylvania, Virginia, West Virginia)
- Region 4, 4 Forsyth Street, SW, Atlanta, GA 30303-3104 (404) 562-9900
  http://www.epa.gov/region04/ (Alabama, Florida, Georgia, Kentucky, Mississippi, North Carolina, South Carolina, Tennessee)
- Region 5, 77 West Jackson Blvd., Chicago, IL 60604-3507 (312) 353-2000
  http://www.epa.gov/region5/ (Illinois, Indiana, Michigan, Minnesota, Ohio, Wisconsin)
- Region 6, 1445 Ross Avenue, Suite 1200, Dallas, TX 75202 (214) 665-2200 (800) 887-6063
  http://www.epa.gov/region06/ (Arkansas, Louisiana, New Mexico, Oklahoma, Texas)
- Region 7, 726 Minnesota Avenue, Kansas City, KS 66101 (913) 551-7003 (800) 223-0425
  http://www.epa.gov/region07/ (Iowa, Kansas, Missouri, Nebraska)
- Region 8, 999 18th Street, Suite 500, Denver, CO 80202-2466 (303) 312-6312 (800) 227-8917
  http://www.epa.gov/region08/ (Colorado, Montana, North Dakota, South Dakota, Utah, Wyoming)

- Region 9, 75 Hawthorne Street, San Francisco, CA 94105 (415) 744-1500
  http://www.epa.gov/ region09/ (Arizona, California, Hawaii, Nevada, American Samoa, Guam, Trust Territories of Pacific Islands, Wake Island)
- Region 10, 1200 Sixth Avenue, Seattle, WA 98101 (206) 553-1200 (800) 424-4372
  http://www.epa.gov/region10/ (Alaska, Idaho, Oregon, Washington)

Federal Energy Regulatory Commission
  825 North Capitol Street, NE, Washington, DC 20426 (202) 208-0000
  http://www.ferc.fed.us

Interstate Commission on Potomac River Basin
  6110 Executive Boulevard, Suite 300, Rockville, MD 20852-3903 (301) 984-1908
  http://www.potomacriver.org

Nuclear Regulatory Commission
  One White Flint North, 11555 Rockville Pike, Rockville, MD 20852-2738 (301) 415-7000
  http://www.nrc.gov

Susquehanna River Basin Commission
  1721 North Front Street, Harrisburg, PA 17102 (717) 238-0422
  http://www.srbc.net

Tennessee Valley Authority
  400 West Summit Hill Drive, Knoxville, TN 37902 (423) 632-2101
  http://www.tva.gov

II. SELECTED STATE, TERRITORIAL, AND CITIZENS' ORGANIZATIONS FOR ENVIRONMENTAL INFORMATION RETRIEVAL

A. Government Agencies

Alabama:
Department of Conservation and Natural Resources
  P.O. Box 301450, Montgomery, AL 36130-1450 (334) 242-3486
  http://www.dcnr.state.al.us

Alaska:
Department of Environmental Conservation
  410 Willoughby Avenue, Suite 105, Juneau, AL 99801-1795 (907) 465-5060
  http://www.state.ak.us

Arizona:
Department of Water Resources
  500 North 3rd Street, Phoenix, AZ 85004-3226 (602) 417-2400
  http://www.adwr.state.az.us

Natural Resources Division
  1616 West Adams Street, Phoenix, AZ 85007 (602) 542-4625

Arkansas:
Department of Pollution Control and Ecology
  8001 National Drive, Little Rock, AR 72209 (501) 682-0744
  http://www.adeq.state.ar.us

Energy Office
  1 State Capitol Mall, Little Rock, AR 72201-1012 (501) 682-7325

California:
Conservation Department Resources Agency
  801 K Street, MS24-01, Sacramento, CA 95814 (916) 322-1080
  http://www.consrv.ca.gov

Environmental Protection Agency
  555 Capital Mall, Suite 525, Sacramento, CA 95814 (916) 445-3846
  http://www.calepa.ca.gov

Colorado:
Department of Natural Resources
  1313 Sherman Street, Room 718, Denver, CO 80203-2239 (303) 866-3311
  http://www.dnr.state.co.us

Connecticut:
Department of Environmental Protection
  State Office Building, 79 Elm Street, Hartford, CT 06106-5127 (860) 424-3000
  http://www.dep.state.ct.us

Delaware:
Natural Resources and Environmental Control Department
  89 Kings Highway, Dover, DE 19903 (302) 739-5823
  http://www.dnrec.state.de.us

District of Columbia:
Environmental Regulation Administration
  2100 Martin Luther King Jr. Avenue, SE, Suite 203, Washington, DC 20020 (202) 404-1167
  http://clean.rti.org/state/washdc.htm

Florida:
Department of Environmental Protection
  3900 Commonwealth Blvd., M.S.10, Tallahassee, FL 32399-3000 (850) 488-1554
  http://www.dep.state.fl.us

Georgia:
Department of Natural Resources
  205 Butler Street, SE, Atlanta, GA 30334-4100 (404) 656-3500
  http://www.ganet.org/dnr/

Guam:
Environmental Protection Agency
  IT&E Harmon Plaza, Complex Unit D-107, 130 Rojas St., Harmon, Guam 96911 (617) 646-9402

Hawaii:
Department of Land and Natural Resources
  P.O. Box 621, Honolulu, HI 96809 (808) 587-0400
  http://www.hawaii.gov/dlnr/Welcome.html

Idaho:
Department of Lands
  P.O. Box 83720, Boise, ID 83720-0050 (208) 334-0200
  http://www2.state.id.us/lands/index.htm

Department of Water Resources
  1301 North Orchard Street, Boise, ID 83706 (208) 327-7900
  http://www.idwr.state.id.us

Illinois:
Department of Natural Resources
524 South 2nd Street, Springfield, IL 62701 (217) 785-0075
http://dnr.state.il.us

Indiana:
Department of Natural Resources
402 W. Washington St., Indianapolis, IN 46204-2212 (317) 232-4020
http://www.state.in.us/dnr/

Iowa:
Department of Natural Resources
Wallace State Office Bldg., Des Moines, IA 50319 (515) 281-4367
http://www.state.ia.us/dnr/

Kansas:
Department of Health and Environment
Landon State Office Bldg., Topeka, KS 66612 (785) 296-1522
http://www.state.ks.us/public/kdhe/

Kentucky:
Natural Resources and Environmental Protection Cabinet
Capital Plaza, Frankfort, KY 40601 (502) 564-3350
http://www.nr.state.ky.us/nrhome.htm

Louisiana:
Department of Environmental Quality
7290 Bluebonnet Blvd., Baton Rouge, LA 70810 70817-4401 (225) 765-0741
http://www.deq.state.la.us
Department of Natural Resources
625 North 4th Street, Baton Rouge, LA 70804-9396 (504) 342-2707
http://www.dnr.state.la.us/index.ssi

Maine:
Department of Environmental Protection
17 State House Station, Augusta, ME 04333-0017 (207) 287-7688 (800) 452-1942
http://www.state.me.us/dep/

Maryland:
Department of Natural Resources
580 Taylor Avenue, Tawes State Office Bldg., Annapolis, MD 21401 (410) 260-8400
http://www.dnr.state.md.us

Massachusetts:
Department of Environmental Management
100 Cambridge Street, 19th Floor, Boston, MA 02202 (617) 727-3163
http://www.state.ma.us/dem/dem.htm

Michigan:
Department of Natural Resources
Box 30028, Lansing, MI 48909 (517) 373-2329
http://www.dnr.state.mi.us

Minnesota:
Department of Natural Resources
500 Lafayette Road, St. Paul, MN 55155-4046 (612) 296-2549
http://www.dnr.state.mn.us

Mississippi:
Department of Environmental Quality
P.O. Box 20305, Jackson, MS 39289 (601) 961-5171
http://www.deq.state.ms.us/domino/deqweb.nsf

Missouri:
Department of Natural Resources
P.O. Box 176, Jefferson City, MO 65102 (800) 334-6946
http://www.dnr.state.mo.us/homednr.htm

Montana:
Department of Natural Resources and Conservation
1625 11th Avenue, P.O. Box 201601, Helena, MT 59620-1601 (406) 444-2074
http://www.dnrc.mt.gov

Nebraska:
Department of Environmental Quality
P.O. Box 98922, Lincoln, NE 68509 (402) 471-2186
http://www.deq.state.ne.us

Nevada:
Department of Conservation and Natural Resources
Capitol Complex, Carson City, NV 89710 (702) 687-5000
http://www.state.nv.us/cnr/

New Hampshire:
Department of Environmental Services
6 Hazen Drive, P.O. Box 95, Concord, NH 03302-0095 (603) 271-3503
http://www.state.nh.us/des/
Department of Resources and Economic Development
P.O. Box 1856, Concord, NH 03302-1856 (603) 271-2411
http://www.dred.state.nh.us

New Jersey:
Department of Environmental Protection
401 E. State Street, P.O. Box 402, Trenton, NJ 08625-0402 (609) 292-2885
http://www.state.nj.us/dep/

New Mexico:
Environmental Department
1190 Saint Francis Drive, Santa Fe, NM 87505 (505) 827-2855 (800) 879-3421
http://www.nmenv.state.nm.us

New York:
Department of Environmental Conservation
50 Wolf Road, Albany, NY 12233 (518) 485-8940
http://www.dec.state.ny.us

North Carolina:
Department of Environment and Natural Resources
P.O. Box 27687, Raleigh, NC 27611 (919) 733-4984
http://www.ehnr.state.nc.us/EHNR/

North Dakota:
Game & Fish Department
100 North Bismarck Expressway, Bismarck, ND 58501 (701) 328-6300
http://www.state.nd.us/gnf/

Ohio:
Department of Natural Resources
1952 Belcher Drive, Building C-1, Columbus, OH 43224 (614) 265-6565
http://www.dnr.state.oh.us

Environmental Protection Agency
P.O. Box 1049, Columbus, OH 43216-1049 (614) 644-3020
http://www.epa.state.oh.us

Oklahoma:
Conservation Commission
2800 North Lincoln Boulevard, Suite 160, Oklahoma City, OK 73105-4210 (405) 521-2384
http://www.oklaosf.state.ok.us/~comscom/
Department of Environmental Quality
707 North Robinson, Oklahoma City, OK 73702 (405) 702-6100
http://www.deq.state.ok.us

Oregon:
Department of Environmental Quality
811 S.W. 6th Avenue, Portland, OR 97204-1390 (503) 229-5696
http://www.deq.state.or.us

Pennsylvania:
Department of Environmental Resources
400 Market Street, Harrisburg, PA 17105 (717) 787-2814
http://www.dep.state.pa.us

Puerto Rico:
Department of Natural Resources
P.O. Box 5887, San Juan, PR 00906 (787) 723-3090

Rhode Island:
Department of Environmental Management
235 Promenade Street, Providence, RI 02908 (401) 222-2771
http://www.state.ri.us

South Carolina:
Department of Health and Environmental Control
2600 Bull Street, Columbia, SC 29201 (803) 734-5000
http://www.state.sc.us/dhec/
Department of Natural Resources
1000 Assembly Street, Columbia 29201 (803) 734-3888
http://water.dnr.state.sc.us

South Dakota:
Department of Environment and Natural Resources
Joe Foss Bldg., 523 East Capitol, Pierre, SD 57501 (605) 773-3151
http://www.state.sd.us/state/executive/ denr/denr.html

Tennessee:
Department of Environment and Conservation
401 Church St., 21st Floor, Nashville, TN 37243 (888) 891-8332
http://www.state.tn.us/environment/

Texas:
Natural Resources Conservation Commission
P.O. Box 13087, Austin, TX 78711 (512) 239-1000
http://www.three.state.tx.us

Utah:
Department of Natural Resources
1594 West North Temple, Suite 3710, Salt Lake City, UT 84114 (801) 538-7200
http://www.nr.state.ut.us

Vermont:
Agency of Natural Resources
103 South Main Street, Waterbury, VT 05671-0301 (802) 241-3614
http://www.anr.state.vt.us

Virgin Islands:
Department of Planning & Natural Resources
396-1 Annas Retreat, Foster Bldg., Charlotte Amalie, U.S. Virgin Islands 00802 (340) 774-3320
http://www.gov.vi/pnr/

Virginia:
Secretary of Natural Resources
P.O. Box 1475, Richmond, VA 23212 (804) 786-0044
http://snr.vipnet.org

Washington:
Department of Ecology
P.O. Box 47600, Olympia, WA 98504-7600 (360) 407-6000
http://www.wa.gov/ecology/
Department of Natural Resources
P.O. Box 47000, Olympia, WA 98504-7000 (360) 902-1000
http://www.wa.gov/dnr/

West Virginia:
Division of Natural Resources
1900 Kanawha Blvd. East, Charleston, WV 25305 (304) 558-2771
http://www.dnr.state.wv.us

Wisconsin:
Department of Natural Resources
Box 7921, Madison, WI 53707 (608) 267-7517
http://www.dnr.state.wi.us

Wyoming:
Department of Environmental Quality
122 West 25th Street, Herschler Bldg., Cheyenne, WY 82002 (307) 777-7758
http://deq.state.wy.us

B. Citizens' Organizations
Advancement of Earth & Environmental Sciences
International Association for, Northeastern Illinois University, Geography and Environmental Studies Department 5500 North St. Louis Avenue, Chicago, Illinois 60625 (312) 794-2628
Air Pollution Control Association
1 Gateway Center, 3rd Floor, Pittsburgh, PA 15222 (412) 232-3444
American Association for the Advancement of Science
1200 New York Avenue, NW, Washington, DC 20005 (202) 326-6400
http://www.aaas.org
American Chemical Society
1155 16th Street, NW, Washington, DC 20036 (202) 872-4600
http://acs.org
American Farm Bureau Federation
225 Touhy Avenue, Park Ridge, IL 60068 (847) 685-8600
http://www.fb.com
American Fisheries Society,
5410 Grosvenor Lane, Suite 110, Bethesda, MD 20014-2199 (301) 897-8616
http://www.fisheries.org

American Forest & Paper Association
    1111 19th Street, NW, Suite 800, Washington, DC 20036
    (202) 463-2438
    http://www.afandpa.org

American Forests
    P.O. Box 2000, Washington, DC 20013 (202) 955-4500
    http://www.amfor.org

American Institute of Biological Sciences
    1444 I Street NW, Washington, DC 20005 (202) 628-1500
    http://www.aibs.org

American Museum of Natural History
    Central Park West at 79th Street, New York, NY 10024-5192 (212) 769-5100
    http://www.amnh.org

American Petroleum Institute
    1220 L Street, NW, Washington, DC 20005-4070 (202) 682-8000
    http://www.api.org

American Rivers
    1025 Vermont Avenue NW, Suite 720, Washington, DC 20005 (202) 347-7550
    http://www.amrivers.org

Association for Conservation Information
    P.O. Box 12559, Charleston, SC 29412 (803) 762-5032

Boone and Crockett Club
    250 Station Dr., Missoula, MT 59801 (406) 542-1888
    http://www.boone-crockett.org

Center for Marine Conservation
    1725 DeSales Street, NW, Suite 600, Washington, DC 20036 (202) 429-5609
    http://www.cmc-ocean.org

Citizens for a Better Environment
    3255 Hennepin Avenue South, Minneapolis, MN 55331 (612) 824-8637
    http://www.cbemw.org

Coastal Conservation Association
    4801 Woodway, Suite 220W, Houston, TX 77056 (713) 626-4222
    http://www.ccatexas.org

Conservation Foundation
    1250 24th Street, NW, Suite 400, Washington, DC 20037 (202) 293-4800

Conservation Fund
    1800 North Kent Street, Suite 1120, Arlington, VA 22209-2156 (703) 525-6300
    http://www.conservationfund.org

Conservation International
    2501 M Street, NW, Suite 200, Washington, DC 20037 (202) 429-5660 (800) 429-5660
    http://www.conservation.org

Defenders of Wildlife
    1101 14th Street, NW, #1400, Washington, DC, 20005 (202) 682-9400
    http://www.defenders.org

Ducks Unlimited, Inc.
    One Waterfowl Way, Memphis, TN (901) 785-3825
    http://www.ducks.org

Earthwatch Institute International
    680 Mt. Auburn Street, Watertown, MA 02471 (800) 776-0188
    http://www.earthwatch.org

Environmental Action Foundation, Inc.
    6930 Carroll Ave., Suite 600, Takoma Park, MD 20912
    (301) 891-1100

Food and Agriculture Organization of the United Nations (FAO)
    Via delle Terme di Caracalla, 00100, Rome, Italy (396) 57051
    http://www.fao.org

Friends of the Earth
    1025 Vermont Avenue, NW, Suite 300, Washington, DC 20005 (202) 783-7400
    http://foe.co.uk

Greenpeace U.S.A.
    1436 U Street, NW, Washington, DC 20009 (202) 462-1177
    http://greenpeaceusa.org

International Association of Fish and Wildlife Agencies
    444 North Capitol Street, NW, Suite 544, Washington, DC 20001 (202) 624-7890
    http://www.teaming.com/iafwa.htm

International Fund for Agricultural Development (IFAD)
    107, Via del Serafico, 00142, Rome, Italy (3906) 54591
    http://www.ifad.org

Keep America Beautiful, Inc.
    1010 Washington Blvd., Stamford, CT 06901 (203) 323-8987
    http://www.kab.org

National Association of Conservation Districts
    509 Capitol Court, NE, Washington, DC 20002 (202) 547-6223
    http://www.nacdnet.org

National Audubon Society
    700 Broadway, New York, NY 10003 (212) 979-3000
    http://www.audubon.org

National Environmental Health Association
    720 South Colorado Boulevard, 970 South Tower, Denver, CO 80246 (303) 756-9090
    http://www.neha.org

National Fisheries Institute
    1901 N. Fort Myer Dr., Suite 700, Arlington, VA 22209 (703) 524-8880
    http://www.nfi.org/main.html

National Geographic Society
    1145 17th Street, NW, Washington, DC 20036 (202) 857-7000
    http://www.nationalgeographic.com

National Parks and Conservation Association
    1717 Massachusetts Avenue, NW, Washington, DC 20036 (202) 223-6722 (800) 628-7275
    http://www.npca.org

National Wildlife Federation
    8925 Leesburg Pike, Vienna, VA 22184 (703) 790-4000
    http://www.nwf.org

Natural Resources Council of America
    801 Pennsylvania Avenue, SE, No. 410, Washington, DC 20003 (202) 333-0411

Nature Conservancy
    1815 North Lynn Street, Arlington, VA 22209 (703) 841-5300
    http://www.tnc.org

Population Association of America

721 Ellsworth Drive, Suite 303, Silver Spring, MD 20910 (301) 565-6710

http://www.jstor.org

Rainforest Alliance

650 Bleecker Street, New York, NY 10012 (212) 677-1900

http://www.rainforest-alliance.org

Save-the-Redwoods League

114 Sansome Street, Room 605, San Francisco, CA 94104 (415) 362-2352

http://www.savetheredwoods.org

Sierra Club

85 Second Street, 2nd Floor, San Francisco, CA 94105 (415) 977-5500

http://www.sierraclub.org

Smithsonian Institution

1000 Jefferson Drive, S.W., Washington, DC 20560 (202) 357-2700

http://smithsonian.org

Society of American Foresters

5400 Grosvenor Lane, Bethesda, MD 20814-2198 (301) 897-8720

http://www.safnet.org

Sport Fishing Institute

1010 Massachusetts Avenue, NW, Suite 320, Washington, DC 20001 (202) 898-0770

United Nations Educational, Scientific, and Cultural Organiza-tion (UNESCO)

UNESCO House, 7, Place de Fontenoy, 75352 Paris 07 SP France, (331) 45 68 10 00

http://www.unesco.org

United Nations Environment Programme/Industry & Environ-ment Centre

Tour Mirabeau 39-43, quai André Citröen 75739 Paris, ce-dex 15, France (331) 44 37 14 50

http://www. unepie.org

Wilderness Society

900 17th Street, NW, Washington, DC 20006-2596 (202) 833-2300

http://www.wilderness.org

World Wildlife Fund

1250 24th Street, NW, Washington, DC 20077 (202) 293-4800

http://www.wwf.org

Zero Population Growth, Inc.

1400 16th Street, NW, Suite 320, Washington, DC 20036 (202) 332-2200

http://www.zpg.org

III. CANADIAN AGENCIES AND CITIZENS' ORGANIZATIONS

A: Government Agencies

Alberta:

Alberta Environmental Protection

23 Legislature Bldg., 10800 97th Avenue, Edmonton, AB T5K 2B6 (403) 427-2391

http://www.gov.ab.ca/env.html

British Columbia:

Ministry of Environment, Lands, and Parks

337, Parliament Bldgs., Victoria, BC V8V 1X4 (250) 387-1187

http://www.env.gov.bc.ca

Manitoba:

Manitoba Environment

344 Legislative Bldg., Winnipeg, MB R3C 0V8 (204) 945-3522

http://www.gov.mb.ca/environ/index.html

Manitoba Natural Resources

Box 22, 200 Saulteaux Crescent, Winnipeg, MB R3J 3W3 (204) 945-6784

http://www.gov.mb.ca/natres/index.html

New Brunswick:

Department of Environment

364 Argyle St., Fredericton, NB E3B 1T9 (506) 453-2558

http://www.gov.nb.ca/environm/

Department of Natural Resources and Energy

P.O. Box 6000, Fredericton, NB E3B 5H1 (506) 453-2614

http://www.gov.nb.ca/dnre/

Newfoundland and Labrador:

Department of Environment and Labour

Confederation Bldg., PO Box 8700, St. John's, NF A1B 4J6 (709) 729-2664

http://www.govt.nf.ca/env/labour/OHS/default.asp

Northwest Territories:

Department of Resources, Wildlife, and Economic Develop-ment, #600 Scotia Centre, Bldg. Box 21, 5102-50 Avenue, Yellowknife, NT X1A 3S8 (867) 669-2366 http:// www.rwed.gov.nt.ca

Nova Scotia:

Department of Natural Resources, P.O. Box 698, Halifax, NS B3J 2T9 (902) 424-59356

http://www.gov.ns.ca/natr/

Department of the Environment

P.O. Box 2107, Halifax, NS B3J 3B7 (902) 424-5300

http://www.gov.ns.ca/envi/

Ontario:

Ministry of Natural Resources

300 Water Street, P.O. Box 7000, Peterborough, ON K9J 8M5 (705) 755-2000 (416) 314-2000

http://www.mnr.gov. on.ca/mnr/

Prince Edward Island:

Department of Technology and Environment

Jones Bldg., 11 Kent Street, 4th Floor, P.O. Box 2000, Charlottetown, PEI C1A 7N8 (902) 892-5000

http://www.gov.pe.ca/te/index.asp

Quebec:

Ministère de l'Environnement et de la Faune

Edifice Marie-Guyart, 675, boulevard René-Lévesque, Est, Québec, PC G1R 5V7 (418) 521-3830 (800) 561-1616

http://www.mef.gouv.qc.ca

Ministère des Ressources Naturelles

#B-302, 5700, 4 Avenue Ouest, Charlesbourg, PQ G1H 6R1 (418) 627-8600

http://www.mrn.gouv.qc.ca

Saskatchewan:

Saskatchewan Environment and Resource Management

3211 Albert Street, Regina, SK S4S 5W6 (306) 787-2700
http://www.gov.sk.ca/govt/environ/

Yukon Territory:

Council on the Economy & the Environment

A-8E, P.O. Box 2703, Whitehorse, YT Y1A 2C6 (867) 667-5811
http://www.gov.yk.ca

Department of Renewable Resources

Box 2703, Whitehorse, YT Y1A 2C6 (867) 667-5811
http://www.gov.yk.ca

B. Citizens' Groups

Alberta Wilderness Association

Box 6398, Station D, Calgary, AB T2P 2E1 (403) 283-2025
http://www.web.net/~awa/

BC Environmental Network (BCEN)

1672 East 10th Avenue, Vancouver, BC V5N 1X5 (604) 879-2272
http://www.bcen.bc.ca

Canadian EarthCare Society

1476 Water Street, Kelowna, BC V1Y 8P2 (604) 861-4788
http://www.earthcare.org

Ducks Unlimited Canada

Oak Hammock Marsh, Stonewall, P.O. Box 1160, Oak Hammock Marsh, MB R0C 2Z0 (204) 467-3000 (800) 665-3825
http://www.ducks.ca

Federation of Ontario Naturalists

355 Lesmill Road, Don Mills, ON M3B 2W8 (416) 444-8419
http://www.ontarionature.org

L'Association des Entrepreneurs de Service en Environnement du Quebec (AESEQ)

911 Jean-Talon, Est 220, Montreal, PQ H2R 1V5 (514) 270-7110

New Brunswick Environment Industry Association

P.O. Box 637, Stn. A, Fredericton, NB E3B 5B3 (506) 455-0212
http://www.nbeia.nb.ca/index.html

Prince Edward Island Environmental Network (PEIEN)

126 Richmond Street, Charleston, PEI C1A 1H9 (902) 566-4170
http://www.isn.net/~network/index.html

Yukon Conservation Society (YCS)

P.O. Box 4163, Whitehorse, YT Y1A 3T3 (403) 668-6637

IV. SELECTED JOURNALS AND PERIODICALS OF ENVIRONMENTAL INTEREST

American Forests

910 17th Street NW, Suite 600, Washington, DC 20006 (202) 955-4500 (800) 368-5748
http://www.amfor.org

American Scientist

Scientific Research Society, P.O. Box 13975, Research Triangle Park, NC 27709-3975 (919) 549-0097
http://www.amsci.org/amsci/amsci.html

Annual Report of the Council on Environmental Quality

Superintendent of Documents, U.S. Government Printing Office, Washington, DC 20401 (202) 512-1800
http://ceq.ch.doc.gov/nepa/reports/reports.htm

Audubon

National Audubon Society, 700 Broadway, New York, NY 10003 (212) 979-3000
http://magazine.audubon.org

BioScience

American Institute of Biological Sciences, 1444 I St. NW, Suite 200, Washington, DC 20005 (202) 628-1500
http://www.aibs.org

California Environmental Directory

California Institute of Public Affairs, Box 189040, Sacramento, CA 95818 (916) 442-2472
http://www.igc.org/cipa/cipa.html#about

The Canadian Field-Naturalist

Box 35069 Westgate, Ottawa, ON, Canada K1Z 1A2 (613) 722-3050
http://www.achilles.net/ofnc/cfn.htm

Conservation Directory

National Wildlife Federation, 8925 Leesburg Pike, Vienna, VA 22184 (703) 790-4000
http://www.nwf.org/nwf/pubs/considir/index.html

Earth First! Journal

P.O. Box 1415, Eugene, OR 97440-1415 (541) 344-8004
http://host.envirolink.org/ef/

E: The Environmental Magazine

Earth Action Network, P.O. Box 5098, Westport, CT 06881 (203) 854-5559
http://www.emagazine.com

Environment

Heldref Publications, 1319 18th Street, NW, Washington, DC 20036-1802 (202) 296-6267
http://www.heldref.org

Environment Reporter

Bureau of National Affairs, Inc. 1231 25th Street, NW, Washington, DC 2037 (202) 452-4200
http://www.bna.com/prodcatalog/desc/ER.html

Environmental Action Magazine

Environment Action, Inc. 6930 Carroll Ave., Suite 600, Takoma Park, MD 20912-4414 (301) 891-1106

Environmental Science and Technology

American Chemical Society Publications Support Services, 1155 16th Street, NW, Washington, DC 20036 (202) 872-4554 (800) 227-5558
http://pubs.acs.org/journals/esthag/

Focus (bimonthly newsletter)

World Wildlife Fund, 1250 24th Street, NW, Washington, DC 20037 (202) 293-4800
http://www.wwf.org

The Futurist

World Future Society, 7910 Woodmont Avenue, Suite 450, Bethesda, MD 20814 (301) 656-8274 (800) 989-8274
http://www.wfs.org/wfs/

Greenpeace Magazine

Greenpeace USA, 1436 U Street, NW, Washington, DC 20009 (202) 462-1177 (800) 326-0959
http://www.greenpeaceusa.org

Journal of Soil and Water Conservation

Soil and Water Conservation Society, 7515 Northeast Ankeny Road, Ankeny, IA 50021-9764 (515) 289-2331
http://www.swcs.org

Journal of Wildlife Management

The Wildlife Society, 5410 Grosvenor Lane, Suite 200, Bethesda, MD 20814-2197 (301) 897-9770
http://www.wildlife.org/journal.html

Mother Earth News

Sussex Publishers Inc., 49 E. 21st Street, 11th Floor, New York, NY 10010 (212) 260-7210
http://www.MotherEarthNews.Com

National Wildlife

National Wildlife Federation, 8925 Leesburg Pike, Vienna, VA 22184 (703) 790-4510
http://www.nwf.org/nwf/natwild/

Natural Resources Journal

University of New Mexico, School of Law, 1117 Stanford, NE, Albuquerque, NM 87131 (505) 277-4820
http://www.unm.edu/~natresj/NRJ/NRJ.html

Nature

Macmillan Publishers Ltd., Porter South, Grinan Street, London N1 9XW, England (44) 0171 833-4000
http://www.nature.com

Nature Canada

Canadian Nature Federation, One Nicholas Street, Suite 606, Ottawa, ON, Canada K1N 787 (613) 562-3447 (800) 267-40880
http://www. cnf.ca/nc_main.html

Nature Conservancy Magazine

1815 North Lynn Street, Arlington, VA 22209-2003 (703) 841-5300 (800) 267-4088
http://www.tnc.org

Omni

Omni Publications International Ltd., 277 Park Ave., New York, NY 10172 (212) 702-6000
http://www.omnimag.com

Pollution Abstracts

Cambridge Scientific Abstracts, 7200 Wisconsin Avenue, Suite 601, Bethesda, MD 20814-4823 (301) 961-6700 (800) 843-7751
http://www.csa.com

Science

American Association for the Advancement of Science, 1200 New York Avenue NW, Washington, DC 20005 (202) 326-6501
http://www.sciencemag.org

Sierra Magazine

Sierra Club, 85 2nd Street, 2nd Floor, San Francisco, CA 94105-3441 (415) 977-5750
http://www.sierraclub.org/sierra/

Smithsonian

900 Jefferson Drive, Washington, DC 20560 (202) 786-2900
http://smithsonianmag.com

Technology Review

201 Vassar Street, Cambridge, MA 02139 (617) 253-8250
http://www.techreview.com

U.S. News and World Report

2400 N Street, NW, Washington, DC 20037-1196 (202) 955-2000
http://www.usnews.com/usnews/home.htm

The World & I

New World Communications,3600 New York Avenue, NE, Washington, DC 20002 (800) 822-2822 (202) 635-4000
http://www.worldandi.com

SOURCES used to compile this list: Canadian Almanac Directory 1997; Carroll's Federal Directory, April 1997; Carroll's State Directory, February 1997; Congressional Quarterly's Washington Information Directory 1997–1998; Encyclopedia of Associations, 32nd Edition, 1997; Gale Directory of Publications and Broadcast Media, 131st edition; The World Almanac, 1999, Web search engines: Google, Metacrawler.

Glossary

This glossary of environmental terms is included to provide you with a convenient and ready reference as you encounter general terms in your study of environment that are unfamiliar or require a review. It is not intended to be comprehensive, but taken together with the many definitions included in the articles themselves, it should prove to be quite useful.

A

Abiotic   Without life; any system characterized by a lack of living organisms.

Absorption   Incorporation of a substance into a solid or liquid body.

Acid   Any compound capable of reacting with a base to form a salt; a substance containing a high hydrogen ion concentration (low pH).

Acid Rain   Precipitation containing a high concentration of acid.

Adaptation   Adjustment of an organism to the conditions of its environment, enabling reproduction and survival.

Additive   A substance added to another in order to impart or improve desirable properties or suppress undesirable ones.

Adsorption   Surface retention of solid, liquid, or gas molecules, atoms, or ions by a solid or liquid.

Aerobic   Environmental conditions where oxygen is present; aerobic organisms require oxygen in order to survive.

Aerosols   Tiny mineral particles in the atmosphere onto which water droplets, crystals, and other chemical compounds may adhere.

Air Quality Standard   A prescribed level of a pollutant in the air that should not be exceeded.

Alcohol Fuels   The processing of sugary or starchy products (such as sugar cane, corn, or potatoes) into fuel.

Allergens   Substances that activate the immune system and cause an allergic response.

Alpha Particle   A positively charged particle given off from the nucleus of some radioactive substances; it is identical to a helium atom that has lost its electrons.

Ammonia   A colorless gas comprised of one atom of nitrogen and three atoms of hydrogen; liquefied ammonia is used as a fertilizer.

Anthropocentric   Considering humans to be the central or most important part of the universe.

Aquaculture   Propagation and/or rearing of any aquatic organism in artificial "wetlands" and/or ponds.

Aquifers   Porous, water-saturated layers of sand, gravel, or bedrock that can yield significant amounts of water economically.

Atom   The smallest particle of an element, composed of electrons moving around an inner core (nucleus) of protons and neutrons. Atoms of elements combine to form molecules and chemical compounds.

Atomic Reactor   A structure fueled by radioactive materials that generates energy usually in the form of electricity; reactors are also utilized for medical and biological research.

Autotrophs   Organisms capable of using chemical elements in the synthesis of larger compounds; green plants are autotrophs.

B

Background Radiation   The normal radioactivity present; coming principally from outer space and naturally occurring radioactive substances on Earth.

Bacteria   One-celled microscopic organisms found in the air, water, and soil. Bacteria cause many diseases of plants and animals; they also are beneficial in agriculture, decay of dead matter, and food and chemical industries.

Benthos   Organisms living on the bottom of bodies of water.

Biocentrism   Belief that all creatures have rights and values and that humans are not superior to other species.

Biochemical Oxygen Demand (BOD)   The oxygen utilized in meeting the metabolic needs of aquatic organisms.

Biodegradable   Capable of being reduced to simple compounds through the action of biological processes.

Biodiversity   Biological diversity in an environment as indicated by numbers of different species of plants and animals.

Biogeochemical Cycles   The cyclical series of transformations of an element through the organisms in a community and their physical environment.

Biological Control   The suppression of reproduction of a pest organism utilizing other organisms rather than chemical means.

Biomass   The weight of all living tissue in a sample.

Biome   A major climax community type covering a specific area on Earth.

Biosphere   The overall ecosystem of Earth. It consists of parts of the atmosphere (troposphere), hydrosphere (surface and ground water), and lithosphere (soil, surface rocks, ocean sediments, and other bodies of water).

Biota   The flora and fauna in a given region.

Biotic   Biological; relating to living elements of an ecosystem.

Biotic Potential   Maximum possible growth rate of living systems under ideal conditions.

Birthrate   Number of live births in one year per 1,000 midyear population.

Breeder Reactor   A nuclear reactor in which the production of fissionable material occurs.

C

Cancer   Invasive, out-of-control cell growth that results in malignant tumors.

Carbon Cycle   Process by which carbon is incorporated into living systems, released to the atmosphere, and returned to living organisms.

Carbon Monoxide (CO)   A gas, poisonous to most living systems, formed when incomplete combustion of fuel occurs.

Carcinogens   Substances capable of producing cancer.

Carrying Capacity   The population that an area will support without deteriorating.

Glossary

Chlorinated Hydrocarbon Insecticide Synthetic organic poisons containing hydrogen, carbon, and chlorine. Because they are fat-soluble, they tend to be recycled through food chains, eventually affecting nontarget systems. Damage is normally done to the organism's nervous system. Examples include DDT, Aldrin, Deildrin, and Chlordane.

Chlorofluorocarbons (CFCs) Any of several simple gaseous compounds that contain carbon, chlorine, fluorine, and sometimes hydrogen; they are suspected of being a major cause of stratospheric ozone depletion.

Circle of Poisons Importation of food contaminated with pesticides banned for use in this country but made here and sold abroad.

Clear-Cutting The practice of removing all trees in a specific area.

Climate Description of the long-term pattern of weather in any particular area.

Climax Community Terminal state of ecological succession in an area; the redwoods are a climax community.

Coal Gasification Process of converting coal to gas; the resultant gas, if used for fuel, sharply reduces sulfur oxide emissions and particulates that result from coal burning.

Commensalism Symbiotic relationship between two different species in which one benefits while the other is neither harmed nor benefited.

Community Ecology Study of interactions of all organisms existing in a specific region.

Competitive Exclusion Resulting from competition; one species forced out of part of an available habitat by a more efficient species.

Conservation The planned management of a natural resource to prevent overexploitation, destruction, or neglect.

Conventional Pollutants Seven substances (sulfur dioxide, carbon monoxide, particulates, hydrocarbons, nitrogen oxides, photochemical oxidants, and lead) that make up the largest volume of air quality degradation, as identified by the Clean Air Act.

Core Dense, intensely hot molten metal mass, thousands of kilometers in diameter, at Earth's center.

Cornucopian Theory The belief that nature is limitless in its abundance and that perpetual growth is both possible and essential.

Corridor Connecting strip of natural habitat that allows migration of organisms from one place to another.

Crankcase Smog Devices (PCV System) A system, used principally in automobiles, designed to prevent discharge of combustion emissions into the external environment.

Critical Factor The environmental factor closest to a tolerance limit for a species at a specific time.

Cultural Eutrophication Increase in biological productivity and ecosystem succession resulting from human activities.

D

Death Rate Number of deaths in one year per 1,000 midyear population.

Decarbonization To remove carbon dioxide or carbonic acid from a substance.

Decomposer Any organism that causes the decay of organic matter; bacteria and fungi are two examples.

Deforestation The action or process of clearing forests without adequate replanting.

Degradation (of water resource) Deterioration in water quality caused by contamination or pollution that makes water unsuitable for many purposes.

Demography The statistical study of principally human populations.

Desert An arid biome characterized by little rainfall, high daily temperatures, and low diversity of animal and plant life.

Desertification Converting arid or semiarid lands into deserts by inappropriate farming practices or overgrazing.

Detergent A synthetic soap-like material that emulsifies fats and oils and holds dirt in suspension; some detergents have caused pollution problems because of certain chemicals used in their formulation.

Detrivores Organisms that consume organic litter, debris, and dung.

Dioxin Any of a family of compounds known chemically as dibenzo-p-dioxins. Concern about them arises from their potential toxicity as contaminants in commercial products. Tests on laboratory animals indicate that it is one of the more toxic anthropogenic (man-made) compounds.

Diversity Number of species present in a community (species richness), as well as the relative abundance of each species.

DNA (Deoxyribonucleic Acid) One of two principal nucleic acids, the other being RNA (Ribonucleic Acid). DNA contains information used for the control of a living cell. Specific segments of DNA are now recognized as genes, those agents controlling evolutionary and hereditary processes.

Dominant Species Any species of plant or animal that is particularly abundant or controls a major portion of the energy flow in a community.

Drip Irrigation Pipe or perforated tubing used to deliver water a drop at a time directly to soil around each plant. Conserves water and reduces soil waterlogging and salinization.

E

Ecological Density The number of a singular species in a geographical area, including the highest concentration points within the defined boundaries.

Ecological Succession Process in which organisms occupy a site and gradually change environmental conditions so that other species can replace the original inhabitants.

Ecology Study of the interrelationships between organisms and their environments.

Ecosystem The organisms of a specific area, together with their functionally related environments; considered as a definitive unit.

Ecotourism Wildlife tourism that could damage ecosystems and disrupt species if strict guidelines governing tours to sensitive areas are not enforced.

Edge Effects Change in ecological factors at the boundary between two ecosystems. Some organisms flourish here; others are harmed.

Effluent A liquid discharged as waste.

El Niño Climatic change marked by shifting of a large warm water pool from the western Pacific Ocean toward the East.

Electron Small, negatively charged particle; normally found in orbit around the nucleus of an atom.

Eminent Domain Superior dominion exerted by a governmental state over all property within its boundaries that authorizes it to appropriate all or any part thereof to a necessary public use, with reasonable compensation being made.

Endangered Species   Species considered to be in imminent danger of extinction.

Endemic Species   Plants or animals that belong or are native to a particular ecosystem.

Environment   Physical and biological aspects of a specific area.

Environmental Impact Statement (EIS)   A study of the probable environmental impact of a development project before federal funding is provided (required by the National Environmental Policy Act of 1968).

Environmental Protection Agency (EPA)   Federal agency responsible for control of air and water pollution, radiation and pesticide problems, ecological research, and solid waste disposal.

Erosion   Progressive destruction or impairment of a geographical area; wind and water are the principal agents involved.

Estuary   Water passage where an ocean tide meets a river current.

Eutrophic   Well nourished; refers to aquatic areas rich in dissolved nutrients.

Evolution   A change in the gene frequency within a population, sometimes involving a visible change in the population's characteristics.

Exhaustible Resources   Earth's geologic endowment of minerals, nonmineral resources, fossil fuels, and other materials present in fixed amounts.

Extinction   Irrevocable elimination of species due to either normal processes of the natural world or through changing environmental conditions.

F

Fallow   Cropland that is plowed but not replanted and is left idle in order to restore productivity mainly through water accumulation, weed control, and buildup of soil nutrients.

Fauna   The animal life of a specified area.

Feral   Refers to animals or plants that have reverted to a noncultivated or wild state.

Fission   The splitting of an atom into smaller parts.

Floodplain   Level land that may be submerged by floodwaters; a plain built up by stream deposition.

Flora   The plant life of an area.

Flyway   Geographic migration route for birds that includes the breeding and wintering areas that it connects.

Food Additive   Substance added to food usually to improve color, flavor, or shelf life.

Food Chain   The sequence of organisms in a community, each of which uses the lower source as its energy supply. Green plants are the ultimate basis for the entire sequence.

Fossil Fuels   Coal, oil, natural gas, and/or lignite; those fuels derived from former living systems; usually called nonrenewable fuels.

Fuel Cell   Manufactured chemical systems capable of producing electrical energy; they usually derive their capabilities via complex reactions involving the sun as the driving energy source.

Fusion   The formation of a heavier atomic complex brought about by the addition of atomic nuclei; during the process there is an attendant release of energy.

G

Gaia Hypothesis   Theory that Earth's biosphere is a living system whose complex interactions between its living organisms and nonliving processes regulate environmental conditions over millions of years so that life continues.

Gamma Ray   A ray given off by the nucleus of some radioactive elements. A form of energy similar to X rays.

Gene   Unit of heredity; segment of DNA nucleus of the cell containing information for the synthesis of a specific protein.

Gene Banks   Storage of seed varieties for future breeding experiments.

Genetic Diversity   Infinite variation of possible genetic combinations among individuals; what enables a species to adapt to ecological change.

Geothermal Energy   Heat derived from the Earth's interior. It is the thermal energy contained in the rock and fluid (that fills the fractures and pores within the rock) in the Earth's crust.

Germ Plasm   Genetic material that may be preserved for future use (plant seeds, animal eggs, sperm, and embryos).

Global Warming   An increase in the near surface temperature of the Earth. Global warming has occurred in the distant past as the result of natural influences, but the term is most often used to refer to the warming predicted to occur as a result of increased emissions of greenhouse gases. Scientists generally agree that the Earth's surface has warmed by about 1 degree Fahrenheit in the past 140 years.

Green Revolution   The great increase in production of food grains (as in rice and wheat) due to the introduction of high-yielding varieties, to the use of pesticides, and to better management techniques.

Greenhouse Effect   The effect noticed in greenhouses when shortwave solar radiation penetrates glass, is converted to longer wavelengths, and is blocked from escaping by the windows. It results in a temperature increase. Earth's atmosphere acts in a similar manner.

Gross National Product (GNP)   The total value of the goods and services produced by the residents of a nation during a specified period (such as a year).

Groundwater   Water found in porous rock and soil below the soil moisture zone and, generally, below the root zone of plants. Groundwater that saturates rock is separated from an unsaturated zone by the water table.

H

Habitat   The natural environment of a plant or animal.

Habitat Fragmentation   Process by which a natural habitat/landscape is broken up into small sections of natural ecosystems, isolated from each other by sections of land dominated by human activities.

Hazardous Waste   Waste that poses a risk to human or ecological health and thus requires special disposal techniques.

Herbicide   Any substance used to kill plants.

Heterotroph   Organism that cannot synthesize its own food and must feed on organic compounds produced by other organisms.

Glossary

Hydrocarbons   Organic compounds containing hydrogen, oxygen, and carbon. Commonly found in petroleum, natural gas, and coal.

Hydrogen   Lightest-known gas; major element found in all living systems.

Hydrogen Sulfide   Compound of hydrogen and sulfur; a toxic air contaminant that smells like rotten eggs.

Hydropower   Electrical energy produced by flowing or falling water.

I

Infiltration   Process of water percolation into soil and pores and hollows of permeable rocks.

Intangible Resources   Open space, beauty, serenity, genius, information, diversity, and satisfaction are a few of these abstract commodities.

Integrated Pest Management (IPM)   Designed to avoid economic loss from pests, this program's methods of pest control strive to minimize the use of environmentally hazardous, synthetic chemicals.

Invasive   Refers to those species that have moved into an area and reproduced so aggressively that they have replaced some of the native species.

Ion   An atom or group of atoms, possessing a charge; brought about by the loss or gain of electrons.

Ionizing Radiation   Energy in the form of rays or particles that have the capacity to dislodge electrons and/or other atomic particles from matter that is irradiated.

Irradiation   Exposure to any form of radiation.

Isotopes   Two or more forms of an element having the same number of protons in the nucleus of each atom but different numbers of neutrons.

K

Keystone Species   Species that are essential to the functioning of many other organisms in an ecosystem.

Kilowatt   Unit of power equal to 1,000 watts.

Leaching   Dissolving out of soluble materials by water percolating through soil.

L

Limnologist   Individual who studies the physical, chemical, and biological conditions of aquatic systems.

M

Malnutrition   Faulty or inadequate nutrition.

Malthusian Theory   The theory that populations tend to increase by geometric progression (1, 2, 4, 8, 16, etc.) while food supplies increase by arithmetic means (1, 2, 3, 4, 5, etc.).

Metabolism   The chemical processes in living tissue through which energy is provided for continuation of the system.

Methane   Often called marsh gas (CH^4); an odorless, flammable gas that is the major constituent of natural gas. In nature it develops from decomposing organic matter.

Migration   Periodic departure and return of organisms to and from a population area.

Monoculture   Cultivation of a single crop, such as wheat or corn, to the exclusion of other land uses.

Mutation   Change in genetic material (gene) that determines species characteristics; can be caused by a number of agents, including radiation and chemicals, called mutagens.

N

Natural Selection   The agent of evolutionary change by which organisms possessing advantageous adaptations leave more offspring than those lacking such adaptations.

Niche   The unique occupation or way of life of a plant or animal species; where it lives and what it does in the community.

Nitrate   A salt of nitric acid. Nitrates are the major source of nitrogen for higher plants. Sodium nitrate and potassium nitrate are used as fertilizers.

Nitrite   Highly toxic compound; salt of nitrous acid.

Nitrogen Oxides   Common air pollutants. Formed by the combination of nitrogen and oxygen; often the products of petroleum combustion in automobiles.

Nonrenewable Resource   Any natural resource that cannot be replaced, regenerated, or brought back to its original state once it has been extracted, for example, coal or crude oil.

Nutrient   Any nutritive substance that an organism must take in from its environment because it cannot produce it as fast as it needs it or, more likely, at all.

O

Oil Shale   Rock impregnated with oil. Regarded as a potential source of future petroleum products.

Oligotrophic   Most often refers to those lakes with a low concentration of organic matter. Usually contain considerable oxygen; Lakes Tahoe and Baikal are examples.

Organic   Living or once living material; compounds containing carbon formed by living organisms.

Organophosphates   A large group of nonpersistent synthetic poisons used in the pesticide industry; include parathion and malathion.

Ozone   Molecule of oxygen containing three oxygen atoms; shields much of Earth from ultraviolet radiation.

P

Particulate   Existing in the form of small separate particles; various atmospheric pollutants are industrially produced particulates.

Peroxyacyl Nitrate (PAN)   Compound making up part of photochemical smog and the major plant toxicant of smog-type injury; levels as low as 0.01 ppm can injure sensitive plants. Also causes eye irritation in people.

Pesticide   Any material used to kill rats, mice, bacteria, fungi, or other pests of humans.

Pesticide Treadmill   A situation in which the cost of using pesticides increases while the effectiveness decreases (because pest species develop genetic resistance to the pesticides).

Petrochemicals   Chemicals derived from petroleum bases.

pH   Scale used to designate the degree of acidity or alkalinity; ranges from 1 to 14; a neutral solution has a pH of 7; low pHs are acid in nature, while pHs above 7 are alkaline.

Phosphate   A phosphorous compound; used in medicine and as fertilizers.

Photochemical Smog   Type of air pollution; results from sunlight acting with hydrocarbons and oxides of nitrogen in the atmosphere.

Photosynthesis   Formation of carbohydrates from carbon dioxide and hydrogen in plants exposed to sunlight; involves a release of oxygen through the decomposition of water.

Photovoltaic Cells   An energy-conversion device that captures solar energy and directly converts it to electrical current.

Physical Half-Life   Time required for half of the atoms of a radioactive substance present at some beginning to become disintegrated and transformed.

Phytoplankton   That portion of the plankton community comprised of tiny plants, e.g., algae, diatoms.

Pioneer Species   Hardy species that are the first to colonize a site in the beginning stage of ecological succession.

Plankton   Microscopic organisms that occupy the upper water layers in both freshwater and marine ecosystems.

Plutonium   Highly toxic, heavy, radioactive, manmade, metallic element. Possesses a very long physical half-life.

Pollution   The process of contaminating air, water, or soil with materials that reduce the quality of the medium.

Polychlorinated Biphenyls (PCBs)   Poisonous compounds similar in chemical structure to DDT. PCBs are found in a wide variety of products ranging from lubricants, waxes, asphalt, and transformers to inks and insecticides. Known to cause liver, spleen, kidney, and heart damage.

Population   All members of a particular species occupying a specific area.

Predator   Any organism that consumes all or part of another system; usually responsible for death of the prey.

Primary Production   The energy accumulated and stored by plants through photosynthesis.

R

Rad (Radiation Absorbed Dose)   Measurement unit relative to the amount of radiation absorbed by a particular target, biotic or abiotic.

Radioactive Waste   Any radioactive by-product of nuclear reactors or nuclear processes.

Radioactivity   The emission of electrons, protons (atomic nuclei), and/or rays from elements capable of emitting radiation.

Rain Forest   Forest with high humidity, small temperature range, and abundant precipitation; can be tropical or temperate.

Recycle   To reuse; usually involves manufactured items, such as aluminum cans, being restructured after use and utilized again.

Red Tide   Population explosion or bloom of minute single-celled marine organisms (dinoflagellates), which can accumulate in protected bays and poison other marine life.

Renewable Resources   Resources normally replaced or replenished by natural processes; not depleted by moderate use.

Riparian Water Right   Legal right of an owner of land bordering a natural lake or stream to remove water from that aquatic system.

S

Salinization   An accumulation of salts in the soil that could eventually make the soil too salty for the growth of plants.

Sanitary Landfill   Land waste disposal site in which solid waste is spread, compacted, and covered.

Scrubber   Antipollution system that uses liquid sprays in removing particulate pollutants from an airstream.

Sediment   Soil particles moved from land into aquatic systems as a result of human activities or natural events, such as material deposited by water or wind.

Seepage   Movement of water through soil.

Selection   The process, either natural or artificial, of selecting or removing the best or less desirable members of a population.

Selective Breeding   Process of selecting and breeding organisms containing traits considered most desirable.

Selective Harvesting   Process of taking specific individuals from a population; the removal of trees in a specific age class would be an example.

Sewage   Any waste material coming from domestic and industrial origins.

Smog   A mixture of smoke and air; now applies to any type of air pollution.

Soil Erosion   Detachment and movement of soil by the action of wind and moving water.

Solid Waste   Unwanted solid materials usually resulting from industrial processes.

Species   A population of morphologically similar organisms, capable of interbreeding and producing viable offspring.

Species Diversity   The number and relative abundance of species present in a community. An ecosystem is said to be more diverse if species present have equal population sizes and less diverse if many species are rare and some are very common.

Strip Mining   Mining in which Earth's surface is removed in order to obtain subsurface materials.

Strontium-90   Radioactive isotope of strontium; it results from nuclear explosions and is dangerous, especially for vertebrates, because it is taken up in the construction of bone.

Succession   Change in the structure and function of an ecosystem; replacement of one system with another through time.

Sulfur Dioxide (SO^2)   Gas produced by burning coal and as a by-product of smelting and other industrial processes. Very toxic to plants.

Sulfur Oxides (SO^x)   Oxides of sulfur produced by the burning of oils and coal that contain small amounts of sulfur. Common air pollutants.

Sulfuric Acid ($H2\ SO^4$)   Very corrosive acid produced from sulfur dioxide and found as a component of acid rain.

Sustainability   Ability of an ecosystem to maintain ecological processes, functions, biodiversity, and productivity over time.

Sustainable Agriculture   Agriculture that maintains the integrity of soil and water resources so that it can continue indefinitely.

T

Technology   Applied science; the application of knowledge for practical use.

Tetraethyl Lead   Major source of lead found in living tissue; it is produced to reduce engine knock in automobiles.

Thermal Inversion   A layer of dense, cool air that is trapped under a layer of less dense warm air (prevents upward flowing air currents from developing).

Glossary

Thermal Pollution   Unwanted heat, the result of ejection of heat from various sources into the environment.

Thermocline   The layer of water in a body of water that separates an upper warm layer from a deeper, colder zone.

Threshold Effect   The situation in which no effect is noticed, physiologically or psychologically, until a certain level or concentration is reached.

Tolerance Limit   The point at which resistance to a poison or drug breaks down.

Total Fertility Rate (TFR)   An estimate of the average number of children that would be born alive to a woman during her reproductive years.

Toxic   Poisonous; capable of producing harm to a living system.

Tragedy of the Commons   Degradation or depletion of a resource to which people have free and unmanaged access.

Trophic   Relating to nutrition; often expressed in trophic pyramids in which organisms feeding on other systems are said to be at a higher trophic level; an example would be carnivores feeding on herbivores, which, in turn, feed on vegetation.

Turbidity   Usually refers to the amount of sediment suspended in an aquatic system.

U

Uranium 235   An isotope of uranium that when bombarded with neutrons undergoes fission, resulting in radiation and energy. Used in atomic reactors for electrical generation.

Z

Zero Population Growth   The condition of a population in which birthrates equal death rates; it results in no growth of the population.

Index

Index

Test Your Knowledge Form

We encourage you to photocopy and use this page as a tool to assess how the articles in *Annual Editions* expand on the information in your textbook. By reflecting on the articles you will gain enhanced text information. You can also access this useful form on a product's book support Web site at *http://www.dushkin.com/online/*.

NAME: _____   DATE: _____

TITLE AND NUMBER OF ARTICLE: _____

BRIEFLY STATE THE MAIN IDEA OF THIS ARTICLE: _____

_____

LIST THREE IMPORTANT FACTS THAT THE AUTHOR USES TO SUPPORT THE MAIN IDEA:

_____

WHAT INFORMATION OR IDEAS DISCUSSED IN THIS ARTICLE ARE ALSO DISCUSSED IN YOUR TEXTBOOK OR OTHER READINGS THAT YOU HAVE DONE? LIST THE TEXTBOOK CHAPTERS AND PAGE NUMBERS:

_____

LIST ANY EXAMPLES OF BIAS OR FAULTY REASONING THAT YOU FOUND IN THE ARTICLE:

_____

LIST ANY NEW TERMS/CONCEPTS THAT WERE DISCUSSED IN THE ARTICLE, AND WRITE A SHORT DEFINITION:

We Want Your Advice

ANNUAL EDITIONS revisions depend on two major opinion sources: one is our Advisory Board, listed in the front of this volume, which works with us in scanning the thousands of articles published in the public press each year; the other is you—the person actually using the book. Please help us and the users of the next edition by completing the prepaid article rating form on this page and returning it to us. Thank you for your help!

ANNUAL EDITIONS: Environment 03/04

ARTICLE RATING FORM

Here is an opportunity for you to have direct input into the next revision of this volume.
We would like you to rate each of the articles listed below, using the following scale:

1. **Excellent: should definitely be retained**
2. **Above average: should probably be retained**
3. **Below average: should probably be deleted**
4. **Poor: should definitely be deleted**

Your ratings will play a vital part in the next revision.
Please mail this prepaid form to us as soon as possible.
Thanks for your help!

RATING	ARTICLE	RATING	ARTICLE
	1. How Many Planets? A Survey of the Global Environment		
	2. Forget Nature. Even Eden Is Engineered		
	3. Crimes of (a) Global Nature		
	4. Population Control Today—and Tomorrow?		
	5. Population and Consumption: What We Know, What We Need to Know		
	6. An Economy for the Earth		
	7. The Eco-Economic Revolution: Getting the Market in Sync With Nature		
	8. Poverty and Environmental Degradation: Challenges Within the Global Economy		
	9. Where the Sidewalks End		
	10. Energy: A Brighter Future?		
	11. Beyond Oil: The Future of Energy		
	12. Renewable Energy: A Viable Choice		
	13. Fossil Fuels and Energy Independence		
	14. Power Struggle: California's Engineered Energy Crisis and the Potential of Public Power		
	15. What Is Nature Worth?		
	16. A Fragile Cornucopia: Assessing the Status of U.S. Biodiversity		
	17. Invasive Species: Pathogens of Globalization		
	18. Where Have All the Farmers Gone?		
	19. All the Wild Rivers		
	20. Growing More Food With Less Water		
	21. Oceans Are on the Critical List		
	22. Feeling the Heat: Life in the Greenhouse		
	23. Three Pollutants and an Emission		
	24. Groundwater Shock: The Polluting of the World's Major Freshwater Stores		
	25. Water Quality: The Issues		
	26. Statehouse and Greenhouse: The States Are Taking the Lead on Climate Change		

(Continued on next page)

ABOUT YOU

Name

Date

Are you a teacher? ☐   A student? ☐
Your school's name

Department

Address   City   State   Zip

School telephone #

YOUR COMMENTS ARE IMPORTANT TO US!

Please fill in the following information:
For which course did you use this book?

Did you use a text with this ANNUAL EDITION?   ☐ yes ☐ no
What was the title of the text?

What are your general reactions to the *Annual Editions* concept?

Have you read any pertinent articles recently that you think should be included in the next edition? Explain.

Are there any articles that you feel should be replaced in the next edition? Why?

Are there any World Wide Web sites that you feel should be included in the next edition? Please annotate.

May we contact you for editorial input?   ☐ yes ☐ no
May we quote your comments? ☐ yes ☐ no